HIGH-ENERGY ASTROPHYSICS IN THE 21ST CENTURY

AIP CONFERENCE PROCEEDINGS 211

HIGH-ENERGY ASTROPHYSICS IN THE 21ST CENTURY
TAOS, NM 1989

EDITOR:
PAUL C. JOSS
MASSACHUSETTS INSTITUTE
OF TECHNOLOGY

American Institute of Physics   New York

Authorization to photocopy items for internal or personal use, beyond the free copying permitted under the 1978 US Copyright Law (see statement below), is granted by the American Institute of Physics for users registered with the Copyright Clearance Center (CCC) Transactional Reporting Service, provided that the base fee of $2.00 per copy is paid directly to CCC, 27 Congress St., Salem, MA 01970. For those organizations that have been granted a photocopy license by CCC, a separate system of payment has been arranged. The fee code for users of the Transactional Reporting Service is: 0094-243X/87 $2.00.

© 1990 American Institute of Physics.

Individual readers of this volume and non-profit libraries, acting for them, are permitted to make fair use of the material in it, such as copying an article for use in teaching or research. Permission is granted to quote from this volume in scientific work with the customary acknowledgment of the source. To reprint a figure, table or other excerpt requires the consent of one of the original authors and notification to AIP. Republication or systematic or multiple reproduction of any material in this volume is permitted only under license from AIP. Address inquiries to Series Editor, AIP Conference Proceedings, AIP, 335 E. 45th St., New York, NY 10017.

L.C. Catalog Card No. 90-55644
ISBN 0-88318-803-1
DOE CONF 891276

Printed in the United States of America.

This book was put into production on 13 June 1990 and was published on 17 August 1990.

CONTENTS

Preface ix

List of Participants xi

I.   INTRODUCTORY REMARKS 3
     Alan N. Bunner

II. NEW SCIENTIFIC CHALLENGES
     Paul C. Joss and Donald Q. Lamb, Chairmen
     X-Ray Emission from Normal Stars 11
         R. Rosner
     Timing Observations of Neutron Stars and Black Holes 23
         K. S. Wood
     Cyclotron Lines: The Next 100 Years 27
         Ira Wasserman
     Gamma Ray Astronomy and Black Hole Astrophysics 32
         Edison P. Liang
     Supernovae and Supernova Remnants at High Energies 38
         Roger A. Chevalier
     Gamma-Rays as a Probe of Type Ia Supernova Physics 48
         Adam Burrows
     The Goals of Gamma-Ray Spectroscopy in High Energy Astrophysics 56
         Richard E. Lingenfelter, James C. Higdon, Marvin Leventhal,
         Reuven Ramaty, and Stanford E. Woosley
     Future X-Ray Observations of "Normal" Galaxies 74
         G. Fabbiano
     X-Ray Observations and the Properties of Active Galactic Nuclei 84
         Richard F. Mushotzky
     The All-Sky X-Ray Foreground: Outstanding Issues 93
         Elihu Boldt
     Scientific Requirements for Studying the Composition
     of the X-Ray Background 99
         R. E. Griffiths

III. TECHNOLOGY FOR NEW INSTRUMENTATION
     Jonathan E. Grindlay and William C. Priedhorsky, Chairmen
     X-Ray Imaging 107
         Richard C. Catura
     An Imaging Photoemission Polarimeter for Soft X-Rays 120
         P. Kaaret, R. Novick, A. Heckler, P. Shaw, G. W. Fraser,
         J. E. Lees, and J. F. Pearson
     The Promise of Replication for X-Ray Astronomy 125
         M. P. Ulmer

High Energy, High Resolution X-Ray Optics..129
  Martin C. Weisskopf, Marshall Joy, and Steven Kahn
Space Astrophysics with Large Structures: CASES and P/OF....................134
  H. S. Hudson and J. M. Davis
Off-Plane Imaging for Milli-Arcsecond X-Ray Observatories....................139
  Webster Cash
Prospects for High Pressure Gas Scintillation Chambers..........................145
  T. K. Edberg, A. Parsons, B. Sadoulet, S. Weiss, J. Wilkerson,
  and G. Smith
X-Ray Spectroscopy Instrumentation for the 21st Century.......................150
  Steven M. Kahn
Low-Temperature Detectors for X-Rays..164
  Gilbert G. Fritz
A New Thermal Sensor for X-Ray Microcalorimetry...................................169
  E. Silver, S. Labov, T. Pfafman, F. Goulding, D. Landis,
  N. Madden, and J. Beeman
Multistep Fluorescence Gated Proportional Counters................................174
  B. D. Ramsey and M. C. Weisskopf
High-Resolution Gamma-Ray Imaging from the Moon................................179
  William A. Mahoney
A Gamma-Ray Imaging Telescope Based on Liquid Xenon.......................189
  Elena Aprile, Reshmi Mukherjee, and Masayo Suzuki
High Resolution Compton Telescope for the 21st Century........................195
  W. N. Johnson, R. A. Kroeger, and J. D. Kurfess
New Drift Chamber Technology for High Energy Gamma-Ray Telescopes..200
  Stanley D. Hunter and Rajani Cuddapah
High Resolution Gamma-Ray Imaging and Spectroscopy..........................205
  G. H. Nakano and J. R. Kilner
Directions in Gamma-Ray Spectroscopy..213
  Neil Gehrels and Robert M. Candey
Imaging Germanium Telescope Array for Gamma-Rays (IGETAGRAY)..224
  Charles J. Hailey, Klaus P. Ziock, Fiona A. Harrison, and
  Judith Fleischmann
A Large Area, Low Cost, Gamma-Ray, Imaging Spectrometer..................229
  K. P. Ziock and C. J. Hailey
A Total Throughput Transient Spectrometer for Gamma-Ray Sources........234
  Kevin Hurley
ASTROGAM: A Magnetic Rigidity Spectrometer for Gamma Ray Astronomy..240
  J. H. Adams, Jr., S. P. Ahlen, L. M. Barbier, J. J. Beatty, P. Carlson,
  H. J. Crawford, R. L. Golden, K. E. Krombel, R. C. Lamb,
  J. Lloyd-Evans, A. A. Marin, J. F. Ormes, M. E. Ozel, G. F. Smoot,
  R. E. Streitmatter, A. J. Tylka, T. C. Weekes, and B. Zhou

Superconducting Transition Detectors for Low-Energy
Gamma-Ray Astrophysics 246
    J. D. Kurfess, W. N. Johnson, G. G. Fritz, M. S. Strickman,
    R. L. Kinzer, G. Jung, A. K. Drukier, and M. Chmielowski
Segmented Ge Detectors and Mechanical Coolers for Future
Gamma-Ray Astronomy Instruments..252
    Larry S. Varnell
Mercuric Iodide Room Temperature Gamma-Ray Detectors.......................257
    Bradley E. Patt, Jeffrey M. Markakis, Vernon M. Gerrish,
    Robert C. Haymes, and Jacob I. Trombka

IV. NASA CAPABILITIES IN THE 21ST CENTURY
Space Capabilities in the 21st Century 265
    Robert C. Rhome

V. MISSION CONCEPTS
    Kevin Hurley and Richard Mushotzky, Chairmen
X-Ray Interferometry for Sub-Milliarcsecond and
Sub-Microarcsecond Imaging...291
    Christopher Martin
High Throughput X-Ray Astronomy 297
    Paul Gorenstein
The X-Ray Large Array (XLA) 314
    K. S. Wood
The Study of Large-Scale Structures with Wide-Field X-Ray Optics 318
    Richard Burg, Christopher J. Burrows, and Riccardo Giacconi
The EXOSS Mission for Hard X-Ray Astronomy 325
    J. Grindlay, T. Prince, M. Weisskopf, and G. Skinner
The Nuclear Astrophysics Explorer 343
    J. L. Matteson, B. J. Teegarden, N. Gehrels, and W. A. Mahoney
Low-Cost Small Satellites for Astrophysical Missions 359
    William C. Priedhorsky
The Energetic Transient Array—ETA—A Network of "Space Buoys"
in Solar Orbit for Observations of Gamma-Ray Bursts............................365
    George R. Ricker

VI. RAPPORTEURS' REPORTS
    Alan N. Bunner, Chairman
X-Ray Astronomy in the 21st Century 381
    Claude R. Canizares
Report on Workshop..391
    S. Rappaport
Gamma-Ray Astronomy—Progress and Instruments...............................395
    L. E. Peterson
Report on Workshop..405
    Richard Epstein

Preface

On December 11–14, 1989, a workshop on the theme of "High-Energy Astrophysics in the 21st Century" was held at the Sagebrush Inn in Taos, New Mexico. This workshop, which drew 71 participants from throughout the high-energy astrophysics community, was the result of an initiative conceived by the NASA High-Energy Astrophysics Management Operations Working Group (HEAMOWG) about eight months earlier.

The workshop had two main purposes. Foremost, the meeting was intended to provide NASA with a sense of the scientific goals, aspirations, and priorities of the high-energy astrophysics community beyond the year 2000, particularly as these goals relate to NASA's "Astrotech 21" initiative. Secondarily, the workshop provided one of several means for informing the Astronomy Survey Committee of the National Academy of Sciences (the "Bahcall Committee") regarding the medium- to long-range interests and needs of our community. These Proceedings represent the collective response of the workshop participants to satisfy these purposes.

Neither the workshop itself, nor these Proceedings, could have been brought into being without the diligent efforts of many people. I am grateful for all of the hard work carried out by the Organizing Committee: Alan Bunner, Jim Cutts, Martin Elvis, Richard Epstein, Lou Kaluzienski, Larry Peterson, Bill Priedhorsky, and George Ricker. Worthy of special note were the efforts of NASA's Alan Bunner, who provided inspiring support and invaluable advice throughout this project. Richard Epstein and Bill Priedhorsky of the Los Alamos National Laboratory performed double duty as our local hosts in New Mexico. Jim Cutts of the Jet Propulsion Laboratory not only provided essential logistical support, but also kindly took on an especially difficult role as a last-minute replacement for our banquet speaker, in which capacity he was both entertaining and informative.

It is a pleasure to thank Louise Blair and the staff of the Sagebrush Inn for providing all of us with a very comfortable and enjoyable stay. Excellent administrative and clerical support for the workshop was provided by Elaine Ruhe of Los Alamos and Karen Blair of the Massachusetts Institute of Technology. A special note of thanks is owed to Christopher Naylor of MIT, who for the past year has worked tirelessly and with great skill on both the organization of the workshop and the preparation of these Proceedings.

I am writing on the eve of the anticipated launch of the Hubble Space Telescope. Within the next decade, several additional major space astrophysics observatories are scheduled to be launched. As a result of what we will learn from these great observatories, our view of the universe ten years hence is likely to be very different in some respects from the one we have today. It is our hope that when that time comes, the future of astrophysics will still look as bright as it does today. This workshop, and these Proceedings, represent one of the beginnings of our efforts to turn this hope into reality.

Paul C. Joss
Cambridge, Massachusetts
April 1990

Participants

James H. Adams
Elene Aprile
Jeffrey Bloch
Elihu Boldt
Alan N. Bunner
Richard Burg
Adam Burrows
Claude Canizares
Webster Cash
Richard Catura
Roger Chevalier
James A. Cutts
Robert F. Doolittle
Timothy Edberg
Martin Elvis
Richard Epstein
Giuseppina Fabbiano
Edward Fenimore
Jess Fordyce
Gilbert G. Fritz
Margaret Geller
Neil Gehrels
Paul Gorenstein
Richard E. Griffiths
Jonathan E. Grindlay
John Grunsfeld
Charles Hailey
Robert Haymes
Hugh Hudson
Stanley D. Hunter
Kevin Hurley
W. Neil Johnson
Paul C. Joss
Philip Kaaret
Steven M. Kahn
Louis Kaluzienski

Richard Kelly
James D. Kurfess
Donald Q. Lamb
Marvin Leventhal
Edison P. Liang
Richard Lingenfelter
William A. Mahoney
Robert Marshall
Christopher Martin
James L. Matteson
Stephen S. Murray
Richard Mushotzky
George H. Nakano
Robert Novick
Laurence E. Peterson
William Priedhorsky
Thomas A. Prince
Brian D. Ramsey
Saul Rappaport
Robert C. Rhome
George Ricker
Robert Rosner
Ethan J. Schreier
Maurice Shapiro
Gerald Share
Eric Silver
Robert Svoboda
Melville P. Ulmer
Larry Varnell
Ira Wasserman
Martin C. Weisskopf
Laura Whitlock
Kent S. Wood
Stanford E. Woosley
Klaus Ziock

I. Introductory Remarks
Alan N. Bunner

INTRODUCTORY REMARKS

Alan N. Bunner

NASA Headquarters,
Washington, D.C.

I'd like to welcome everyone to what I expect will be a very important workshop for the future of high energy astrophysics. It's been about 10 years since we have had a workshop or conference devoted to thinking about the next steps in this field. In November 1979, NASA's High Energy Astrophysics Management Operations Working Group (HEAMOWG) submitted a recommended "Program of High Energy Astrophysics for the 1980's" (Opp et al., 1979).

This workshop has grown out of discussions at meetings of the HEAMOWG, in which it became clear that, in view of the fact that the 21st Century is nearly here, and that in view of the 15-20 years that are often required to turn a mission concept into reality, we need to begin planning for the mission ideas of the next century and the technology advances that are needed to pave the way.

I would just like to make a few introductory remarks on the several purposes for this workshop, from a NASA point of view. As we all know, the National Academy of Sciences has just embarked on its decadal "Astronomy and Astrophysics Survey Committee" activity, chaired by Prof. John Bahcall. This survey has for its goal identifying the most important projects and research activities to be undertaken over the next 15 years or so. This committee, and in particular, the panel on High Energy Astrophysics, chaired by Bruce Margon, is looking to this workshop and others like it, to identify scientific needs and the key areas of technology development required to assist future space-based research programs.

Quite independently from the Bahcall Committee activities, the Astrophysics Division at NASA Headquarters has also been developing plans for a budget initiative based on the advanced technology needs of the 21st century. It is our intent to propose to the Administration and Congress for a budget augmentation in Fiscal Year 1992 for this so-called "Technology-21" initiative.

Finally, President Bush has recently directed the National Space Council to investigate the potential of a program including a manned lunar base and a manned journey to Mars, all called the "Human Exploration Initiative". In turn, the National Space Council has asked NASA to consider the scientific program that could be accomplished at a lunar base. Thus a possible lunar base program has now emerged with our planned Technology 21 program as unifying themes that we may look forward to, for astrophysical research in the 21st century.

Incidentally, for those of you that are particularly interested in NASA's plans for the lunar program, I can recommend another workshop planned to be held February 5-7 in Annapolis, Maryland. Those interested should contact Dr. Mike Mumma, at Goddard Space Flight Center.

So, we have a rather rich set of planning activities for the future taking place: the Bahcall Committee, the NASA Technology-21 initiative, and the possibility of a major manned exploration program. But I should emphasize an important point: that with all the recent activity at NASA responding to calls for rapid answers to complex questions, there has not been time for the appropriate dialogue to take place within the scientific community on the scientific needs, the priorities, and the alternatives. And that is the function of this week's workshop, to hear the ideas and arguments, to think, to debate and to discuss, with emphasis on what directions research in the fields of x-ray and gamma-ray astronomy should take over the next two decades. To let the dialogue begin.

Figures 1 through 5, prepared by Dr. James A. Cutts of JPL, illustrate the goal and approach to the Astrophysics Technology 21 program. The primary philosophy is a recognition that over the next 10 years, we will be launching, operating, and analyzing data from the Great Observatories, COBE, ROSAT, EUVE, XTE and so forth. Our next task is to develop the technology advances needed to enable space astronomy beyond these missions.

Figures 2 and 4 illustrate the logical progression of steps that must take place: From identifying a science area with a potential for a major breakthrough (some examples are shown in Figure 4), to developing the needed technologies such as new detectors or cryogenic techniques. The status today with respect to step 1, the identification of science ideas, is illustrated in Figure 5: (1) a report of NASA's Gamma Ray Astronomy Program Working Group has been published, which is available from NASA HQ on request, (2) an X-ray Astronomy Program Working Group has also been chartered and has met twice to date, (3) a special HEAMOWG meeting was held in January 1989 focussing on technology objectives, (4) the Astronomy and Astrophysics Survey Committee and its subcommittees have been formed, and (5) we are now at this workshop. The intention is that the recommended scientific and technological thrusts resulting from this workshop will be reviewed and absorbed by the Margon subcommittee,

ASTROPHYSICS TECHNOLOGY 21

THE PRIMARY THRUST OF THE ASTROPHYSICS TECHNOLOGY 21 PROGRAM IS TO DEVELOP THE TECHNOLOGICAL BASE FOR THE ASTROPHYSICS MISSIONS OF THE 21ST CENTURY. THE PROGRAM MUST START NOW TO PROVIDE THE ASTROPHYSICS COMMUNITY WITH THE RANGE OF CHOICES TO DEVELOP ITS SCIENCE STRATEGY AND TO IMPLEMENT ITS MISSION PLAN.

Figure 1

ASTROPHYSICS TECHNOLOGY 21
PROGRAM INTRODUCTION

- GOAL:
  - TO PROVIDE THE ENABLING TECHNOLOGIES FOR SPACE ASTRONOMY BEYOND THE GREAT OBSERVATORIES
- ASSUMPTION:
  - MAJOR INCREASES IN OBSERVATIONAL CAPABILITY BRING MAJOR SCIENTIFIC REWARDS
- APPROACH:
  - STEP 1: IDENTIFY FUTURE MISSION CONCEPTS WITH OUTSTANDING SCIENCE POTENTIAL (WITH MOWGS)
  - STEP 2: REVIEW MISSION CONCEPTS WITH NAS ASTRONOMY SURVEY
  - STEP 3: DEVELOP THE ASTROPHYSICS TECHNOLOGY 21 PROGRAM PLAN
    A: USE FUTURE MISSION CONCEPTS TO DERIVE ASTROPHYSICS PROGRAM TECHNOLOGY REQUIREMENTS
    B: USE LDR WORKSHOP FORMAT TO IDENTIFY AND REFINE TECHNOLOGY REQUIREMENTS
    C: CONDUCT FEASIBILITY/PROOF-OF-CONCEPT TECHNOLOGY STUDIES IN ASSOCIATION WITH THE WORKSHOPS
  - STEP 4: ITERATE MISSION REQUIREMENTS AND TECHNOLOGICAL SOLUTIONS TO DETERMINE THE COMMUNITY'S CHOICES FOR THE NEXT GENERATION OF ASTROPHYSICS MISSIONS

Figure 2

Figure 3

Figure 4

ASTROPHYSICS TECHNOLOGY 21
APPROACH
STEP 1 IDENTIFICATION OF OUTSTANDING SCIENCE POTENTIAL

- THE MOWG ARE DEVELOPING PLANS FOR FOUR DISCIPLINE AREAS
- STATUS OF THESE PLANS IS AS FOLLOWS:

  - HIGH ENERGY ASTROPHYSICS
    - GAMMA RAY PROGRAM WORKING GROUP REPORT PUBLISHED 10/88
    - HEAMOWG HOLDS PLANNING MEETING FOR TECHNOLOGY WORKSHOP (7/07/89)
    - X-RAY PROGRAM WORKING GROUP MEETING (7/12/89)
    - TECHNOLOGY WORKSHOP (DECEMBER, 89)

  - INFRARED RADIO ASTROPHYSICS
    - STRATEGIC PLAN COMPLETE (6/24/89)
    - LIST OF MISSION CONCEPTS WILL BE DEVELOPED AT THE JUNE 26-27 IRMOWG AND REFINED FOR THE FALL MEETING

  - UV/VISIBLE AND RELATIVITY
    - DRAFT OF UV/VISIBLE PLAN COMPLETE 9/1/89 (AFTER AUG 15-16 MOWG)
    - GRAVITATIONAL PHYSICS PROGRAM PLAN WILL BE COMPLETE 2/1/90

  - SCIENCE OPERATIONS
    - THREE MISSION CONCEPTS DEFINED, STRATEGIC PLAN IN PREPARATION

Figure 5

Introductory Remarks

the X-ray Program Working Group, and in our own planning at NASA.

Table 1, at left, lists the NASA future mission plans that are already in some stage of planning or definition or development for the near future. This list includes international collaborative missions to which the U.S. is contributing, and Space Station Freedom attached payloads currently under study.

Following the initial announcement for this workshop, the Presidential directive was issued to NASA regarding the formulation of a long-term strategy of space exploration culminating in a manned mission to Mars, including an intermediate step involving establishment of a manned lunar base. Participants are urged to consider the possibility of instrumentation on the lunar surface.

The workshop organizers have structured the sessions this week into three general categories: (1) the scientific needs of high energy astrophysics in the 21st century, (2) emerging, high-potential technologies that may be important for the next century, and finally (3) mission concepts, for those authors who may wish to bring scientific needs and technology together in a concept proposal deserving of further consideration.

Table 1

HIGH ENERGY ASTROPHYSICS MISSION PLANS

PROJECT	LAUNCH	LIFETIME
BBXRT/ASTRO BROAD BAND X-RAY TELESCOPE on ASTRO-1	4/1990	7 DAYS
ROSAT ROENTGEN SATELLIT (FRG)	5/1990	3 YEARS
GRO GAMMA RAY OBSERVATORY	6/1990	10 YEARS
SHEAL-2/DXS SHUTTLE HIGH ENERGY ASTROPHYSICS LABORATORY/DIFFUSE X-RAY SPECTROMETER	11/1992	7 DAYS
ASTRO-D SPECTROSCOPIC X-RAY OBSERVATORY (JAPAN)	2/1993	3 YEARS
TGRS TRANSIENT GAMMA RAY SPECTROMETER, on WIND SPACECRAFT	1/1993	4 YEARS
ASTRO-2/BBXRT BROAD-BAND X-RAY TELESCOPE ON ASTRO-2	1993	7 DAYS
SPECTRUM-X STELLAR X-RAY POLARIMETER + X-RAY ALL SKY MONITOR (USSR)	1994	10 YEARS
MARS OBSERVER GAMMA-RAY BURST DETECTOR ON MARS OBSERVER	1994	3 YEARS
XTE X-RAY TIMING EXPLORER	9/1994	3 YEARS
HETE HIGH ENERGY TRANSIENT EXPERIMENT	1995	2 YEARS
XBSS X-RAY BACKGROUND SURVEY SPECTROMETER	1996	2 YEARS
AXAF ADVANCED X-RAY ASTROPHYSICS FACILITY	1997	15 YEARS
NAE NUCLEAR ASTROPHYSICS EXPLORER	1997	3 YEARS
XMM X-RAY MULTI-MIRROR MISSION (ESA)	1998	15 YEARS
LAMAR LARGE AREA MODULAR ARRAY OF REFLECTORS, on SPACE STATION	1998	3 YEARS
EXOSS ENERGETIC X-RAY OBSERVATORY ON SPACE STATION	1999	3 YEARS

We also have scheduled, on Wednesday, a presentation by Mr. Bob Rhome of NASA's Office of Space Science and Applications, on the capabilities and infrastructure that NASA plans to have in place by the 21st century. Near the end of the workshop, a series of rapporteurs' reports, by Saul Rappaport, Claude Canizares, Richard Epstein, and Larry Peterson, will serve to highlight for discussion some of the key themes of the week's presentations.

REFERENCES

A. Opp et al., 1979, A Program in High Energy Astrophysics for the 1980's, by the High Energy Astrophysics Management Operations Working Group.

DISCUSSION

Neil Johnson

Is the Technology 21 initiative to be an increment to the SR&T program or a separate program?

Bunner

It's too early to say. We're talking about plans for Fiscal Year 1992. It's likely that, depending on the technology development involved, a combination of Astrophysics Division (OSSA) and Office of Aeronautics and Space Technology (OAST) support would be sought.

II. New Scientific Challenges
Paul C. Joss and Donald Q. Lamb, Chairmen

X-RAY EMISSION FROM NORMAL STARS

R. ROSNER

Enrico Fermi Institute
and Department of Astronomy and Astrophysics
The University of Chicago

ABSTRACT. With the closing of the *Einstein* and *EXOSAT* eras, the characteristics of stellar x-ray emission are now fairly well understood, but the more fundamental reasons why there is any x-ray emission at all is not at all well understood. Future x-ray missions will have to play a key role in unraveling this mystery.

1. Introduction

Given the brevity of space for this report, I have chosen to focus my discussion on the core problem of stellar activity studies: the ultimate cause of stellar x-ray emission, and its relation to the so-called "activity-rotation connection" and other observed correlations between activity level and stellar parameters. This topic indeed well-illustrates both the breadth of results now available and the depth of analysis now possible; furthermore, its observational elucidation defines many of the key experimental constraints on future x-ray missions aimed at stellar x-ray astronomy, and therefore provides a convenient point-of-reference for the purpose of this talk.

Now, the connection between stellar activity levels and stellar rotation rate was recognized early on to be central to our understanding of how activity for stars with outer convection zones comes about. Indeed, the idea that rotation and activity level were connected for such stars first derived from theoretical ideas about the nature of stellar magnetic activity (e.g., that the drivers of stellar magnetism are a combination of stellar rotation and surface convective motions); this idea, in combination with observations of the Sun during the *Skylab* era which showed a clear connection between the level of surface magnetic activity and the level of coronal activity (as measured by the level of coronal x-ray emission), naturally suggested a connection between stellar rotation and the level of stellar x-ray emission (see a fuller discussion of the history of these ideas in Vaiana and Rosner 1978 and in Rosner, Golub, and Vaiana 1985). This anticipated connection is of course eminently testable; and in fact was tested, starting with the early *Einstein* Observatory stellar surveys (cf. Vaiana *et al.* 1981; Pallavicini *et al.* 1981). This sort of study is not readily carried out for stars which are clearly very different from the Sun (such as early-type stars, evolved stars, and pre-main sequence stars); for such stars, our understanding of the origins of the activity seen from them is at a much more primitive level, and it is much more difficult to construct observational tests that can meaningfully constrain the available possible models. In the following, I will discuss some of the most recent results from such studies, and suggest future directions.

Of necessity, a substantial number of other interesting topics will either not be discussed here, or will only be briefly alluded to without the benefit of detailed discussion: These topics revolve about the nature of the coronal atmosphere itself — that is, is the solar analogy really an apt description of the coronae of other stars; what is the geometry or structure of these coronae; are there normal stars with x-ray emission, but no coronae (e.g., the early-type stars)? What occurs to stellar activity on the surface of evolved stars? It is in these areas that stellar x-ray spectroscopy comes to the fore, and forms an essential tool in distinguishing between competing models. This relative neglect is regrettable, but unavoidable; although I will mention these and related issues only in passing in the following, I will nevertheless take into account the observational desirata which follow from these studies in order to define future observational capabilities.

This paper is structured as follows: I first outline the main scientific issues, using recent key studies of stellar x-ray emission to exemplify the kind of work being done, and to highlight the unresolved issues; and then discuss the desired observational constraints, and the instrumental capabilities which follow from them. In writing this article, I have drawn extensively on previous reviews of this subject, principally from Rosner, Golub, and Vaiana (1985), Vaiana and Sciortino (1986, 1987), Rosner (1988), and Pallavicini (1989).

2. What are THE Determinants of Stellar X-ray Emission?

Starting from the early papers of Pallavicini *et al.* (1981), Walter (1982), and others, it was realized that the connection between rotation and activity level which standard stellar magnetic dynamo theory (cf. Parker 1979) had predicted in fact did exist — the x-ray luminosity scales as a power law with the stellar rotation rate. This connection holds, however, only for the so-called solar-type stars, that is, the late-type stars which show substantial evidence for surface convection zones. Thus, as shown by Pallavicini *et al.*, the early-type stars show no such relation between x-ray luminosity and rotation. The key questions for the late-type stars are then: What exactly is the universal scaling of activity on rotation rate, and is there any evidence for departure from this scaling at the extremes of stellar attributes on the main sequence (e.g., at the dA-dF spectral type boundary, where surface convection zones first appear, and at the terminus of nuclear burning at the low-mass end of the main sequence) and at the extremes of stellar rotation rate (or stellar age).

In contrast, the key question for early-type stars is at a much more elementary level: Why is there any x-ray emission at all? That is, in the case of late-type stars, the Sun forms a kind of Rosetta stone which allowed us to fix on a physical model for the origins of stellar x-ray emission, a model which I believe could not have easily been developed in the absence of the spatially-resolved solar observations. Such a Rosetta stone does not exist for the early-type stars; and hence we lack an immediate physical picture of the circumstances under which x-ray emission arises in the outer atmospheres of these stars. The only strong hint we have is the observed correlation between the level of x-ray emission and the bolometric luminosity of these massive stars (Pallavicini *et al.* 1981), which suggested a deep connection between the presence of keV-temperature matter and possible instabilities in the powerful winds known to emanate from these stars (winds whose mass loss rate has been known for some time to be related to the bolometric luminosity; cf. Lamers 1981, Lucy and White 1980, and Lucy 1982).

3. Is there a "universal" rotation-activity connection?

In a recent paper, Fleming, Gioia, and Maccacaro (1989) addressed one of these issues directly, namely the nature of the "universal" activity-rotation scaling law for late-type main sequence stars. The particulars are as follows: Using the Extended Medium Sensitivity Survey (EMSS) from *Einstein*, they constructed an x-ray-selected sample of 128 F-M dwarf stars, and combined this sample with optical observations (including spectroscopy) to establish the relations between spectral type, rotation rate, x-ray luminosity, etc. among these sample stars.

The key result is shown in their Figure 1, which on cursory inspection indicates that unlike the classical result of Pallavicini *et al.* (1981) and others, stellar x-ray luminosity L_x for this sample scales like $(v_e \sin i)^1$, rather than $(v_e \sin i)^2$, where v_e is the equatorial stellar rotation rate and i is the inclination angle of the stellar rotation axis to the line-of-sight. The question of exactly how L_x scales with the rotation rate Ω has of course been a rather hotly debated topic since the early results of Pallavicini *et al.* and Walter (1982); and it is therefore crucial to understand the present result, which emerges from the first x-ray-selected sample constructed to date for this purpose.

To begin with, it is important to note that since an x-ray-selected sample of the kind examined here by its very nature preferentially picks out the x-ray-bright sources, one way of reconciling the apparently conflicting results is if the intrisically brighter sources (e.g., the ones in the present sample) are simply showing the effect of stellar activity "saturation"; that is, it may be that at the upper end of the stellar luminosity function, the dependence of activity on rotation rate is weaker than it is for less-active stars, so that the "universal" scaling does not in fact hold over the entire range of observed rotation rates. This argument has been made most recently by Vilhu and Walter (1987). Can we determine whether this simple and intuitively-appealing interpretation is correct?

To answer this question, Fleming *et al.* considered the correlations between L_x, Ω, and stellar radius R; and found that (1) L_x is uncorrelated with $\Omega \sin i$, while (2) the x-ray luminosity instead scales with R^2. These latter results seem to deepen the mystery: It appears that for this well-determined sample of late-type dwarf stars, it is the stellar radius (or surface area) which sets the x-ray emission level, not the stellar rotation rate. Is there a simple explanation for the discrepancy between this result and earlier studies?

The following argument seems to resolve the difficulty (Rosner 1988). Since the sample we are considering is x-ray-selected, and indeed represents the upper tail of the L_x distribution, one might expect that the stars in this sample have mean surface fluxes which are comparable to those in solar active regions; this is in fact borne out by comparing L_x of stars in this sample with the expected L_x if these stars were fully covered by solar-like active regions (cf. Vaiana and Rosner 1978). In that case, the x-ray luminosity of stars in this sample can be 'predicted' simply by determining the stellar surface area for each star, and multiplying this area by the mean solar active region x-ray flux; that is, one expects a strong correlation between L_x and the square of the stellar radius for stars in this sample. This is exactly what is found by Fleming *et al.*

However, recall that $v_e = \Omega R$, and that Fleming *et al.* found L_x to be uncorrelated with $\Omega \sin i$ for this sample; the implication is then immediate that $L_x \approx v_e^2$, contrary to what is in fact found — Fleming *et al.* find instead a linear dependence of L_x on v_e. Is there then an internal inconsistency in the data analysis? I believe not. Careful reading of Fleming *et al.*'s analysis shows that the power law fit connecting L_x to $v_e \sin i$ is obtained by simply assuming that the upper limits on rotation rates $v_e \sin i\ <\ 10$ km/s can be regarded as 'true detections', with fractionally-large errors, and applying a least-squares fit to all of the data. The effect is to overweight (with respect to the upper bounds) the few 'detections' at large rotation rates and relatively low luminosities, and thus to underestimate the exponent in the power law. This can be avoided by instead applying the by-now standard detection-and-bounds correlation analysis (cf. Schmitt *et al.* 1985). Indeed, if one uses the data presented in Table 1 of Fleming *et al.* to compute the power law index, one finds that within the statistical errors of the data, the power law index connecting v_e to L_x for this sample cannot be distinguished from 2! This is remarkable: the exponent is 2 despite the fact that L_x for this sample is demonstrably *uncorrelated* with stellar rotation. I suspect that if the $L_x \approx v_e^2$ relation had been found first, then there would have been little reason to probe further, since this result would have confirmed the 'well-known' result that activity and rotation are correlated in precisely this manner; it is to Fleming *et al.*'s credit that they did in fact probe further, and now provide the first good evidence for saturation of stellar coronal activity.

This discussion points out the great benefit of analyzing results from x-ray selected samples of stars. Such samples are non-trivial to construct from the total *Einstein* data base; and a key objective of the *ROSAT* mission in the stellar area is going to be the construction of large samples of stars with reasonable uniform sensitivity limits.

4. The rotation-activity connection for evolved stars

A rather different perspective on the activity-rotation problem is provided by the recent completion of the *Einstein* survey of late-type giants and supergiants (Maggio *et al.* 1988). The total sample encompassed 380 stars, and was used in the first instance to resolve the problem of locating the "dividing line" separating coronally-active from inactive evolved stars in the H-R diagram. Having confirmed earlier work placing this boundary at $\approx$ spectral type K, Maggio *et al.* proceeded to consider the x-ray luminosity-rotation rate correlation for this stellar sample; it is of some considerable surprise that the sought-for correlation is remarkable weak.

The reasons for this surprise are of course the fact that for late-type dwarf stars, x-ray luminosity and rotation are well-correlated; and that there is a rather commonly-accepted and intuitively-appealing picture of how this correlation might come about. However, the data for giants serve to remind us that in some sense the data for dwarf stars are too 'neat'. That is, the common explanation for the activity-rotation connection is based on the idea that in standard α–ω dynamo theory, the rate of magnetic flux production depends on the rate of differential rotation, and hence (for fixed convection zone depth, for example) should scale with the rotation rate. What is commonly forgotten is this latter caveat: namely, in order for this correlation to be evident, it must be the case that the depth of the region over which the differential rotation field acts either does not vary with stellar type, or varies randomly with stellar type. Thus, the x-ray data from evolved stars tell us quite directly that, given that we are wedded to standard α–ω dynamo theory, the depth over which the ω-dynamo must act within the convection zone of these stars must be correlated with stellar type. This would seem to be a rather stringent constraint for dynamos for evolved stars; and is confirmed by Maggio *et al.*'s initial success in correlating L_x with the Rossby number (at least for F and early G giants).

This subject area is again one in which the data available from *Einstein* and *EXOSAT* are very sparse, especially if one wishes to construct samples with uniform x-ray sensitivity, or volume-limited samples. Thus, as in the case of the main sequence stars, a key task for the future will be to simple vastly enlarge the available set of detected objects, a task which will require imaging telescopes with considerable sensitivity.

5. The "Decay" of Stellar Activity at the Low-mass End of the Main Sequence

The possibility that the nature of stellar activity may change in some dramatic way for very low-mass stars has been the focus of detailed studies for a number of years, and plays an important role in subjects ranging from the galactic contribution to the diffuse soft x-ray background and stellar dynamo theory to the genesis of the "period gap" in cataclysmic variables (see review by Rosner, Golub, and Vaiana 1985). Very recently, the *Einstein* Extended Medium Sensitivity Survey (EMSS), already alluded to above, has been used by Fleming, Liebert, Gioia, and Maccacaro (1988) to attack this problem as well.

Table 1. Number of predicted and observed stars as a function of detection threshold [1]

	Volume [pc^3]	Predicted [4] # [Sp>M5]	Observed # [Sp>M5]
$\log <L_x> = 28.73$ [2]	3,870	91	0
$\log <L_x> = 27.98$ [3]	265	6	0

[1] Table adapted from Fleming *et al.* (1988).
[2] Mean L_x for young disk population stars (Bookbinder 1985), dM only.
[3] Mean L_x for old disk population stars (Bookbinder 1985), dM only.
[4] Using stellar mass function of Reid (1987).

Fleming *et al.* obtain two basic results from their study of the EMSS x-ray selected sample (consisting of some 31 dM stars, all optically identified). First, they show that the sample as a whole has a tendency for $<L_x/L_{bol}>$ to be roughly constant; second, they pointed out that no dM stars later than M5 were either detected or identified. The above table (extracted from Fleming *et al.*) summarizes the second result quantitatively, and shows quite explicitly that the absence of dM stars later than M5 is not likely to be a statistical fluke. Since this sample is x-ray-selected, it is clearly the most explicit demonstration of an effect first noticed in the dM star analysis of Bookbinder (1985), namely that there simply are no x-ray-bright stars later than M5 (there are certainly x-ray-emitting stars later than M5, but *all* of these stars detected to date show very modest (solar-like or less) x-ray emission levels.

The fundamental reason for this deficit of x-ray-bright stars remains a mystery. Rosner and Vaiana (1980) some time ago pointed out that for stars later than roughly M5, the presumptive change in stellar structure as a result of convection dominating energy transport everywhere within the star ought to be reflected in the efficiency with which such stars produce magnetic fields, and hence in how active they were; this would be especially the case if the dominant dynamo process for main sequence stars later than M5 were a "shell" dynamo located at the interface between the outer convection zone and the radiative core. This expectation is apparently borne out by the x-ray data; but unfortunately, we are no closer to understanding exactly in what way this change in stellar structure has brought about a *decrease* in stellar magnetic field production efficiency. That is, magnetic field production has certainly not ceased for the very low mass stars (since we still observe coronal emission), so that it cannot be simply a matter of 'shutting down' field production. Instead, one presumes that the basic dynamo process has been modified; this remains a problem for the future.

The key observational task here is to develop a much more substantial sample of dwarf stars with spectral type later that M5. In particular, a substantially larger sample will be needed in order to definitively test the idea that there are essentially no bright $(L_x > 10^{28.5}$ $ergs$ $s^{-1})$ M dwarfs later than M5. As in the case of the giants, the primary observational requisite to accomplish this goal is sensitivity.

6. Some other key questions

The above discussions only highlight the vigorous level of research activity in stellar x-ray astronomy today, and one cannot end this presentation without at least alluding to some of the other central unresolved questions. For this reason, I have summarized what are in my opinion the major remaining problems in Table 2, organized according to the type of star involved.

Table 2: *Some of the General Unanswered Questions*

Object Class	Question
Solar-like Stars	Why the strong correlation between L_x and Ω?
OB Stars	Coronae or shocked winds? Why is $<T>$ high, but $\frac{1}{4}$ keV extinction low?
RS CVn Stars	Do close binaries have intrinsically different magnetic activity cycles?
Evolved Stars	Why is there an "x-ray dividing line"?
Low-mass Stars	What happens to magnetic activity for fully-convective stars?
Pre-main sequence Stars	Is the observed activity "solar-like"?
dA Stars	Do they really not emit x-rays?

7. The Observational Requirements for Future Missions

We now turn to defining the instrumental capabilities which are needed in order to resolve the issues discussed above. For the sake of definiteness, I have set myself the task of trying to answer the questions posed in Table 2; and in the main, my calculations of the required instrumental characteristics are based on the assumption that the instrumentation must be capable of addressing these questions, and be within the realistic capability of the next generation of x-ray telescopes. In order to be as clear as possible, I have placed the information into two tables, the first (Table 3) organized by the type of star to be studied (as in Table 2), and the second organized by instrumental characteristics. In the main, the contents of these tables are self-explanatory.

Table 3: Summary of Desired Instrumental Capabilities

Object Class	Question
Solar-like Stars	Why the strong correlation between L_x and Ω?
OB Stars	Coronae or shocked winds? Why are the "coronal" temperatures as high as 1-1.5 keV? Why is the extinction below 0.5 keV so low in many OB x-ray sources? How extended are the emission regions?
RS CVn Stars	Do close binaries have intrinsically different magnetic activity cycles? What is the spatial relation between the low and high temperature emission components?
Evolved Stars	Why is there an "x-ray dividing line"? What is its relation to other dividing lines?
Low-mass Stars	What happens to magnetic activity for fully-convective stars?
Pre-main sequence Stars	Is the observed activity "solar-like"? Are flares an important component of coronal heating for these stars?
dA Stars	Do they really not emit x-rays? Can a Jupiter-like auroral model work if x-ray emission is detected?

Table 4: Instrumental Characteristics

Characteristic	Capability	Key Defining Task
Sensitivity	10^{-15} ergs s^{-1} cm^{-2}	Luminosity function tails at low L_x
Field-of-view	$> (1\ \text{degree})^2$	Stellar associations
Angular resolution	$< 2"$	Source confusion in associations
Time resolution	$< 1\text{-}5$ minutes	Flares in dK-dM stars
Energy band	0.05 - 10 keV	X-ray emission from dA stars (very soft) to RS CVn's (hardest)
Energy resolution	< 0.5	2-temperature coronal models
	< 0.01	Detailed spectroscopy

Finally, it is amusing and of some interest to consider what the observational desirata would be if one were basically unconstrained by technology. In my view, it is *angular resolution* which is the singular instrumental characteristic whose improvement could change our understanding of stellar x-ray emission in a fundamental, qualitative, way; this is illustrated in the following Table 5.

Table 5: Angular Resolution

Resolution (")	Science Impact
0.1	Resolve many dM binaries, allowing one to distinguish activity properties of single and binary stars
0.01	Resolve the wind "bubble" surrounding nearby early-type stars, and determine the gross geometry of the x-ray emission region
0.001	Resolve the coronae of nearest stars; separate distinct emission measure components

Many of the items on these "wish lists" will be provided by the new generation of x-ray telescopes, starting with *ROSAT*; and as suggested by W. Cash at this meeting, even the more extreme objectives listed in Table 5 may be attainable with technological developments over the next decade. This of course bodes extremely well for stellar x-ray astronomy.

Acknowledgments: The work reported here was largely supported by NASA grants, including several *Einstein* Observatory Guest Observer grants to the University of Chicago.

References

Bookbinder, J. 1985, PhD Thesis, Harvard University.
Fleming, T.A., Gioia, I.M., and Maccacaro, T. 1989, *Ap. J.*, in press.
Fleming, T.A., Liebert, J., Gioia, I.M., and Maccacaro, T. 1988, *Ap. J.*, **331**, 958.
Lamers, H.J.G.L.M. 1981, *Ap. J.*, **245**, 593.
Lucy, L.B. 1982, *Ap. J.*, **255**, 286.
Lucy, L.B., and White, R.L. 1981, *Ap. J.*, **241**, 300.
Maggio, A., Vaiana, G.S., Haisch, B., Stern, R.A., Bookbinder, J., Harnden, Jr., F.R., and Rosner, R. 1990, *Ap. J.*, in press.
Pallavicini, R., *et al.* 1981, *Ap. J.*, **248**, 279.
Pallavicini, R. 1989, *Astron. Ap. Rev.*, **1**, 177-207.
Parker, E.N. 1979, *Cosmical Magnetic Fields* (Oxford: Oxford Univ. Press).
Reid, I.N. 1987, *M.N.R.A.S.*, **225**, 873.
Rosner, R. 1988, in *IAU Highlights*, ed. E.R. Priest.
Rosner, R., Golub, L., and Vaiana, G.S. 1985, *Ann. Rev. Astron. Ap.*, **23**, 413-52.
Rosner, R., and Vaiana, G.S. 1980, in R. Giacconi and G. Setti (eds.), *X-ray Astronomy* (Dordrecht: D. Reidel Publ.), pp. 129-51.
Schmitt, J.H.M.M., *et al.* 1985, *Ap. J.*, **290**, 307.
Walter, F.M. 1982, *Ap. J.*, **253**, 745.
Vaiana, G.S., *et al.* 1981, *Ap. J.*, **245**, 163.
Vaiana, G.S., and Rosner, R. 1978, *Ann. Rev. Astron. Ap.*, **16**, 393-428.
Vaiana, G.S., and Sciortino, S. 1986, *Adv. Space Res.*, **6**(8), 99.
Vaiana, G.S., and Sciortino, S. 1987, in *Circumstellar Matter*, eds. I. Appenzeller and C. Jordan (Dordrecht: D. Reidel), p. 333.
Vilhu, O., and Walter, F.M. 1987, *Ap. J.*, **321**, 958.

DISCUSSION

Steven Kahn

If you could eventually have instrumentation with unlimited area, resolution, etc., what sort of measurements should be made to really determine the relation between dynamo activity and the X-ray emission?

Rosner

The key point of looking at *stellar* activity from this point of view is that this allows us to "vary" stellar parameters which cannot be varied for the Sun: rotation rate, effective surface gravity, color, age, etc. The difficulty to date has been that the available data are difficult to analyze from a statistical point of view: The statistically complete samples are small, even through the actual number of detected stars is large — the reason is that most stars seen by EINSTEIN, for example, were serendipitous detections, and hence the sensitivity threshold for the overall EINSTEIN stellar survey is extremely non-uniform. The resulting small statistically complete samples are then really too small to allow definitive conclusions regarding the correlation of X-ray luminosity with parameters of late-type dwarf stars *other* than rotation.

Gerald Share

Could some of the scatter in the X-ray emission vs. rotation rate be explained by flares?

Rosner

If you mean individual flares, rather than a superposition of (individually unresolved) small flares, the answer is no, one cannot appeal to flares in order to explain the scatter in L_x seen in the L_x vs. $vsini$ diagram. The reason is that typical (viz., median) exposure times for the stars in such plots are sufficiently long — e.g., typically 1000-2000 sec — that if a flare occured during the observing interval, one would have seen the variation explicitly. That is, flare decay times for *observed* stellar flares are of the order of the observing intervals, and thus one should recognize flares even if they are observed during the period in which they show the slowest variation, namely the decay phase.

Kent Wood

To return to the question of flares, we know their number increases as we go to smaller flare sizes. Could the excess X-ray emission be from the merged contribution of many smaller flares?

Rosner

The effect you speak of has been known of in solar X-ray astronomy since the early 1970's, and indeed Bob Liu (UCB) suggested, on the basis of his balloon data, that such microflares may constitute an important component of coronal heating. Similar arguments have been made over the past five years in the stellar case, but careful statistical analyses by Sciortino et al. of EINSTEIN stellar data show that the stellar X-ray data cannot in fact be used to argue for microflare coronal heating. Hugh Hudson also has reminded me that the same authors have also shown that extrapolation of Bob Lin's flare frequency spectrum to lower flare energies leads to an insufficient coronal energy input. Thus, in the absence of a steepening of the frequency spectrum below present sensitivity limits, there is no good evidence for an important role of (small) flares in coronal heating.

(Unknown)

Do you view X-ray polarimetry as being much of a help on these questions?

Rosner

Polarimetry is a terrific tool for understanding the source geometry. As such, it has a particularly useful role to play in solar observations, especially in unravelling the spatial structure of solar flares. In the stellar case, I'm not so certain. This is principally because polarimetry data is most useful when one also has X-ray images — a spatially-resolved picture of the flaring region — which are not going to be available in the foreseeable future. The one exception I can imagine at this point is the case of RS CVn stars: in this case, the use of eclipse data for the high and low temperature components, together with polarimetry data gathered as a function of orbital phase, might well improve our understanding of RS CVn coronae. Perhaps there are other such exceptions.... I'm sure it's well worth thinking about.

Webster Cash

If X-ray imaging were sufficiently good to resolve the coronal structure of normal stars, would this help solve the listed fundamental problems?

Rosner

My first instinct is to say no, but upon further reflection, I think that you may be onto something. What I have in mind is that the high-temperature (e.g. "active sun") component of coronal emission typically occurs in a restricted latitude band, whose position varies systematically during the course of the solar cycle. This positional variation is just the X-ray counterpart of the classic sunspot "Butterfly Diagram," i.e., the systematic eruption of solar-active region magnetic complexes closer and closer to the solar equator as the activity cycle passes through its maximum phase. Now, the pattern of flux emergence *is* a diagnostic for the nature of magnetic dynamos, and hence the answer to your question is really, yes! I note in this connection that some diagnostic capability would also be called for, since this would help in disentangling the (hotter) active component from the (relatively cooler) quiescent coronal X-ray background.

Claude Canizares

I am puzzled by your comment that the dynamo theory will be sorted out primarily by surveys that add more points to the L_x vs. Ω plot. Don't you need additional parameters from the X-ray data (such as T_x) as well as UV, etc., to sort out the scatter in this plot?

Rosner

Perhaps the best way of answering your question is to turn it around: What additional constraints on dynamo theory would we add by, for example, knowing the coronal temperature? With one crucial exception, I believe that such data do not add any constraints whatever. This exception refers to the fact that for the Sun, there is a quite noticeable difference in coronal temperature between the "Quiet Sun" (which is present throughout the solar activity cycle) and the "Active Sun" (which is due essentially to emission from active regions, and thus waxes and wanes with the solar cycle). For this reason, relatively low-energy resolution ($\Delta E/E \lesssim 0.5$) imaging *is* an important desideratum. Why do further details not matter? The reason is simply that the X-ray luminosity turns out to be such an excellant proxy for measuring the rate of magnetic flux emergence, better indeed than any other measuring tool other than direct flux detection. Why this is so is of course *the* unanswered question — the point of my talk.

TIMING OBSERVATIONS OF NEUTRON STARS AND BLACK HOLES

K.S. Wood
E.O. Hulburt Center for Space Research
Naval Research Laboratory
Washington DC 20375

ABSTRACT

X-ray timing studies on neutron stars and black holes need large collecting areas, >> 10 m2, to permit millisecond and sub-millisecond studies. Physical processes near neutron star surfaces and general relativistic effects could be seen. The key requirement is collecting many photons rapidly, by viewing bright sources with a large aperture. Lack of a requirement for X-ray mirrors permits rapid realization of such a capability.

Neutron star and black hole studies based on X-ray timing can make dramatic progress in coming decades given appropriate instrumentation. Timing applied to sources such as Her X-1, Cyg X-1, the Rapid Burster has produced major discoveries during the past two decades of X-ray astronomy. Further increases in instrumental power will extend studies of known phenomena and bring additional discoveries. The principal benefits will be realized simply by making a large increase in collecting area. We may be on the threshold of a quantum jump in the size of space platforms, hence would be foolish to miss the scientific potential of having instrumental apertures keep pace with that anticipated growth or to ignore the fact that it will soon become possible to build instruments too large to have been carried to orbit fully assembled.

Engineering and mission-specific aspects are covered in a companion paper in this volume (Wood, 1990). Here we consider how a very large X-ray-sensitive area (>> 10 m^2) would assist neutron star and black holes studies. These objects have great importance to astrophysics, deriving largely from their role as testbeds or laboratories for physics. Extreme densities, high temperatures and strong magnetic fields are inferred in neutron stars. Some conditions found there are found nowhere else in nature. Together, neutron stars and black hole candidates give a unique look at regions of strong-field gravity. Dynamical timescales for physical processes in near surfaces of neutron stars and or near the Schwarzschild radius in disks around black holes are very short, milliseconds or shorter. Effective probing of processes in these regions will result from gathering many photons on those timescales. This means collecting tens of photons in milliseconds to microseconds, which in turn requires large areas. It is crucial to understand that (i) this is not an field where seeing great numbers sources at lower flux levels is necessarily fruitful, and (ii) in many of the most important scientific objectives integration time is no substitute for collecting area. The most exciting future possibilities come from achieving photon-rich X-ray astronomy, by viewing the brightest objects in the sky with a very large aperture.

Once the merit of large aperture for timing is appreciated, there is seen to exist a programmatic gap. Timing is poorly covered in present plans. After XTE there is no timing mission planned by anyone, to the end of the century. We should not let XTE become the endpoint of X-ray timing, but instead plan a follow-on mission that would have area >> 10 m^2.

TIMESCALES AND LEVELS OF MODULATION

Development through the 1970s and 1980s of X-ray timing can be understood as discovery of effects at ever-shorter timescales and lower levels of modulation, most of them not anticipated in advance. There has also been increasing emphasis on aperiodic effects and variability patterns that are not coherent over long times. Great physical interest attaches to these subtle and complex variability signatures.

Some highlights deserve mention. At the start of the 1970s the only known X-ray pulsar was the Crab. The period from Uhuru through HEAO-1 saw discovery of the binary pulsars and the use of pulse timing to determine neutron star masses. Rapid flickering in the black hole candidate Cyg X-1 was found in this period and X-ray bursts caused by thermonuclear flashes on neutron stars were discovered. Timing discoveries continued impressively through the decade of the 1980s, with many results from Exosat and Ginga. Prior to 1980, ~ 50 periods associated with binary X-ray sources were known (i.e., spin periods of neutron stars or white dwarfs, orbital periods, and longer periods that might be disk precession). Today the number stands near 200. Orbital periods for sources without pulsations began to be determined using dips and occasionally eclipses. The most common neutron star binaries -- bright objects seen near the center of our galaxy -- began to be understood for the first time in the latter half of the decade. The relationship to improvement in methodology is perhaps most clearly seen in the fact that quasiperiodic oscillations (QPOs) in Sco X-1, the brightest X-ray source, were first seen only in 1985, whereas the source had been extensively studied since 1962 by many observers. Other phenomena first found in the 1970s were studied with new levels of refinement. There were studies of the fluctuations in the periods of accreting neutron stars and new methods for studying rapid variability in black hole candidates.

Future goals in X-ray timing can be quantified largely in terms of timescales and modulation depths, which readily relate to instrumental capabilities. Two categories of frontier phenomena come up for discussion, (i) those that represent extrapolation of known phenomena into parameter space domains that cannot now be reached in practice and (ii) future discoveries, about which we have only theoretical speculations and which likewise fall in inaccessible domains. The following table compares some neutron star phenomena from both categories, illustrating timing parameter space issues as well as some aspects of the associated astrophysics.

TABLE 1

Phenomenon	Timescale	"Altitude"	modulation depth
Binary pulsar spin	1-100 s	> 1000 km	50%
QPO	50 ms	100 km	few %
Millisecond pulsar	1 ms	< 1 km	$<10^{-3}$
NS vibration	~1 ms	surface	?
Photon bubble	. 1 ms?	1-10 km?	30%?

The spin period of a binary pulsar has a timescale characteristic of the orbital period at the Alfven radii of highly magnetic objects, 1000 km or more. The high level of modulation and high coherence of the periodicity mean this phenomenon is relatively easy to discover. Neither advantage exits for

QPOs. Seeing them requires gathering the photons in a substantially shorter timescale, in numbers sufficient to permit work at a much lower level of modulation. Large area and improved telemetry (in pointed instruments) made possible their detection in the 1980s. The QPO is also associated with modulation introduced at a lower altitude, ~100 km above the surface.

Seeing processes still closer to the surface will often require work at yet shorter timescales and lower levels of modulation. That generalization encompasses a number of phenomena that have been described theoretically and for which only upper limits or tantalizing suggestions exist observationally. Millisecond binary X-ray pulsars are one example. They have been predicted on the basis of several different lines of argument. In each case the level of modulation expected is small; it is quite difficult to produce a millisecond pulsation in an accreting neutron star with high modulation depth. But the integration time of the observation used for discovery purposes must be kept short enough that the period is not smeared by the frequency modulation (Doppler shift) associated with binary orbital motion. Using the largest available areas it has at best been possible to reach modulation depths of at best several x 10^{-3}. The level we need to reach is very likely 10^{-3} or 10^{-4}. Neutron star vibrations are another possible effect whose detection is hampered by the same problems, because the expected periods are similar and the Doppler issue remains the same.

NEW PHYSICS

New phenomena can attract interest from other scientific communities. The relativistic gravitation community is a prime candidate, because strong-field gravitational phenomena can appear as central regions of compact sources are probed. One example is described here and others may be found in Wood and Michelson (1988).

Rapidly rotating neutron stars become secularly unstable when subjected to viscous dissipative forces or gravitational radiation reaction (Chandrasekhar 1970, Friedman and Schutz 1978). Modes that grow via gravitational radiation are damped by viscosity and vice versa. This process, often referred to as the Chandrasekhar-Friedman-Schutz or CFS instability can be excited when a neutron star with a weak magnetic field is spun up by accretion from a binary companion. Accretion torque drives the star into the unstable regime, whereupon it becomes a gravity wave source. (Wagoner 1984). It reaches an equilibrium wherein angular momentum lost by gravitational radiation equals that gained from accretion. Most of the energy release associated with accretion comes in X-rays. Sco X-1 and most other low-mass X-ray binaries are candidate sites for this effect. The X-ray flux is expected to be weakly modulated at the same frequency as the gravitational radiation. This situation constitutes one of the several possibilities for millisecond pulsations mentioned above. General reasons given there for the need for large areas to detect millisecond pulsations apply here in particular. Detection of such a source in X-rays facilitates detection in gravitational waves, since a gravity wave detector can be built to be resonantly responsive at the frequency found in X-ray. Dual channel detection would provide two measures of the neutron star distortion and would be tremendously important.

Another community that would benefit from large areas consists of numerical physicists modeling flows and bursts with radiation hydrodynamics codes. Such

codes exist now and will develop over the next decade with improved
computational power. They could be used to model the rise of an X-ray burst
predicting how the spectrum evolves as the released energy interacts with the
stellar envelope. Observations needed to test such models would consist of
many repeated spectral determinations during burst rise, one spectrum every
few milliseconds. This calls for apertures $\gg 10$ m^2. Another radiation
hydrodynamics example was presented by R. Klein and J. Arons at the recent
23rd ESLAB symposium (Klein and Arons, 1989). They describe how super-
Eddington accretion near neutron star polar caps can give rise to "photon
bubbles", regions of low matter density with a high radiation density of 10
keV photons. Bubbles become buoyant and may rise in the accretion column or
move laterally to the edge of the flow to become observable. Calculations to
date suggest ~30% modulation at timescales 10^{-4} s might be seen. Only a 10^6
cm^2 array could see such an effect, and then only in the brightest pulsars.

Millisecond pulsars and photon bubbles are two examples of possible future
discoveries that lie in presently unobservable domains of short timescale or
low modulation depth. There are also many well-established phenomena where
area limitations are recognized. A list would include many kinds of QPO
studies (trying to see strong QPOs in the time domain, or faint QPOs near
transition points where the QPO is lost and then appears to change character),
period fluctuation studies in binary pulsars, time-resolved spectra of X-ray
bursts, temporal and spectral resolution of 3-millisecond flickering seen by
HEAO-1 in Cyg X-1 (Meekins et al., 1984), and time-resolved studies of dips
and eclipses. (See Wood and Michelson (1988) for further discussion.)

CONCLUDING REMARKS

The potential of millisecond and submillisecond X-ray timing should be
recognized and the gap in future plans should be addressed. Neutron star and
black hole candidate sources from our galaxy and the Magellanic Clouds are so
bright that background considerations are no concern. Just as in the timing
missions in the 1 m^2 class (HEAO-1, Ginga, XTE), collimators can provide
adequate source isolation. Mirrors bring no great benefit and lose important
higher energy photons that an array without mirrors can easily detect. If we
do not insist on mirrors then a large array can be essentially a 2-dimensional
planar structure covered with some appropriate detector element. It is
important to initiate serious discussions and feasibility studies to look into
the possibility that areas $\gg 10$ m^2 could be achieved at the Space Station or
at some other large platform, sooner rather than much later.

REFERENCES

Chandrasekhar, S. 1970, Phys. Rev. Lett., **24**, 611.
Friedman, J.L., and Schutz, B.F. 1978, Ap.J., **222**, 281.
Klein, R.I, and Arons, J., 1989, preprint.
Meekins, J.F., et al., 1984, Ap. J., **278**, 288.
Wagoner, R.V., 1984, Ap. J., **278**, 345.
Wood, K.S, 1990, these proceedings.
Wood, K.S., and Michelson, P.F., 1988 in _International Symposium on Experimental Gravitational Physics_, eds. P.F. Michelson, Hu Enke, and G. Pizella (World Scientific Publishing: Singapore), p. 475.

CYCLOTRON LINES: THE NEXT 100 YEARS

Ira Wasserman

Center for Radiophysics and Space Research
Cornell University, Ithaca, N.Y.

Cyclotron lines have now been observed, with varying degrees of reliability, in the spectra of a number of X-ray pulsars (Trumper et al. 1978, Voges et al. 1982, Wheaton et al. 1979, White, Swank and Holt 1983, Clark et al. 1989, Mihara et al. 1990) and γ-ray bursts (Mazets et al. 1981, Hueter 1984, Murakami et al. 1988, Fenimore et al. 1988, Wang et al. 1989). Obviously, cyclotron lines furnish the most direct evidence for the association of strong magnetic fields, $B \sim 10^{12}$ Gauss, with these phenomena, reinforcing the identification of X- and γ-ray sources with high energy emission in the vicinity of neutron stars. Moreover, cyclotron lines, like other spectral features, may provide additional diagnostic information on the physical conditions in the line-forming region, thereby imposing significant constraints on models for the overall emission process.

Unfortunately, the extraction of line properties from broad band X- and γ-ray spectra is highly model dependent (e.g. Holt and McCray 1982, Fenimore et al. 1988). Conventionally, various analytic models are compared (after convolution with instrumental response functions that may be only incompletely known) with the raw data. (For an alternative approach, see Loredo and Epstein 1989.) Although reasonable, this procedure can result in potentially inaccurate (and even inappropriate) impressions of the physical characteristics of the line-forming region. The problem is most acute when only a single cyclotron line feature is detected, in which case emission and absorption line interpretations may fit the data equally well, but with different fieldstrengths, linewidths and underlying continua (e.g. Voges et al. 1982).

Higher harmonic features can remove the ambiguity somewhat, because the nearly harmonic spacings of cyclotron transition energies tightly restrict the set of acceptable line center positions. Moreover, because electric dipole transitions dominate the radiative decay of excited Landau levels for $B \ll m_e^2 c^3/e\hbar \approx 4.4 \times 10^{13}$ Gauss, higher harmonic photons are effectively destroyed after one or two resonant scatters. As a result, higher harmonic features closely resemble "true absorption" lines whose properties depend only on the line absorption profiles (e.g. Fenimore et al. 1988, Wang et al. 1989 and references therein), but it is impossible to model the cyclotron fundamental in this way. There is, as yet, no convenient analytic description for the appearance of the cyclotron fundamental valid for a broad range of physical conditions, and very likely there will never be such a description. Truly self-consistent modelling of X- and γ-ray spectra with cyclotron line features requires numerical construction of theoretical spectra on the one hand and numerical comparison of the simulated and actual data on the other.

With the advent of fast processors and supercomputing, this procedure can now be carried out. Recently, Wang et al. (1989) published fits of a particular physical model for cyclotron line formation to data from GB880205 (Murakami et al. 1988). The model assumed that the γ-ray burst continuum is, by and large, formed "below" the line-forming region, in which a cool, quasi-thermal population of electrons resides. The electrons are kept cool by resonant Compton scattering, which tends to maintain an electron temperature $T \approx 3.1 B_{12}$ keV (Lamb, Wang and Wasserman 1990). Thus the temperature of the line-forming region was not a free parameter in this model, but the column depth, fieldstrength and viewing angle were. The model successfully accounted for the approximately equal strengths of the fundamental and first harmonic cyclotron features detected in GB880205, a nontrivial consequence of the detailed radiative transfer that cannot be understood in any simple absorption line picture. In spite of its success, the Wang et al. (1989) model is not "proven" true; distinct physical models may be able to reproduce the observed line strengths equally well. However, any uncertainty in the applicability of the fits is now in the basic *physical* assumptions of the model, not in *mathematical* choices for (analytic) line shapes adopted for computational convenience. Thus, the uncertainty is moved to the basic physics at the heart of the line-forming process, which is what we aim to study in the first place!

Given the finite human lifespan, it is tempting to anticipate somewhat the directions in which such a procedure, and the insights it engenders, will propel the study of X-ray sources and γ-ray bursts. To do so let us focus on γ-ray bursts. As a straw man, let us adopt the Wang et al. (1989) model, in which continuum and line formation are presumably physically distinct problems. This point of view is reasonable, at first sight, given the fact that the narrowness of the observed lines requires relatively low temperature scattering electrons ($T \lesssim 10$ keV) that cannot be responsible for continuum emission up to photon energies $\gtrsim 1$ MeV. To proceed further, we shall need data. Instead of inventing numbers, let us look some up in the literature, specifically the old Mazets et al. (1981) data, which furnished the earliest (and much-criticized) evidence for cyclotron lines in γ-ray burst spectra.

To use the Mazets et al. (1981) data, let us first introduce three (perhaps justified) assumptions. First, let us assume that the line features claimed by Mazets et al. (1981) were real. Second, let us assume that the line features were actually first harmonic ($n = 2$), not fundamental ($n = 1$), cyclotron lines. Since the Konus experiment was only sensitive above ~ 30 keV, Mazets et al. (1981) may have been unable to detect fundamental cyclotron lines at modest fieldstrengths ($B_{12} \lesssim 3$). Moreover, Lamb, Wang and Wasserman

Table I

Date	B_{12}	$D_{max}(pc)$	L_{max}/L_E	T_N(keV)
7 Mar 79	1.9	220	0.033	0.42
1 Nov 79	2.8	240	0.027	0.38
12 Jun 79	2.0	340	0.036	0.41
12 Apr 79	2.2	360	0.031	0.41
10 Jun 79	1.7	290	0.043	0.43
29 Mar 79	2.2	260	0.031	0.41
24 May 79	1.2	490	0.057	0.49
31 Jul 79	2.4	310	0.027	0.41
6 Oct 79	2.7	500	0.033	0.38
22 Jun 79	2.2	350	0.031	0.41

(1990) argue that it may be natural to detect two and perhaps three cyclotron lines in γ-ray bursts since, they estimate, the thickness of the cool line-forming region is likely to be ~ 10^{21-22} electrons-cm^{-2}, corresponding to first harmonic optical depths $\tau_2 \sim 1 - 10$. (The line equivalent widths given by Mazets et al. 1981 are generally consistent with $\tau_2 \sim 1 - 10$.) Third, let us adopt the continuum spectral fits published by Mazets et al. (1981), assuming in addition that it is correct to extrapolate the fitted continua to energies in the neighborhood of the (unseen) first harmonic.

Table 1 summarizes information that can be derived from the Mazets et al. (1981) data in the Wang et. al. (1989) model. Limiting distances can be derived from the requirement that the resonant line radiation pressure be insufficient to expel the line forming region. Assuming that the lines are produced in a layer near the surface of a neutron star that is gravitationally bound, Lamb, Wang and Wasserman (1990) show that the distance D is bounded by

$$D(pc) \lesssim 770 B_{12}^{-1/2} \left[\frac{F_{obs}(E_B) E_B}{1 \, cm^{-2} s^{-1}} \right]^{-1/2} \frac{\tau_2^{1/2}}{[\ln \tau_{eff}]^{1/4}} \left(\frac{\Delta\Omega}{4\pi} \right)^{1/2} \quad (1)$$

where $\tau_{eff} \sim \tau_1$, the (average) optical depth of the cyclotron fundamental, $F_{obs}(E_B)$ is the observed photon flux per unit energy at the cyclotron fundamental, and it is assumed that both the continuum and line forming regions occupy a total solid angle $\Delta\Omega$ on the neutron star surface. Equation (1) assumes that the mass of the layer is dominated by protons, and holds for neutron star mass and radius $1.4 M_\odot$ and 10 km, respectively. Given the distance bound, equation (1), we can compute an upper limit to the ratio of the continuum luminosity to the Eddington luminosity L_E (or more appropriately, the continuum flux at emission to the Eddington flux). In addition, we can define an effective "number temperature",

$$T_N(keV) \approx 0.074 \left[\left(\frac{A_0}{1 \, cm^{-2} s^{-1}} \right) D_{100}^2 \left(\frac{\Delta\Omega}{4\pi} \right)^{-1} \ln \left(\frac{E_F}{E_0} \right) \right]^{1/3} \quad (2),$$

which is the temperature of a blackbody emitting photons at the observed number flux at distance $D = 100 D_{100}$ pc, assuming a continuum spectrum of the form $F_{obs}(E) = A_0 E^{-1} \exp(-E/E_F)$ above a low energy cutoff E_0. To fill in the entries in Table 1, the arbitrary assignments $\tau_2^{1/2}/(\ln \tau_{eff})^{1/4} = E_0 (keV) = \Delta\Omega/4\pi = 1$ have been made; in practice, it should be possible to fit for these parameters.

In general, the numbers in Table 1 support the notion that γ-ray bursts are relatively local (e.g. Hartmann, Epstein and Woosley 1989, Paczynski 1989 and van Paradijs 1989). From data like these, a global picture of the neutron star distribution in the Galaxy is likely to emerge, and a deeper understanding of the evolution of neutron star magnetic fields ought to result. In addition, the relatively low values of T_N suggest that γ-ray bursts may also manifest themselves at rather low energies ($\lesssim 1$ keV), and that the hard X-ray spectra of γ-ray bursts may result from Compton upscattering of soft, ambient photons by energetic electrons.

At first glance, it seems likely that observations of cyclotron lines in γ-ray burst spectra may significantly advance our understanding of neutron star astrophysics. It is entirely possible that γ-ray burst activity is more common to neutron stars than either radio pulsation (which requires rapid rotation and/or strong fields) or X-ray emission (which requires membership in a [close] binary). Moreover, the formation of cyclotron line features *may* be somewhat decoupled from the formation of the hard X-ray continuum in γ-ray bursts, as in the Wang et al. (1989) picture.

By contrast, cyclotron line formation in accreting X-ray sources is a broad-band phenomenon, as line equivalent widths may be comparable to line center energies for characteristic atmospheric stopping depths (e.g. Wasserman and Salpeter 1980, Miller, Salpeter and Wasserman 1989). Indeed, it is probably impossible to adequately model the continuum spectra of strongly magnetized, accreting neutron stars without including the effects of cyclotron resonances (Pravdo et al. 1979, Meszaros and Nagel 1985, Wang, Wasserman and Salpeter 1989, Clark et al. 1989 and Mihara et al. 1989). Detailed spectra of X-ray sources (such as Her X-1 and 4U1538-52) in which *single* line features—attributed to the fundamental cyclotron transition—have been observed should reveal higher harmonic features at energies $\lesssim 100$ keV. Since $n > 1$ cyclotron features resemble true absorption lines, the observation of higher harmonics would yield valuable information for modelling the X-ray emission from these sources.

This research was supported in part by NASA Grant No. NAGW-666 and NSF Grant No. AST89-13112 to Cornell University.

REFERENCES

Clark, G.W. et al. 1989, preprint.
Fenimore, E.E. et al. 1989, *Ap. J. Lett.*, **335**, L71.
Hartmann, D., Epstein, R.I. and Woosley, S.E. 1989, preprint.
Holt, S.S. and McCray, R. 1982, *Ann. Rev. Astr. Ap.*, **20**, 323.
Hueter, G.J. 1984, in *High Energy Transients in Astrophysics*, S.E. Woosley, ed. (New York, A.I.P.), p. 373.
Lamb, D.Q., Wang, J.C.L.W. and Wasserman, I. 1990, in preparation.
Loredo, T.J. and Epstein, R.I. 1989, *Ap. J.*, **336**, 896.
Mazets, E.P.S. et al. 1981, *Nature*, **290**, 378.
Meszaros, P. and Nagel, W. 1985, *Ap. J.*, **298**, 147.
Mihara, T. et al. 1989, preprint.
Miller, G.S., Salpeter, E.E., and Wasserman, I. 1989, *Ap. J.*, **346**, 405.
Murakami, T. et al. 1988, *Nature*, **335**, 234.
Paczynski, B. 1989, preprint.
van Paradijs, J. 1989, preprint.
Pravdo, S.H. et al. 1978, *Ap. J.*, **225**, 988.
Trumper, J. et al. 1978, *Ap. J. Lett.*, **219**, L105.
Voges, W. et al. 1982, *Ap. J.*, **263**, 803.
Wang, J.C.L.W. et al. 1989, *Phys. Rev. Lett.*, **63**, 1550.
Wang, J.C.L.W., Wasserman, I. and Salpeter, E.E. 1989, *Ap. J.*, **338**, 343.
Wasserman, I. and Salpeter, E.E. 1980, *Ap. J.*, **241**, 1107.
Wheaton, W.A. et al. 1979, *Nature*, **282**, 240.
White, N.E., Swank, J.H. and Holt, S.S. 1983, *Ap. J.*, **270**, 711.

DISCUSSION

Paul Joss

You didn't give a precise definition for T_{eff}, but the expression you displayed seemed to have a formal (though gradual) divergence in the limit as T_{eff} approached infinity. This seems anti-intuitive. Would you explain?

Wasserman

(1) $T_{\text{eff}} \sim T_1$, the optical depth of the cyclotron fundamental. (2) The formula only holds for $T_{\text{eff}} > 1$ but line wings optically thin. (3) At large enough depths so the wings are optically thick, the line force might still increase, at first, but in practice continuum scattering takes over eventually, so the line force levels off. Therefore, the line force is finite as $T_{\text{eff}} \to \infty$.

Neil Gehrels

What are the line widths in your models?

Wasserman

The line widths for the X-ray sources are large, because the line wings are optically thick. Equivalent widths of order of the line center energies are reasonable, so interaction of line and continuum transfer is likely. Line widths for γ-ray bursts are smaller.

GAMMA RAY ASTRONOMY AND BLACK HOLE ASTROPHYSICS

Edison P. Liang

Lawrence Livermore National Laboratory,
University of California
and
Stanford University

One of the most exciting recent development in black hole astrophysics is the discovery by the HEAO-3 experiment of transient enhanced MeV gamma ray emissions from Cygnus X-1 and the Galactic Center, and the potentially correlated appearance of the 511 keV annihilation line from the Galactic Center (Ling et al 1987, Riegler et al 1985). If these soft gamma ray bumps and annihilation lines can be proven to be universal signatures of black hole accretions, then they will provide a new window on the overall black hole phenomenon from stellar mass black holes to AGNs. Hence it is important for future high energy astrophysics space missions to incorporate the study of soft gamma emissions from black hole candidates as a major component of their scientific goals.

Fig. 1 reproduces the spectra of Cyg X-1 as detected by HEAO-3 in 1979-80. The MeV gamma ray bump is apparent only during the gamma-1 state lasting less than two weeks when the hard x-ray intensity is lowest (cf. Ling et al 1987). Note that the bump is extremely hard (photon spectral index ~ 0) with a very sharp cutoff above ~2 MeV. This is in contrast to the nonthermal power-law gamma ray tails reported for many compact objects and gamma ray bursts. It strongly hints at a thermal origin. To date, the most natural and simplest model for the emission of this bump is that proposed by Liang and Dermer (1988). This model invokes the swelling of the innermost accretion disk into a quasi-spherical pair-dominated hot cloud of temperature ~ mc^2 where m is the electron rest mass, and the bump is due to a combination of Comptonized bremsstrahlung and annihilation radiation. First principle calculations of the spectral output by Liang and Dermer (1988) show that the best fit cloud parameters are: radius ~ 300 km (~ 10 Schwarzschild radii for a 10 $M_\odot$ black hole), temperature ~ 0.8 mc^2, radial Thompson depth τ_T ~ 2, pair density $n_\pm$ ~ $10^{17} cm^{-3}$ and compactness (cf. Svensson 1984) ℓ ~ 12.

The Galactic Center was observed in 1979 by HEAO-3 to have a similar behavior (Riegler et al 1985, Fig. 2). Using the same spectral modeling techniques Dermer and Liang (1990) find that the best fit cloud parameters are: radius ~ 5000 km (~16 Schwarzschild radii for a 100 $M_\odot$ black hole), temperature ~ 1.1 mc^2, radial Thompson depth ~ 1.5, and compactness ℓ ~ 8 assuming a source distance of 8.5 kpc.

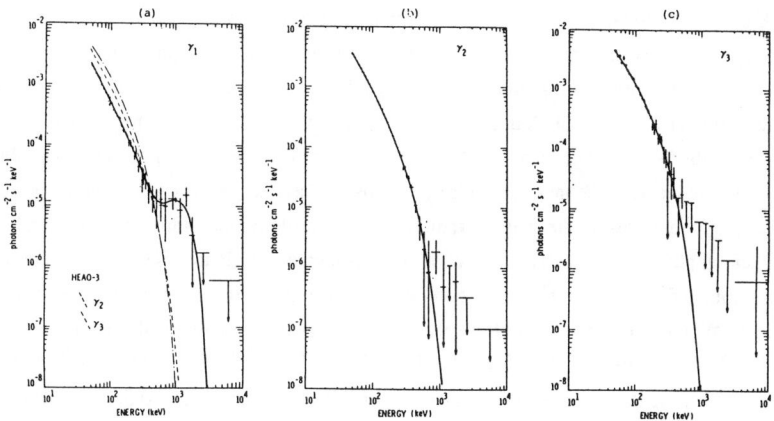

Fig. 1.   HEAO-3 1979-80 spectra of Cyg X-1. Only the γ-1 state during which the hard x-ray intensity is superlow shows a strong MeV bump (from Ling et al. 1987).

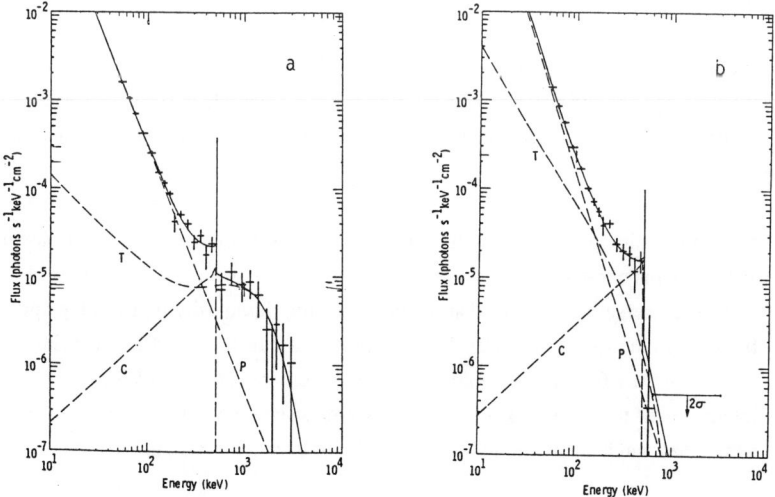

Fig. 2.   HEAO-3 1979-80 spectra of the Galactic Center. Only the Fall 79 spectrum (a) shows a strong MeV bump with correlated increase in the 511 KeV line flux (from Riegler et al. 1985).

What could be causing the quiescent x-ray disk emitting at tens of keV temperature (e.g. Steinle et al 1982) to heat up to a temperature of $\sim mc^2$ and produce a hard spectrum? Note that the lack of soft photons, whether external or internal synchrotron photons, is crucial to obtaining a hard spectrum. We speculate that the quenching of the soft photon source that normally cools the inner disk via inverse Compton (Shapiro et al 1976) is the culprit. A likely, but not necessarily unique scenario for the depletion of soft photons and the heating of the inner disk is recently suggested by Wandel and Liang (1989) in the context of AGNs but can be directly applied to Cyg X-1 and the Galactic Center. To understand this we have to go back to the model of Thorne and Price (1975) who postulate that the radiation pressure-dominated inner disk is unstable and can thicken up physically to become optically thin. It then cools via inverse Compton upscattering of soft photons from the external blackbody disk. This is the foundation of the Shapiro et al (1976) model. Wandel and Liang (1989) go one step further and show that if the Thorne-Price thinning radius moves further out to say, ≥ 50 Schwarschild radii, then the external soft photon will no longer penetrate the inner part of the disk, and the disk interior to, say 10 Schwarschild radii, can only cool by Comptonized bremsstrahlung and pair production. First principle calculations show that indeed a relativistically hot cloud and an MeV spectral bump could result (Fig. 3). The exact temperature is regulated by the viscosity parameter. For Cyg X-1 and Galactic Center the derived viscosity $\bar{\alpha}$ lie in the range 0.1-0.01 (cf. Liang 1989). Since the magnetic field must be $\lesssim 1\%$ of equipartition value in order for the synchrotron soft photons not to destroy the hard spectrum (Dermer 1989), we conclude that magnetic reconnection cannot be a major contributor to the azimuthal stress.

If we accept the thermal pair cloud picture as the correct baseline model, then MeV gamma ray observations can provide new clean diagnostics of black holes which x-ray observations fail to. Fig. 4 summarizes the direct observables and the derivable parameters of the hole and the inner accretion flow.

The thermal pair cloud model predicts a number of observational tests. The following is a list of some of the major ones:
a) Energy conservation requires that the gamma bump luminosity replaces the appropriate fraction of the quiescent x-ray luminosity. This can be tested by studying the pair temperature as a function of gamma to total luminosity ratio (Liang 1989).
b) Interaction of the gamma rays with the disk x-rays should produce an escaping pair wind, which may annihilate in the cool circumstellar or interstellar medium to produce a narrow

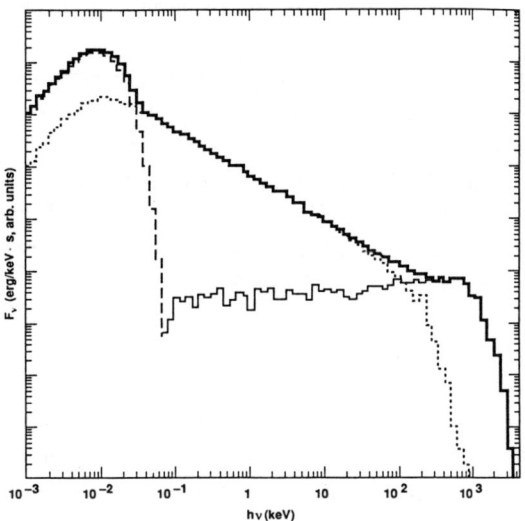

Fig. 3. Composite Monte Carlo model spectrum of a 10^8 M$_\odot$ AGN disk with the thinning radius at 100 GM/c^2. The disk interior to 20 GM/c^2 is not cooled by external soft photons and heats up to produce an MeV bump. Details see Wandel and Liang (1989).

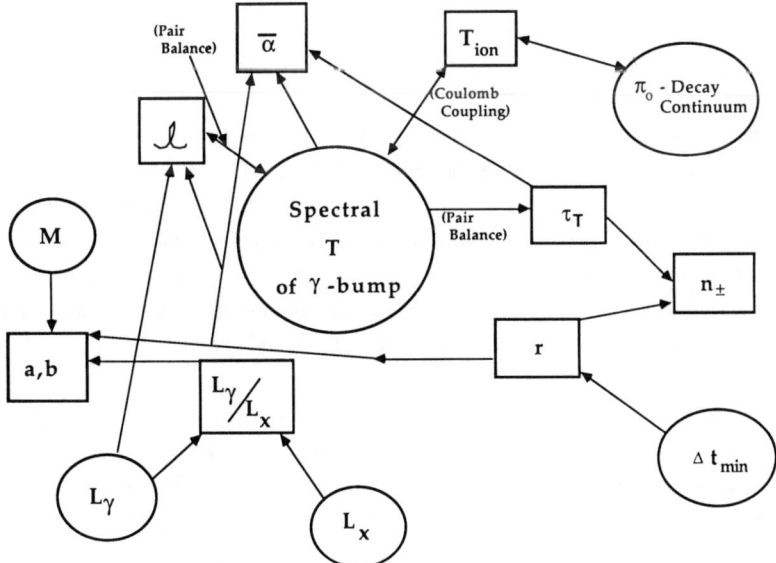

Fig. 4. Flow chart showing the observable (circles) and derived parameters (squares) of the pair cloud around a black hole. a is the angular momentum of the hole and b the Keplerianess of the disk inner edge (cf. Liang 1989). See text for definition of the other parameters.

511 keV line. This can be tested by observing the correlation of the narrow 511 line with the MeV continuum and measuring its absolute intensity (e.g. Lingenfelter and Ramaty 1989)

c) Based on the picture of Wandel and Liang (1989) we expect an anticorrelation between the intensity of the MeV gamma bump and the soft (UV or soft x-rays) continuum. Future simultaneous multiwavelength observations can test this.

d) If rapid time variability in gamma rays (say down to 10's of ms) can one day be measured, then the compactness of the source ~ $L/c\Delta t$ may become an observable. Its correlation or anti-correlation with temperature could provide a clean test of the theory of pair-balanced plasmas (Svensson 1984, Zdziarski 1984).

From the space observation points of view what are the near term priorities? We list a few objectives which should be achievable with observations by GRO and GRANAT:

a) The most important one is the confirmation of the MeV bump in Cyg X-1 and the Galactic Center with higher significance and spectral quality then that obtained by HEAO-3 and balloon flights, (e.g. Baker et al. 1973, McConnell et al. 1987) especially the details of the spectrum above ~ MeV.

b) Time profile for the rise and decay of the bump and its long term duty cycle.

c) Confirmation of the narrow 511 keV line and its time correlation with the MeV continuum.

d) Search for similar bumps in other prime black hole candidates, including A0620, LMC X-3, GX339-4, and the nearby AGNs brightest in the hard x-rays. The best time to look for this bump is when the hard x-rays is lowest.

e) Simultaneous multiwavelength observations from UV to gamma rays whenever an MeV excess is detected.

f) Search for the π_0-decay continuum near 70 MeV which may exist if the hot cloud ion temperature approaches the virial temperature of ~ 100 MeV (Dermer 1989).

g) Check on the gamma ray spectra of LMXB neutron stars to see if they might exhibit similar bumps. Based on the above thermal model we expect neutron stars not to exhibit such bumps due to the preponderance of soft x-rays from the central star. But if we find that neutron stars exhibit similar bumps than we can conclude that they are no longer a unique signature of black holes and the thermal pair cloud model must be revised.

What are the future challenges, especially observations requiring new technologies projected for the XXIst century? One obvious major challenge is the measurement of rapid time variability down to say the natural time scales of the inner region of the accretion flow. For Cyg X-1 type masses these are 3-30 ms. Let us say we want to detect at least 10 photons of the MeV continuum over a 1 (10) ms integration time. The flux of Fig. 1 says that we need a detector area of at least 10^6 (10^5)cm^2. This is far-fetched for now but conceivable for a space station or lunar base platform. The time resolution requirement scales inversely with the mass of the hole. Hence for the Galactic Center (say 100 M$_\odot$) the required detector area would be ten times smaller, etc. Detection of the narrow 511 keV line from Cyg X-1 is expected to be feasible for GRO (cf. Dermer and Liang 1988), but for other sources we would probably need at least the NAE sensitivity or better. Also GRO would not be able to resolve the line profile which would tell us much about the annihilation site. Even the ~ 1 keV resolution at 511 keV projected for NAE would not be able to distinguish between a cold ISM versus a warm (10^4 K) circumstellar wind. We would like to get down to ~ 100 eV spectral resolution at 511 keV in the XXIst century if possible. High spatial resolution ($\leq$ arc min) imaging of the Galactic Center region both in hard x-rays and gamma rays is also needed to separate out the absolute contribution of different sources. Finally, gamma ray polarization measurement poses another useful challenge for the future. A detection of even ~ 1% linear polarization in the MeV continuum would cast strong doubt on the quasi-spherical geometry picture.

This work was performed under the auspices of the U.S. DOE by the Lawrence Livermore National Laboratory under contract #W-7405-ENG-48. and at Stanford this work is partially supported by NASA grant NGR 05-020-668.

References

Baker, R.E. et al 1973, Nature, 245, 18.
Dermer, C.D. 1989 Proc. 14th Texas Symp. on Rel. Astrophys., ed. E. Fenyves (N.Y. Acad Sci.N.Y.).
Dermer, C.D. and Liang, E.P. 1988, AIP Conf. Proc. No. 170, 326.
Dermer, C.D. and Liang, E.P. 1990, to be submitted to Ap. J. Lett.
Liang, E.P. and Dermer, C.D. 1988, Ap. J. Lett. 325, L39.
Liang, E.P. 1989, Ast. Ap. 224, to appear.
Ling, J.C., Mahoney, W.A., Wheaton, W.A. and Jacobson, A.S., 1987, Ap. J. Lett. 321, L117.
Lingenfelter, R. and Ramaty, R. 1989, Ap. J. to appear.
McConnell, M.L. et al. 1987, XX ICRC Conf. Proc. 1, 58 (Moscow).
Riegler, G.R. et al. 1985, Ap. J. Lett. 294, L13.
Shapiro, S., Lightman, A.P. and Eardley, D.M. 1976, Ap. J. 204, 187.
Steinle, H. et al. 1982, Ast. Ap. 107, 350.
Svensson, R. 1984, Mon. Not. Roy. Ast. Soc. 209, 175.
Thorne, K.S. and Price, R.H. 1975, Ap. J. Lett. 195, L101.
Wandel, A. and Liang, E.P. 1989, submitted to Ap. J.
Zdziarski, A. 1984, Ap. J. 283, 842.

SUPERNOVAE AND SUPERNOVA REMNANTS AT HIGH ENERGIES

Roger A. Chevalier

Department of Astronomy
University of Virginia

1. Introduction

Supernovae are among the most dramatic events in the universe and they can give rise to a variety of types of high energy emission. These include early photospheric emission, radioactivity, shock interaction, and neutron star effects. Except for the first one, these topics are also relevant to supernova remnants. The X-ray and γ-ray emission from supernovae has been recently reviewed by Sutherland (1990), with an emphasis on radioactivity. Chevalier (1990) has reviewed the shock interactions around supernovae. Shock interactions are particularly important for supernova remnants, where the spatial structure of the interaction becomes observable.

The emphasis here is on physical phenomena which have not yet been well observed and which can be elucidated by future high energy astrophysics experiments. The time when the experiment becomes possible depends to a large extent on the unpredictable occurrence of nearby supernovae.

2. Prompt Photospheric Emission and its Echo

The shock front created in a supernova explosion is expected to accelerate through the outer layers of the progenitor star, where it produces high radiation energy densities. The emission of this radiation is the first electromagnetic signal of the stellar explosion. Colgate (1974) discussed the γ-ray radiation that would be produced in this way by the explosion of a compact star. He found that high energy γ-rays might be produced in a compact Type I supernova, but that the total radiated energy was only $\sim 10^{44}$ ergs because of the small initial radius. Type II supernovae are expected to have extended progenitor stars with significant prompt emission, and have been modeled in some detail.

Research on supernova light curves during the 1970's showed that red supergiant stars, with radii of a few $\times 10^{13}$ cm, were the likely progenitors of most Type II supernovae. Accurate computer models for the prompt emission required fine zoning in the outer atmospheric layers and were undertaken by Klein and Chevalier (1978), Falk (1978), and Lasher and Chan (1979). Klein and Chevalier (1978) modeled a 10^{51} erg explosion in a star with radius 3.3×10^{13} cm. They found a peak luminosity of 1.9×10^{45} erg s^{-1} with an effective temperature $T_e = 2.3 \times 10^5$ K. They found, as pointed out by Imshennik and Utrobin (1977), that the scattering opacity dominates the absorption opacity so that the spectrum is formed in deeper, hotter layers. This has the effect of increasing the typical photon energies by a factor of 2-3, which is a crucial factor for producing observable X-ray emission. The duration of the pulse, including the light travel time across the star, is about 30 minutes. Table 1 shows some of the pulse properties. The absorption corrected flux at 0.2-0.5 keV assumes a 5×10^{20} cm^{-2} column density in our Galaxy and in the supernova galaxy for a total factor 25 flux reduction. The 0.5-3 keV flux has no reduction.

TABLE 1

Soft X-ray Burst

Band (keV)	Mean Peak Luminosity (~30 min) (erg s^{-1})	Mean Photon Energy (keV)	Absorption Corrected Flux at Earth from 100 Mpc (Photon cm^{-2} s^{-1} keV^{-1})
0.2-0.5	1×10^{44}	0.3	2×10^{-2}
0.5-3	6×10^{42}	1.0	1×10^{-3}

Falk (1978) assumed an initial radius of 7.4×10^{13} cm for the progenitor star and obtained a peak luminosity of 2×10^{45} ergs s^{-1} with $T_e = 1.5 \times 10^5$K. He estimated a factor 2 increase in the effective photon temperature due to non-LTE effects and obtained a total energy in ionizing radiation of 10^{48} ergs, very similar to Klein and Chevalier (1978). Lasher and Chan (1979) carried out semi-analytic calculations for a number of models and obtained general agreement with the above results. They argued that expansion opacity effects (Karp et al. 1977) might increase the absorption opacity so that the LTE temperature is the appropriate one, but it is not clear whether bound-bound transitions are a significant opacity source at the high temperatures present at the time of shock break out.

The occurrence of SN1987A has generated a new set of finely zoned supernova models. In this case, we know that the progenitor star radius was about 3×10^{12} cm, considerably smaller than that used in previous calculations. Model 11E1Y6 of Shigeyama, Nomoto, and Hashimoto (1988) reached a peak $T_e = 4.3 \times 10^5$K, while models 10L and 10H of Woosley (1988) reached a peak $T_e \approx 2 \times 10^5$K. These models give a number of photons with energies >100 eV of 1.3×10^{56} and 8.9×10^{54}, respectively (Fransson and Lundqvist 1989). Efforts to model the non-LTE spectrum are underway. Riffert and Begelman (1988) pointed out that first-order Fermi acceleration of photons across the shock front can give rise to a high energy power-law tail to the emitted spectrum. However, the luminosity in this component is small.

The soft X-ray burst from a supernova has not yet been directly observed. Klein et al. (1979) searched the HEAO-1 data set for a burst, with negative results. Considering the limited sensitivity and sky coverage of the satellite, no significant constraints were placed on burst properties. Indirect evidence for the prompt burst from SN1987A has come from the IUE observations of ultraviolet lines of NV, NIV, NIII, CIII, and HeII (Fransson et al. 1989). The initial photospheric emission is the only plausible ionization source for these ions in the circumstellar gas. Ionization models by Fransson and Lundqvist (1989) show that the required ionizing source has a temperature of $(4-8) \times 10^5$K, or at least 10^{56} photons with energies >100 eV ($\sim 2 \times 10^{46}$ ergs). The LTE calculations by Shigeyama et al. (1988) just satisfy these conditions, while those of Woosley (1988) are too cool. This may be evidence for hardening of the photospheric spectrum by non-LTE effects. The burst is weaker than for a typical Type II supernova because of the smaller initial stellar radius.

The X-ray burst properties of Type II supernovae are moderately well understood and the Type II supernova rate is out to 100 Mpc is roughly 80 (H_0/55 km s^{-1} Mpc^{-1})^3 yr^{-1} (Klein and Chevalier 1978). The properties of a telescope needed to observe these bursts include a) wide sky coverage; b) ability to detect a 30 minute burst; c) positional accuracy to about $2°$, so the supernova can be found on a Schmidt plate; d) sensitivity of a few 10^{-3} photon cm^{-2} s^{-1} keV^{-1}; e) rapid data analysis for variable sources. The benefits of detection include information on the physics of shock break-out, the initial stellar radius, and the early evolution of supernovae. This information should allow more detailed modeling of individual supernovae, which will be helpful in determining their distances. Ultimately, detection of the soft X-ray burst may be an efficient means to discover Type II supernovae.

Even if it is not possible to discover supernovae by their soft X-ray burst, the dust scattered X-ray echo of the initial burst may be observable after a supernova has been discovered optically. Here, we follow Xu, McCray, and Kelley (1986) for the theory of the echo. We assume silicate grains with a radius range $0.01 \mu m \leq a \leq 0.25 \mu m$ and distribution $dn_g/da \propto a^{-3.5}$. In the Rayleigh-Gans approximation, the optical depth to dust scattering is $\tau_s \approx 0.5 A_v$ $(E/0.5 \text{ keV})^{-2}$, where A_v is the visual extinction and E is the photon energy. The photoabsorption optical depth is $\tau_a = 2.5 A_v (E/0.5 \text{ keV})^{-8/3}$. The dominance of the photoabsorption opacity implies that the scattered energy is 0.1 or less of the initial radiated energy.

In this model the echo luminosity remains constant until time

$$t_1 = 0.6 \left(\frac{E}{0.5 \text{ keV}}\right)^{-2} \left(\frac{d}{100 \text{pc}}\right) \text{ day}$$

where d is the distance of the cloud from the supernova, after which the luminosity decreases as $t^{-0.75}$ because the scattered light from the larger grains is no longer observed. After a time $t_2 = 620 t_1$, the echo rapidly fades because the light from even the smallest grains is not scattered into the field of view. As an example, if the scattered energy in a band near 0.5 keV is 10^{44} erg and d = 100 pc, the initial luminosity in the band is 1.1×10^{38} erg s^{-1} and the luminosity between time t_1 and t_2 is 1.4×10^{37} $(t/10 \text{ days})^{-0.75}$ erg s^{-1}. The angular diameter of the X-ray echo is $0\rlap{.}''2$ $(D/\text{Mpc})^{-1}(d/100\text{pc})^{1/2}(t/\text{day})^{1/2}$, where D is the distance to the supernova.

The X-ray echo will not be easy to detect, but it is mentioned here because it may dominate the early X-ray emission from a Type II supernova. Requirements for detection are the early discovery of the supernova and an X-ray telescope with excellent sensitivity and spatial resolution.

3. Radioactivity

The synthesis of radioactive isotopes is believed to be a widespread process in supernova explosions. The decays can give rise to γ-ray lines directly and to a hard X-ray continuum through scattering. Observations of SN1987A have recently focussed attention on these processes (e.g. Kumagai et al. 1989; Woosley, Pinto, and Hartmann 1989). Table 2 lists the major decay chains expected to be present and some of the important γ-ray lines. In SN1987A, about 0.075 $M_\odot$ of ^{56}Ni was synthesized in the explosion and the flux in the 847 and 1238 keV lines peaked at somewhat less than 10^{-3} photons cm^{-2} s^{-1} and an age of several 100 days. Models with mixing of heavy elements into outer, faster layers have been able to reproduce the observed line evolution, including the unexpectedly early turn-on of the γ-ray lines. In the models of Kumagai et al. (1989), 0.0043$M_\odot$ of ^{57}Ni produces a flux of 10^{-4} photons cm^{-2} s^{-1} in the ^{57}Co 122 keV line at an age of about 10^3 days and the 1159 keV ^{44}Sc line from $1.2 \times$

TABLE 2

γ-Ray Lines

Decay	Half-Life	Energies of Important Lines (keV)
$^{56}Ni \rightarrow {}^{56}Co$	6.1 days	750, 812
$^{56}Co \rightarrow {}^{56}Fe$	77 days	847, 1238
$^{57}Ni \rightarrow {}^{57}Co \rightarrow {}^{57}Fe$	36 hrs, 271 days	122 + 14, 136 (^{57}Co decays)
$^{44}Ti \rightarrow {}^{44}Sc \rightarrow {}^{44}Ca$	48 yrs, rapid	78.4, 67.9, 511(x2), 1159 (^{44}Sc decays)

10^{-4}–10^{-6} $M_\odot$ of ^{44}Ti becomes a dominant line after day 1600 at a flux of about 5 x 10^{-6} photons cm^{-2} s^{-1}. These fluxes may be representative of a Type II supernova at a distance of 50 kpc. In the 21st century, ^{44}Ti decay lines will be observable from SN1987A, but other Type II supernovae will be difficult to observe. Supernovae are even a distance of 5 Mpc (10^4 flux reduction) are rare.

There will be a better opportunity to observe γ-ray lines from Type I supernovae because of their larger mass of ^{56}Ni (0.5-1.0 $M_\odot$), small ejected mass, and larger ejection velocities. In models described by Sutherland (1990), the ^{56}Co γ-ray lines peak at 90 days at a flux level about 10^3 above that of SN1987A. Because of the rapid development of transparency, the ^{56}Ni lines (see Table 2) may be observable from a Type I supernova during the first two weeks (Burrows, this meeting). ^{44}Ti has a sufficiently long life that γ-ray lines from its decay may be observable in galactic supernova remnants. In a Type I model discussed by Woosley, Weaver, and Taam (1980), 5.2 x 10^{-4} $M_\odot$ of ^{44}Ti is synthesized, which would lead to a 1159 keV line flux of 4 x 10^{-6} photons cm^{-2} s^{-1} in the year 2000 from SN1572 at a distance of 2.5 kpc.

Measurements of the profiles of the γ-ray lines can give additional information on the density structure of the nearby synthesized matter. Detailed models for the profiles of the ^{56}Co lines have been developed for SN1987A (Bussard, Burrows, and The 1989 and refs. therein). At late times, the optical depth effects are small and a symmetrical line is expected with a width that reflects the velocity distribution of the ^{56}Co. For SN1987A, this width is about 20 keV. At early times, the emission from the far (redshifted) side of the ^{56}Co region is preferentially scattered out of the line, so that the line is observed with a net blueshift. Line profile measurements for SN1987A have been on the borderline of present technology. Teegarden et al. (1989) did observe the 1238 keV line with the interesting result that the line profile did not show opacity effects although the line flux required substantial opacity to be present. The implication is that the ^{56}Co was not uniformly mixed. Accurate line profiles can give information on the clumping of the ^{56}Co. Line profile observations of Type II supernovae will be difficult even in the 21st century unless there is an unusually close event; Type I supernovae are more favorable in this regard.

While the optical depth to γ-ray scattering is still substantial, the downscattered photons may escape at lower energies. McCray, Shull, and Sutherland (1987) predicted the emergence of a hard X-ray continuum from SN1987A. The peak was expected somewhat before the peak in the γ-ray line flux,

when the scattering optical depth was of order 10. The photons are scattered
down in energy to the point where they are photoabsorbed, so that the cutoff
energy does depend on the heavy element abundances in the supernova gas. For
solar abundances, the cutoff is at about 20-30 keV; for the hydrogen poor matter
in a Type I supernova, the cutoff may be closer to 50 keV. As with the γ-ray
line emission, the X-ray emission observed from SN1987A turned on earlier and
had a flatter peak than initially expected (Arnett et al. 1989). The reason is
the same as for the γ-ray emission properties - the outward mixing of the
freshly synthesized ^{56}Co. Thus the evolution of the X-ray emission provides
another constraint on the distribution of the heavy element matter. In
addition, the fact that there is scattered emission at smaller energies than
that of the γ-ray line but not longer can provide evidence for a particular line
even though the line is not directly observed. For example, recent hard X-ray
observations of SN1987A have given evidence for the presence of the ^{57}Co 122 keV
line in SN1987A (Sunyaev et al. 1989).

4. Circumstellar and Interstellar Interaction

Supernovae are expected to be surrounded by gas which has been lost in a
wind by the progenitor star. The densest winds are expected around the red
supergiant progenitors of normal Type II supernovae. The nonthermal radio
emission from supernovae gives excellent evidence for the presence of such a
wind (Chevalier 1982b; Weiler et al. 1986). SN1987A was a weak radio and
thermal X-ray emitter because the blue supergiant progenitor is expected to have
a low density wind (Chevalier and Fransson 1987).

The emission of X-rays depends on the formation of a hot ion shock wave at
the interface between the fast moving supernova ejecta and the slow moving
circumstellar gas. It is not yet clear when the shock wave forms. Chevalier
and Klein (1979) found a shock wave forming at the time of shock break-out, with
an early hard X-ray burst. However, this result may be related to the treatment
of the flux limited diffusion (Epstein 1981). The shock wave probably forms at
least on the doubling time for the initial explosion (Fransson 1982), but a
detailed calculation has not yet been carried out. Even for an extended star,
the doubling time is about one day.

Once the shock front forms, a double shock structure is expected with a
reverse shock on the inside where the supernova gas catches up with the
decelerating shocked layer. Because the outer supernova density profile can
normally be approximated by a steep power law in radius, the structure of the
interaction region is described by a self-similar solution (Chevalier 1982a;
Nadyozhin 1985). Hot gas produced at the shock fronts may initially cool not by
X-ray emission, but by Compton scattering with photospheric photons. This
results in an ultraviolet tail to the supernova energy distribution (Fransson
1982) and may last for tens of days. Another impediment to X-ray emission is
absorption by circumstellar gas in front of the shock front. Although the
initial soft X-ray burst is able to escape because it highly ionizes the
surrounding gas, the gas subsequently recombines. For a dense wind, the
absorption optical depth can be large for tens of days. A measurement of the
absorption would be useful for giving an estimate of the column density of gas
in front of the shock front.

Once the X-rays are able to escape from the shocked region, the emission is
dominated by the inner layer of the shell which contains shocked supernova gas

because it is at a higher density and lower temperature (about 10^7K) than the shocked circumstellar gas (about 10^9K). If the supernova density gradient is steep and the circumstellar density is high, the inner shock front may be in a radiative phase for about 100 days. During this phase, the X-ray luminosity is roughly constant and is 10^{41} erg s^{-1} for a red supergiant mass loss rate of 5 x 10^{-5} M$_\odot$ yr^{-1}. After the radiative phase, the shell expansion is energy-conserving and the X-ray luminosity declines as t^{-1}.

The thermal X-ray properties of a supernova depend sensitively on the density of the presupernova mass loss. Einstein Observatory observations of SN1980K provided the first X-ray detection of an extragalactic supernova (Canizares, Kriss, and Feigelson 1982) and the emission can be interpreted as thermal radiation from the shocked supernova gas (Chevalier 1982b). At a distance of 10Mpc, the X-ray luminosity about 40 days after discovery of the supernova was 2 x 10^{39} erg s^{-1}. Over the next 50 days, the flux declined by a factor of 2 or more, indicating that the shell was in the energy-conserving phase. Inverse Compton scattering of the photospheric photons with relativistic electrons is another possible source of X-ray emission (Beall 1979; Canizares et al. 1982). It would be useful to observe the supernovae with sufficient spectral resolution to distinguish between these mechanisms. Detailed X-ray observations, together with radio observations, should provide information on the density structure of the surrounding gas and on the outer parts of the supernova ejecta. For Type II supernovae, we will obtain a knowledge of the range of wind densities. For Type Ia and Ib supernovae, there are still uncertainties on the stellar evolution leading to the explosion. X-ray observations may help to constrain the possible models.

For nearby supernovae and supernova remnants, high spectral resolution observations of thermal X-ray emission are of particular importance. The spectral region from 200 eV to a few keV is very rich in spectral lines and observations can lead to the ionization state, temperature, composition, and Doppler velocities of the gas (Canizares 1990). Velocity measurements will require a resolution E/ΔE ~ 10^3. For galactic supernova remnants, spatially resolved spectroscopy will yield the composition structure of the remnants. Recent supernova modeling and observations of SN1987A have shown that supernovae are likely to have a clumpy, mixed structure. Optical observations of the remnant Cas A have already shown this type of structure in the ejecta (Chevalier and Kirshner 1979). While spherically symmetric models have had some success in reproducing low resolution observations of supernova remnants (e.g. Hamilton et al. 1985), it is likely that the actual situation is more complex.

Synchrotron emission from supernovae shows that relativistic electrons are accelerated in the interaction region. Relativistic protons are also likely to be present, especially because theories of shock acceleration predict a high efficiency for proton acceleration (Ellison and Eichler 1985). Berezinsky and Ptuskin (1989) have described in some detail how protons may be accelerated in supernova shocks. The relativistic protons interact with dense gas to produce pions which decay into γ-rays with an energy of about 500 MeV. Assuming efficient proton acceleration and a distance of 23 Mpc, Chevalier (1983) calculated a γ-ray flux of 1 x 10^{-9} (t/20 days)$^{-6/5}$ photons cm^{-2} s^{-1} for SN1979C. A galactic supernova with a dense circumstellar medium should be detectable in high-energy γ-rays. If the circumstellar medium is constrained by observations at other wavelengths, this will provide a test of acceleration mechanisms for heavy particles. The same statement applies to galactic supernova remnants. Chevalier (1983) estimated the pion γ-ray flux from energy-conserving remnants to be at best about 10^{-7} photons cm^{-2} s^{-1}. However, the situation is more favorable for remnants with radiative shock waves because relativistic particles and dense thermal gas become concentrated in the same region. Chevalier (1977) and Blandford and Cowie (1982) estimate that the Vela remnant should have a γ-ray flux (> 100 MeV) close to 10^{-6} photons cm^{-2} s^{-1}, which is the lower limit for detection by the COS-B satellite. Another source of high-energy radiation is bremsstrahlung from relativistic electrons. Cowsik and Sarkar (1980) have used the radio synchrotron flux and the COS-B limit on the γ-ray flux from Cas A to set a lower limit of 8×10^{-5} G on the magnetic field.

A supernova remnant of particular interest for the 21st century is SN1987A. Ultraviolet observations have shown that there is a shell at about 1 lt.-yr. from the supernova with an electron density of $(1-3) \times 10^4$ cm^{-3} (Fransson et al. 1989). Chevalier and Liang (1989) estimate that the supernova shock wave will interact with the shell at an age of about 15-20 years, when the thermal X-ray luminosity will be $\geq 10^{38}$ ergs s^{-1}. These estimates are uncertain, but the opportunity to observe a remnant in which we have detailed information on the supernova and its surroundings before the interaction is unprecedented. The angular radius of the remnant will be about 1 arcsec.

5. Neutron Star Effects

Type II supernovae (and possibly Type Ib supernovae) are the natural birthplace for pulsars. The galactic pulsar birthrate is consistent with the Type II supernova rate (Gunn and Ostriker 1970; Narayan 1987). The Crab pulsar is associated with SN1054 and the neutrino signature from SN1987A is consistent with the formation of a neutron star (Burrows 1988). However, there is no unambiguous evidence for pulsar or neutron star activity in any extragalactic supernova and the interactions between neutron stars and supernovae remain unclear.

One effect that can reduce neutron star activity is fall-back of supernova matter to the neutron star (Colgate 1971; Michel 1988; Chevalier 1989). Chevalier (1989) estimates that fall-back is the dominant process for at least the first 7 months. The mass accretion rate is super Eddington, but the radiative efficiency is small. After 7 months, the neutron star accretion can provide an Eddington luminosity and it is possible that the radiative pressure cuts off the fall-back. Woosley, Hartmann, and Pinto (1989) have calculated the X-ray emission resulting from placing a typical binary X-ray source inside SN1987A. However, while mass transfer and mass loss can naturally produce a mass accretion rate close to the Eddington rate in a binary system, it is not clear whether long-lived accretion at this rate occurs inside of a supernova.

If the fall-back is cut off, the minimum emission from the neutron star is expected to be the thermal radiation from the newly formed object. At an age of a few years, the effective temperature is about 3×10^6 K, giving a soft X-ray luminosity of 6×10^{34} erg s^{-1} (e.g. Nomoto and Tsuruta 1987). The supernova envelope is likely to be opaque at soft X-rays for decades. The detection of thermal neutron emission from SN1987A is a problem for the 21st century. The detection of such emission is important because it is possible that fall-back of matter turned a central neutron star into a black hole (Chevalier 1989). In this case, super-Eddington accretion is expected to continue for more than a decade with a low radiative efficiency (e.g. Blondin 1986). The source would emit only at low energies with a luminosity of about 10^{35} erg s^{-1}.

From the arguments given above, it is likely that the central object in a Type II supernova is a pulsar, i.e. a rotating, magnetic neutron star. Statistics of the general population of observed pulsars suggest that they are born with a fairly narrow range of magnetic fields around 10^{12} Gauss (Gunn and Ostriker 1970; Lyne, Manchester, and Taylor 1985). However, the power output of the pulsar depends sensitively on the initial rotation rate. Pulsar statistics suggest that many pulsars are "injected" with slow rotation periods (> 0.5 sec) (Narayan 1987; Emmering and Chevalier 1989). It is plausible that the "injection" period is the period near the time of birth. The magnetic dipole power of a pulsar with a 10^{12} Gauss field is ~10^{34} (P/0.5 sec)$^{-4}$ ergs s^{-1}, so that most pulsars may be weak energy sources. However, the Crab Nebula, with its X-ray luminosity of about 10^{38} ergs s^{-1}, shows that some Type II supernovae do produce moderately rapidly rotating pulsars. Woosley, Pinto, and Hartmann (1989) show how the Crab Nebula would appear inside of SN1987A. The 20 keV flux becomes substantial at an age of about 3 years. If a powerful pulsar is present, the acceleration of protons and heavy particles may occur; Gaisser, Harding and Stanev (1989) have suggested that such acceleration occurs at the shock front where the relativistic pulsar wind is slowed by its interaction with the surrounding supernova envelope. The interaction of the relativistic particles with the supernova gas can produce very high energy radiation, but the luminosity depends on the amount of mixing of the relativistic particles with the gas. The high energy luminosity cannot be reliably predicted because of the many uncertainties.

A better determined program would be a search of nearby galaxies for supernova remnants like the Crab Nebula. The energy range 2-10 keV, where interstellar absorption is small, would be suitable for the search. Statistics in this area would give valuable information on the rate of formation of pulsar-dominated nebulae.

6. Summary

The study of supernovae and their remnants has matured in the past 20 years to the point where there are predictions for high energy observations in a number of areas. First, a soft X-ray burst is expected at the time of shock breakout from a normal Type II supernova with a red supergiant progenitor. The main uncertainties are the details of hardening of the spectrum by non-LTE effects and the degree to which the radiation will be absorbed. Detection of the bursts would give information on the supernova progenitor radii and the early light curve evolution. Second, radioactivity is expected to lead to γ-ray lines and continuum emission in most supernovae, except perhaps low mass Type II events like SN1054. Type Ia supernovae should be the most luminous, Type II's the least luminous, and Type Ib's, which are thought to synthesize about 0.2 M$_\odot$ of ^{56}Ni, at an intermediate level. The evolution of the emission yields information on the amount and distribution of the newly synthesized elements in the supernovae. Gamma-ray line profiles will give more detailed information on the distribution and on clumping of the gas. Lines from the remnants of Type Ia supernovae in our Galaxy should be detectable. Third, thermal X-ray emission is expected from the shock interaction with the circumstellar and interstellar medium. A wide range of luminosities is expected from circumstellar interaction, with normal Type II supernova with dense red supergiant progenitor winds being the brightest. High resolution spectroscopy of galactic supernova remnants will give detailed information on the ionization, temperature, composition, and velocities of the X-ray emitting gas. For young remnants, we will learn about the structure of the ejecta and for old remnants, about the initial mixing process. The shock interactions can also accelerate relativistic protons which give high-energy γ-rays through pion decay interactions. Observations of the γ-rays will tell us about the efficiency of particle acceleration. Finally, neutron stars are thought to be born in supernovae, but they probably have a very wide range of initial power outputs. A useful initial project would be a search of nearby galaxies for objects like the Crab Nebula.

This work was supported in part by NASA grant NAGW-764.

REFERENCES

Arnett, W. D., Bahcall, J. N., Kirshner, R. P., and Woosley, S. E. 1989, Ann. Rev. Astr. Ap., 27, 629.
Beall, J. H. 1979, Ap. J., 230, 713.
Berezinsky, V. S. and Ptuskin, V. S. 1989, Ap. J., 340, 351.
Blandford, R. D. and Cowie, L. L. 1982, Ap. J., 260, 625.
Blondin, J. M. 1986, Ap. J., 308, 755.
Burrows, A. 1988, Ap. J., 334, 89.
Bussard, R. W., Burrows, A., and The, L. S. 1989, Ap. J., 341, 401.
Canizares, C. R. 1990, in Imaging X-ray Astronomy, ed. M. Elvis (Cambridge: Cambridge Univ. Press), in press.
Canizares, C. R., Kriss, G. A., and Feigelson, E. D. 1982, Ap. J. (Letters), 253, L17.
Chevalier, R. A. 1977, Ap. J., 213, 52.
Chevalier, R. A. 1982a, Ap. J., 258, 790.
Chevalier, R. A. 1982b, Ap. J., 259, 302.
Chevalier, R. A. 1983, Ap. J., 272, 765.
Chevalier, R. A. 1989, Ap. J., 346, 847.
Chevalier, R. A. 1990, in Supernovae, ed. A. Petschek (Berlin: Springer-Verlag), in press.
Chevalier, R. A. and Fransson, C. 1987, Nature, 328, 44.
Chevalier, R. A. and Kirshner, R. P. 1979, Ap. J., 233, 154.
Chevalier, R. A. and Klein, R. I. 1979, Ap. J., 234, 597.
Chevalier, R. A. and Liang, E. P. 1989, Ap. J., 344, 332.
Colgate, S. A. 1971, Ap. J., 163, 221.
Colgate, S. A. 1974, Ap. J., 187, 333.
Cowsik, R. and Sarkar, S. 1980, M.N.R.A.S., 191, 855.
Ellison, D. C. and Eichler, D. 1985, Phys. Rev. Lett., 55, 2735.
Emmering, R. T. and Chevalier, R. A. 1989, Ap. J., 345, 931.
Epstein, R. I. 1981, Ap. J. (Letters), 244, L89.
Falk, S. W. 1978, Ap. J. (Letters), 225, L133.
Fransson, C. 1982, Astr. Ap., 111, 140.
Fransson, C., Cassatella, A., Gilmozzi, R., Kirshner, R. P., Panagia, N., Sonneborn, G., and Wamsteker, W. 1989, Ap. J., 336, 429.
Fransson, C. and Lundqvist, P. 1989, Ap. J. (Letters), 341, L59.
Gaisser, T. K., Harding, A. K., and Stanev, T. 1989, Ap. J., 345, 423.

Gunn, J. E. and Ostriker, J. P. 1970, Ap. J., 160, 979.
Hamilton, A. J. S., Sarazin, C. L., Szymkowiak, A. E., and Vartanian, H. 1985, Ap. J. (Letters), 297, L5.
Imshennik, V. S. and Utrobin, V. P. 1977, Pisma Astr. Zh., 3, 68 (English Transl. in Soviet Astr. Letters, 3, 34).
Karp, A. H., Lasher, G., Chan, K. L., and Salpeter, E. E. 1977, Ap. J., 214, 161.
Klein, R. I. and Chevalier, R. A. 1978, Ap. J. (Letters), 223, L109.
Klein, R. I., Chevalier, R. A., Charles, P. A., and Bowyer, S. 1979, Ap. J., 234, 566.
Kumagai, S., Shigeyama, T., Nomoto, K., Itoh, M., Nishimura, J., and Tsuruta, S. 1989, Ap. J., 345, 412.
Lasher, G. and Chan, K. L. 1979, Ap. J., 230, 742.
Lyne, A. G., Manchester, R. N., and Taylor, J. H. 1985, M.N.R.A.S., 213, 613.
McCray, R., Shull, J. M., and Sutherland, P. 1987, Ap. J. (Letters), 317, L73.
Michel, F. C. 1988, Nature, 333, 644.
Nadyozhin, D. K. 1985, Ap. and Sp. Sci., 112, 225.
Narayan, R. 1987, Ap. J., 319, 162.
Nomoto, K. and Tsuruta, S. 1987, Ap. J., 312, 711.
Riffert, H. and Begelman, M. C. 1988, Bull. A. A. S., 20, 961.
Shigeyama, T., Nomoto, K., and Hashimoto, M. 1988, Astr. Ap., 196, 141.
Sunyaev, R. A. et al. 1989, Sov. Astr. Lett., 15, 125.
Sutherland, P. G. 1990, in Supernovae, ed. A. Petschek (Berlin: Springer-Verlag), in press.
Teegarden, B. J., Barthelmy, S. D., Gehrels, N., Tueller, J., Leventhal, M. and MacCallum, C. J. 1989, Nature, 339, 122.
Weiler, K. W., Sramek, R. A., Panagia, N., van der Hulst, J. M., and Salvati, M. 1986, Ap. J., 301, 790.
Woosley, S. E. 1988, Ap. J., 330, 218.
Woosley, S. E., Pinto, P. A. and Hartmann, D. 1989, Ap. J., 346, 395.
Woosley, S. E., Weaver, T. A., and Taam, R. E. 1980, in Type I Supernovae, ed. J. C. Wheeler (Austin: Univ. Texas), p. 96.
Xu, Y., McCray, R., and Kelley, R. 1986, Nature, 319, 652.

DISCUSSION

Richard Mushotsky

Since you were not able to get to the interstellar medium, I would like to add that (1) high spectral resolution absorption studies can give total column densities in the line of sight to sources, which is very useful for understanding the structure of the cold phase of the interstellar medium; and (2) X-ray studies of dust halos allow measurement of dust-to-gas ratios.

GAMMA-RAYS AS A PROBE OF TYPE Ia SUPERNOVA PHYSICS

Adam Burrows

Departments of Astronomy and Physics
University of Arizona
Tucson, AZ 85721

I. INTRODUCTION

The detection of gamma-ray lines and a hard X-ray continuum from SN1987A in the Large Magellanic Cloud has reminded the astrophysics community that supernovae are profoundly radioactive explosions (Cook et al. 1988; Leising 1988; Sunyaev et al. 1987; Matz et al. 1988; Woosley and Pinto 1988; Nomoto et al. 1988). All supernovae should produce ^{56}Ni at some level, (Woosley and Weaver 1986), along with other radioactive isotopes (e.g. ^{57}Ni and ^{44}Ti), and should have gamma-ray line signatures characteristic of the corresponding decay sequences. However, Type Ia supernovae are thought to be thermonuclear explosions of carbon/oxygen white dwarfs near the Chandrasekhar mass ($\sim 1.4\,M_\odot$) and to involve the production of copious quantities of ^{56}Ni (~ 0.5–$0.7\,M_\odot$). The decay sequence, ^{56}Ni $\rightarrow$ ^{56}Co $\rightarrow$ ^{56}Fe, is thought to power the entire classical light curve (Branch et al. 1983; Arnett 1982). When the debris expands to gamma-ray transparency, the gamma-ray lines of this sequence escape and can be detected by terrestrial technologies. Eventually, all supernovae become "gamma-ray supernovae," just as SN1987A in fact now is. Those gammas that first Compton-scatter off of the debris and escape before being photoelectrically absorbed do so as part of a continuum that is down-scattered to tens of keV from primary energies that might have been as high as 3.0 MeV. The combination of the large ^{56}Ni yield, the small progenitor mass, and the rapid expansion should make Type Ia supernovae ~ 1000 times brighter in gamma-lines than SN1987A. The hard X-ray spectrum, with its peak, low-energy cutoff, continuum shelves, and time behavior, the luminosity evolution of the gamma-ray lines at 750 keV and 812 keV (^{56}Ni) and 847 keV and 1238 keV (^{56}Co) (and at other energies), and the line ratios and profiles all bear the stamp of the explosion model. The ^{56}Ni yield, the production of intermediate-mass elements (Ca, Si, S, etc.), the velocity and mass density profiles, and the degree of explosive mixing influence the hard X-ray and gamma-ray line signals in characteristic, distinctive, and diagnostically useful ways. The detection of the hard photon emissions from a Type Ia supernova should pin down the crucial global and structural parameters of the explosion and differentiate one Type Ia model from another. The use of the gamma-ray line and hard X-ray data anticipated from OSSE on the GRO (Kurfess 1988) and the NAE (Matteson 1989), among other future space-based instruments, to probe Type Ia physics has motivated the series of Monte Carlo investigations (Burrows and The 1989, hereafter BT) we summarize in this communication.

II. MODEL W7A, CORRELATIONS, AND RESULTS

Due to the enforced brevity of this paper, we focus on the emissions of only one model from the work of BT on Type Ia hard photon emissions: model W7A. In this model, we employ the mixing prescription of Branch et al. (1985) for model W7 (Nomoto, Thielemann, and Yokoi 1984) and completely homogenize the elements between an enclosed mass of $0.7\,M_\odot$ and the inner edge of the outer C/O mantle at $1.25\,M_\odot$. Branch et al. (1985) and Harkness (1986) have concluded that the best spectral syntheses of Type Ia optical spectra indicate mixing of the intermediate-mass elements (Ca, Si, S, Ne, Mg). These elements are thermonuclearly generated as the deflagration wave that is incinerating the doomed star stalls. Interior to this region, $0.58\,M_\odot$ of fresh ^{56}Ni has been produced. At a given epoch in the explosion, the nucleosynthesis and the velocity and mass-density profiles directly influence the continuum and line emissions through their effect on the photoelectric and Compton opacities. Importantly, these quantities are correlated in a given model of the explosion wave. A weak deflagration wave will produce less ^{56}Ni and fewer intermediate mass elements, will result in lower velocities and, hence, will march more slowly towards transparency, and will stall deeper in the disassembling star. As pointed out in BT, the models cdtg7 (WW2 in BT) and W7 show just such correlated features. The major fundamental difference between these models is the speed of the deflagration wave that in WW2 is a tad slower than in W7. From this one difference do all the manifold differences described above arise to affect the gamma-line ratios, line profiles, low-energy continuum cut-offs, spectral peaks, etc. that mark a model. These structural and elemental characteristics will be imprinted on the hard photon emissions whose observations will provide an excellent and crucial check of the white-dwarf deflagration model of Type Ia supernovae.

Figure 1 summarizes the behavior of the integral luminosities for model W7A between 10 and 300 days. "Total decay rate" is the total amount of energy from gamma-rays and positrons (including the positron kinetic energy) that is released in the ^{56}Ni $\rightarrow$ ^{56}Co $\rightarrow$ ^{56}Fe decay chain. The steep slope in the first few weeks reflects the fact that ^{56}Ni has a shorter mean life (8.8 days) than ^{56}Co (111.3 days). Though this total decay luminosity exceeds 10^{43} ergs/s in the first 20 days, the expanding debris are still opaque enough to trap most of the radiation, degrade it by Compton scattering, and absorb it by the photoelectric effect. At 20 days, the total photoelectric opacity from the center to the surface at 10 keV is $\sim 10^5$ and even at 100 keV is $\sim 10^2$. Nevertheless, after 15 days, the hard radiation starts to emerge, first as the degraded continuum ("Continuum X-rays"), and only a bit later as direct gamma-ray lines. For a given ^{56}Ni mass, it is the hard emission of the early explosion phase ($\sim$ the first month) that most clearly bears the stamp of the mass and abundance profiles of the model. Though the gamma-ray line shapes do at all times reflect the distinctive ^{56}Ni distribution in velocity-space, the progressive Compton transparency at late times forces the gamma-line luminosities themselves to converge to a roughly model-independent behavior. Therefore, hard-photon observations in the first month of a Type Ia supernova exoplosion should be the most revealing.

As Figure 1 shows, in the first 20 days the total energy radiated in Comptonized photons in model W7A exceeds that radiated directly in gamma-ray lines. Furthermore, for the first 50 days, the energy trapped in the debris, indicated in the figure by "energy deposition," exceeds that radiated in hard photons to infinity. Interestingly, the integrated X-ray luminosity peaks at 35–40 days, while the total gamma-line luminosity peaks near 80 days.

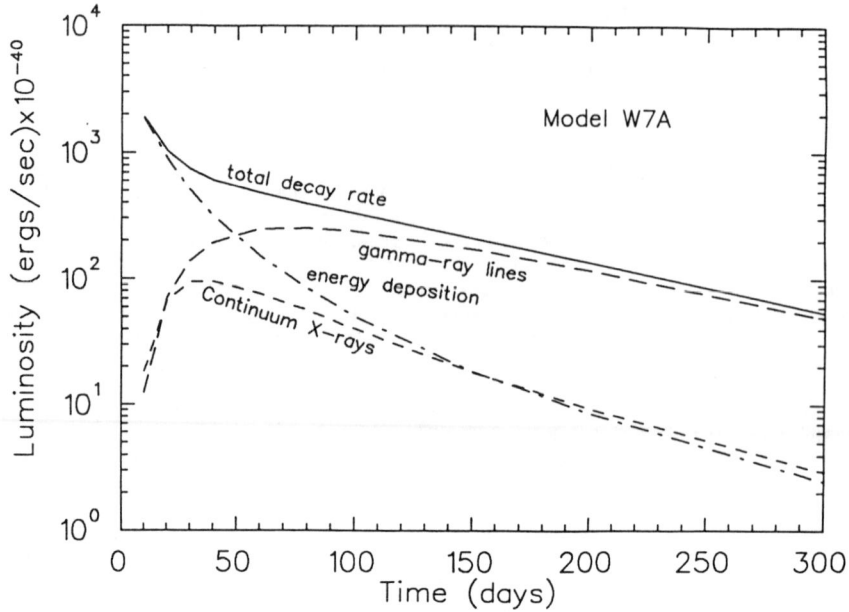

Figure 1. Various "luminosities" (in units of 10^{40} ergs/s) versus time (in days) for model W7A in the first 300 days. "Total decay rate" is the total energy production rate, including positron kinetic energy and annihilation, available from the ^{56}Ni $\to$ ^{56}Co $\to$ ^{56}Fe decay sequence. "Gamma-ray lines" is the total luminosity of all escaping γ-ray lines. "Energy deposition" is the total rate of energy (including positron kinetic energy) absorption in the debris. "Continuum X-rays" is the energy escape rate via all continuum X-rays.

At late times (>100 days), the gamma-ray line luminosity starts to track the total decay rate (minus the contribution due to the e^+ kinetic energy) quite well. In addition, not too curiously, between 100 and 300 days, the energy deposition rate, that is in principle re-radiated in the "optical," tracks the continuum X-ray rate to better than 30%. When good data are available, one should be a check on the other.

Figure 2 depicts the number flux at 1 megaparsec (Mpc) of the dominant ^{56}Ni and ^{56}Co gamma-ray lines at 812 keV, 750 keV, 847 keV, and 1238 keV. Though the decay of ^{56}Ni is quick, the debris do expand rapidly enough to liberate a significant flux of the ^{56}Ni lines at 812 and 750 keV before the ^{56}Ni becomes ^{56}Co. Indeed, the ^{56}Ni lines dominate the gamma-line flux during the first 20 days, in distinct contrast with massive star explosions, such as SN1987A, in which opacity is maintained until almost all the ^{56}Ni is gone. The peak 812 keV flux is $\sim 6 \times 10^{-4}$ cm^{-2}s^{-1} at ~ 30 days, which should

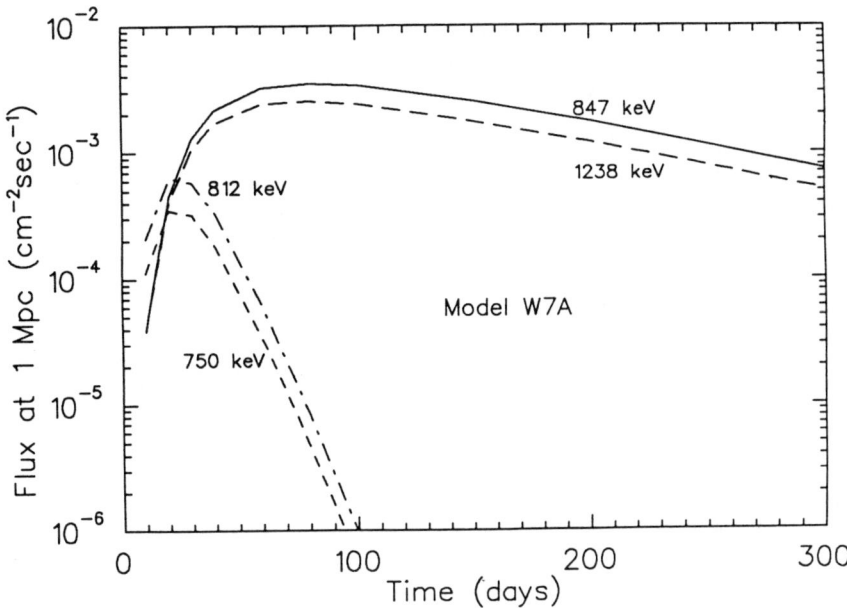

Figure 2. The number flux (in units of $cm^{-2}s^{-1}$) at 1 megaparsec (Mpc) of the escaping gamma-ray lines at 750, 812, 847, and 1238 keV versus time (in days) for model W7A in the first 300 days. Notice the distinctively different temporal behavior of the ^{56}Ni and ^{56}Co lines.

be visible out to ~4 Mpc by the GRO, while the peak 847 keV flux of ~3 ×10^{-5} $cm^{-2}s^{-1}$ at 80 days should be visible out to ~10 Mpc on the GRO. Had the GRO been up during SN1986G in CenA (~4 Mpc), it would have seen the 812 keV line of ^{56}Ni, in addition to the standard ^{56}Co lines.

As Figure 3 indicates, the hard X-ray continuum turns on quickly, rising from a peak at 10 days of $1 \times 10^{-5}\,cm^{-2}s^{-1}keV^{-1}$ at ~100 keV, to a peak at 30 days of $3 \times 10^{-5}\,cm^{-2}s^{-1}keV^{-1}$ at ~150 keV. The systematic shift of the spectral peak in 20 days is a consequence of the rapidly increasing Compton transparency of the expanding ejectum. Progressively higher energy photons can escape without being photoelectrically absorbed. The shift of this peak is accompanied by a shift of the low-energy X-ray photoelectric cutoff. Defining this cutoff as the X-ray energy at which the flux is 10% of the peak flux, we see that the cutoff moves from 30 keV at 10 days, through 60 keV at 30 days. These cutoff energies are significantly higher than the corresponding cutoffs in SN1987A because of the substantially lower heavy-element fraction in the latter (McCray, Shull, and Sutherland 1987). The high photoelectric opacities of the heavy-element-rich

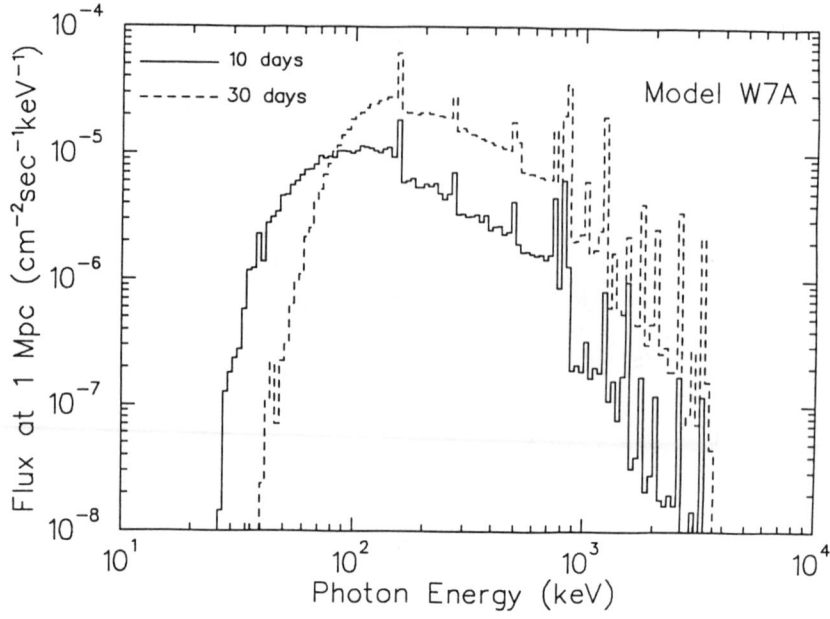

Figure 3. The spectral flux (in units of $cm^{-2}s^{-1}keV^{-1}$) at 1 Mpc of the continuum X-rays between 10 keV and 4.0 MeV for model W7A at 10 days (solid) and 30 days (dashed).

incinerated white dwarf of model W7A shift the observable energy range up beyond both the Ginga (Tanaka 1988) and the AXAF bands ($\lesssim 10$ keV) for all but close galactic Type Ia explosions. However, the >100 keV flux from Type Ia's beyond the local group of galaxies will be detectable for many months by OSSE on the GRO (to $\sim$10 Mpc), the Sigma/GRANAT detector (Durouchoux 1989) (to $\sim$3-4 Mpc), and HEXE on the Mir Space Station (Sunyaev et al. 1987) (to $\sim$3 Mpc), among other instruments.

We close our discussion of the X-ray spectra of model W7 by pointing out that after as early as day 40 (not seen in Figure 3 but in the more comprehensive work of BT), at 511 keV a clear positron line and a Compton ledge begin to emerge. The line flux peaks near 80 days at $\sim 4 \times 10^{-4} cm^{-2}s^{-1}$ at 1 Mpc. After only 150 days, the ledge, which comes predominantly from the Compton degradation of gamma-rays from the 2- and 3-photon decay of positronium, achieves a contrast of an order of magnitude. The appearance of a strong positron line and the early emergence of a "positron ledge" are distinctive features of the white dwarf model of Type Ia supernovae. A positron ledge should be visible in SN1987A, in particular, and massive star supernovae, in general, but only after $\sim$500 days (The, Burrows, and Bussard 1989), and at a significantly lower flux.

III. CONCLUSION

There are many characteristics expected in the X-ray and γ-ray line radiations of Type Ia supernovae (as well as in all other types of supernovae) that can aid in the thorough diagnosis of such explosions. Some of these features are correlated in straightforward ways. Measurements of the ^{56}Ni and ^{56}Co γ-ray line light curves, the line ratios, the line profiles, the X-ray cutoffs and spectral peaks, the X- and γ-ray turn-on times, and the discontinuity in the X-ray spectrum at 511 keV would provide rigorous tests of the current theory of Type Ia progenitors and explosions. Such data would directly reveal the energetics, nucleosynthesis, instabilities, and total mass of the debris and would open a new chapter in nuclear astrophysics. Perhaps, if need be, the γ-ray and X-ray signatures of these explosions will enable future astronomers to type supernovae by their high-energy emissions alone.

Acknowledgements:

The authors would like to thank D. Clayton and L.-S. The for cordial discussions and K. Nomoto and S. Woosley for kindly sending us their models in machine-readable form. This work was supported in part by the NSF Grants no. AST 87-14176 and AST 89-14346.

REFERENCES

Arnett, W. D. 1982, *Ap. J.*, **253**, 785.

Branch, D., Lacy, C. J., McCall, M. L., Sutherland, P. G., Uomoto, A., Wheeler, J. C., and Willis, B. J. 1983, *Ap. J.*, **270**, 123.

Branch, D., Doggett, J. B., Nomoto, K., and Thielemann, F.-K. 1985, *Ap. J.*, **294**, 619.

Burrows, A., and The, L.-S. 1989, *Ap. J.*, submitted (BT).

Cook, W. R., Palmer, D. M., Prince, T. A., Schindler, S. M., Starr, C. H., and Stone, E. C. 1988, *Ap. J. (Letters)*, **334**, L87.

Durouchoux, P. 1989, in the *Proceedings of the Xth Santa Cruz Workshop, "Supernovae"*, ed. S. E. Woosley (Springer-Verlag).

Harkness, R. P. 1986, in *Radiation Hydrodynamics in Stars and Compact Objects*, ed. D. Mihalas and K.-H. A. Winkler (Berlin: Springer), p. 166.

Kurfess, J. D. 1988, in *AIP Proc. 170, the Workshop on Nuclear Spectroscopy of Astrophysical Sources*, ed. N. Gehrels and G. H. Share (Washington, DC).

Leising, M. D. 1988, *Nature*, **332**, 516.

Matteson, J. 1989, in *The Proceedings of the Xth Santa Cruz Summer Workshop, Supernovae*, ed. S. E. Woosley, Santa Cruz, CA, July 10-21.

Matz, S. M., Share, G. H., Leising, M. D., Chupp, E. L., Vestrand, W. T, Purcell, W. R., Strickman, M. S., and Reppin, C. 1988, *Nature*, **331**, 416.

McCray, R., Shull, J. M., and Sutherland, P. 1987, *Ap. J. (Letters)*, **317**, L69.

Nomoto, K., Thielemann, F.-K., and Yokoi, K. 1984, *Ap. J.*, **286**, 664 (model W7).

Nomoto, K., Shigeyama, T., Kumagai, S., and Hashimoto, M. 1988, in *Proc. Astron. Soc. Aus.*, ed. K. M. Proust and W. J. Couch (Sydney: Astron. Soc. Aust.), p. 490.

Sunyaev, R., et al. 1987, *Nature*, **330**, 227.

The, L.-S., Burrows, A., and Bussard, R. W. 1989, *Ap. J.*, in press.

Woosley, S. E., and Weaver, T. A. 1986, *Ann. Rev. Astr. Ap.*, **24**, 205.

Woosley, S. E. and Pinto P. A. 1988, in *Nuclear Spectroscopy of Astrophysical Sources*, AIP Conf. Proc. 170, ed. N. Gehrels and G. H. Share (Washington, DC: AIP), p. 98.

DISCUSSION

Gerald Share

I'd like to make a brief comment and then ask a question. SMM has obtained a limit of about 0.6 $M_\odot$ of ^{56}Ni for SN 1986G (for an assumed distance of 3 MPC). This may be an underluminous Type 1a supernova. Do your models consider such underluminous Type 1a's? What type(s) of experiments are critical for distinguishing the different models which you described?

Burrows

The ^{56}Ni mass limit you quote for SN 1986G is certainly still consistent with the preferred mass for Type Ia's: $0.5 - 0.6$ $M_\odot$. Note, however, that the distance to Cen A may be closer to 4.0 Mpc and, therefore, that your upper limit would be closer to 1 $M_\odot$. A wide range of models has been considered for Types Ia, Ib, etc., and Woosley, in particular, has emphasized that a low ^{56}Ni yield (~ 0.2 $M_\odot$) in a white-dwarf explosion might be connected with Type Ib's. One problem in using SMM data for SN 1986G to limit Type Ia physics is the fact that this supernova was not caught near peak, but months later. The data we will need to clearly distinguish different supernova models are good hard continuum spectra from 10 keV to 2.6 MeV, high sensitivity to γ-ray lines from the 68 keV line of ^{44}Ti to the 2.6 MeV line of ^{56}Co, spectral resolution of at least 500 ($= E/\Delta E$), and time resolution of about one day for the rapidly evolving Type Ia's. Sensitivity is more important than angular resolution, as long as source confusion is minimal (a worry at late times). An effective area of at least $\sim 10^4$ cm^2 is desirable. The general distinguishing features for different Type Ia supernova models are the low-energy hard X-ray cutoff, the evolution of the hard X-ray peak the development of the 511 keV line and continuum shelf, the γ-ray line ratios, and the line profiles and their evolution. Most important for the bright Type Ia's is good data in the first month, since at late times when the debris is transparent all models merge in behavior (except perhaps in their line shapes). Also important is sensitivity to the ^{56}Ni lines at 750 keV, 812 keV, etc., since they dominate over the standard ^{56}Co lines in the first weeks in most models.

Neil Gehrels

How might clumping and fragmentation of the ejecta change the line shapes? GRIS balloon observations of SN 1987A do not show the blueshifted line shapes predicted by unclumped models. The line shapes are reasonably symmetric about the rest energy for these data.

Burrows

Clumping and mixing in supernova ejecta should be a function of depth. While there is no evidence in the optical spectra from SN 1987A of deviations from sphericity during the first months, the "Bochum" event in Hα, etc., the early emergence of X- and γ-rays, the fits to the broad UVOIR peak (to $\sim$ 150 days), and the modeling of the dust and the effects of the dust on the oxygen, etc., lines after 450 days, all point to instabilities, mixing, and clumping near the hydrogen-helium or helium-heavy element boundary in the exploding progenitor. Therefore, I would expect that the first gamma-ray lines we would see would be blueshifted in the classical way, but that as the debris thins, the reduction in the effective opacity casued by holes and fragmentation in the interior would diminish the asymmetry of the lines predicted in purely spherical models. The widths of the lines, however, should not be significantly increased or decreased from those consistent with the UVOIR, X-ray, and gamma-ray light curves and a complete theory must reconcile the gamma-line profiles with the infrared Co II lines seen from the Kuiper Observatory. The philosophy of your question is apt: there are hydrodynamic instabilities in most types and sub-types of supernovae, and we should be prepared to digest and use this probability.

THE GOALS OF GAMMA-RAY SPECTROSCOPY IN HIGH ENERGY ASTROPHYSICS

Richard E. Lingenfelter
Center for Astrophysics & Space Sciences
University of California, San Diego
La Jolla, CA 92093

James C. Higdon
Joint Science Center
The Claremont Colleges
Claremont, CA 91711

Marvin Leventhal
AT&T Bell Labs 1E-349
Murray Hill, NJ 07974

Reuven Ramaty
Laboratory for High Energy Astrophysics
NASA Goddard Space Flight Center
Greenbelt, MD 20771

Stanford E. Woosley
Board of Study Astronomy & Astrophysics
University of California, Santa Cruz
Santa Cruz, CA 95064

ABSTRACT

We review here some of the goals of high resolution gamma-ray spectroscopy in astrophysics for the beginning of the 21st Century, and we discuss how the proposed Nuclear Astrophysics Explorer will address these goals.

INTRODUCTION

High resolution gamma-ray spectroscopy explores the most energetic phenomena that occur in nature and addresses some of the most fundamental problems in physics and astrophysics. It embraces a great variety of processes – nuclear deexcitation, radiative capture, positron annihilation, Compton scattering, bremsstrahlung, and synchrotron emission – and an even greater diversity of astrophysical sources – solar flares, gamma-ray bursts, nova and supernova explosions, cosmic ray interactions and sources, neutron stars, black holes, active galactic nuclei and the cosmic gamma-ray background. Not only do gamma rays allow us to see deeper into these objects, but the bulk of the power radiated by them is often at gamma-ray energies.

High resolution gamma ray spectroscopy, in fact, provides the most direct means of studying nuclear processes in explosive nucleosynthesis, in accreting compact objects and in many other astrophysical sites. Since the gamma ray lines are emitted in nuclear, rather than atomic transitions, their interpretation is not dependent on the modeling of the nonequilibrium ionization processes as is the case with X-ray and other lower energy observations. High resolution gamma-ray spectroscopy thus enables us to directly extract the unique astrophysical information encoded in the energies, shapes, and intensities of the gamma-ray line emission from these sources. Not only do lines indicate the presence of specific nuclei or electron-positron pairs, but the line parameters, i.e. intensities, centroid shifts, widths and profiles, contain information on abundances, bulk velocities, gravitational potentials, densities, temperatures, and accelerated particle spectra. Furthermore, the high transparency of matter to gamma rays allows them to be used as tracers of high-energy processes almost anywhere in the Galaxy.

The high resolution gamma ray spectrometer on HEAO-3 has already made important advances with the discovery (Mahoney et al. 1984) of diffuse galactic 1.809 MeV line emission from nucleosynthetic ^{26}Al and the discovery (Riegler et al. 1981) of the variable, compact nature of the 0.511 MeV positron annihilation line source at or near the Galactic Center. Recent observations with a high resolution Ge detectors on balloons (Leventhal et al. 1989, Matteson et al. 1990) have shown that this compact source, after not being seen for 9 years, has become active again. Recent balloon-borne high resolution spectrometers have also made very valuable measurements of the gamma-ray lines from ^{56}Co in the nearby Supernova 1987A. These and many other gamma-ray lines observed from a wide range of sources are summarized in Table 1. The Nuclear Astrophysics Explorer (Matteson et al. these proceedings) with a sensitivity 100 times better than HEAO-3 can follow up on these discoveries and explore a rich variety of other astrophysical problems, discussed below.

NUCLEOSYNTHESIS AND GALACTIC MIXING

Observations of gamma-ray lines from nuclear transitions in the decay of radionuclei and the annihilation of positrons, produced in various explosive nucleosynthetic processes, provide the most direct method of studying current sites, rates and models of nucleosynthesis (Clayton, Colgate, and Fishman 1969). In addition, these gamma-ray lines can be used as tracers to investigate galactic structure, the mixing of interstellar gas, and the properties of the interstellar plasma.

DIFFUSE EMISSION — Observations at high spectral resolution and few-degree angular resolution of the diffuse galactic emission in gamma ray lines from ^{26}Al and from positron annihilation are of primary importance. These lines can give the first direct information on the current rates and spatial distribution of galactic nucleosynthesis. In addition, high resolution observations of cosmologically red-shifted lines in the diffuse extragalactic gamma-ray continuum could also provide a direct measure of the time-dependent rate of ^{56}Fe nucleosynthesis in the universe.

Galactic ^{26}Al — The search for the very narrow (FWHM < 3 keV) diffuse galactic 1.809 MeV line from the radioactive decay of ^{26}Al produced by supernovae was first suggested by both Arnett (1977) and Ramaty and Lingenfelter (1977). The discovery of this line with the high resolution spectrometer on the HEAO 3 (Mahoney et al. 1984, see Table 1), with an intensity of about about a factor of 5 higher than predicted, stimulated much

Table 1. ASTROPHYSICAL GAMMA-RAY LINE OBSERVATIONS

Process	Observed Energy	Source	Flux, ph/cm^2-s	Ref
e$^\pm$ Annihilation	511	Galactic Center	up to 1.8×10^{-3}	1
Radiation	511	Interstellar Gas	1.5×10^{-3}/rad	2
	511	Solar Flares	up to ~ 0.1	3
(Redshifted)	400-460	Gamma Ray Bursters	up to 70	4
(Redshifted)	~ 400	CrabPulsar Transient	$2-7 \times 10^{-3}$	5
(Redshifted)	~ 413	10June74 Transient	7×10^{-3}	6
	500-2000	Cygnus X-1	up to 2×10^{-2}	7
Radioactive Decay				
^{56}Co$(\epsilon\gamma, \beta^+\gamma)^{56}$Fe	847	Supernova 1987A	$\sim 10^{-3}$	8
	1238	" "	$\sim 10^{-3}$	8
	2598	" "	$\sim 10^{-3}$	9
^{26}Al$(\beta^+\gamma)^{26}$Mg	1809	Interstellar Gas	4.8×10^{-4}/rad	10
Nuclear Excitation				
^{4}He$(\alpha,n)^7$Be*	429	Solar Flares	up to ~ 0.05	11
^{4}He$(\alpha,p)^7$Li*	478	" "	up to ~ 0.05	11
^{56}Fe $(p,p'\gamma)$	847	" "	up to ~ 0.05	3
^{24}Mg $(p,p'\gamma)$	1369	" "	up to ~ 0.08	3
^{20}Ne $(p,p'\gamma)$	1634	" "	up to ~ 0.1	3
^{28}Si $(p,p'\gamma)$	1779	" "	up to ~ 0.08	3
^{12}C $(p,p'\gamma)$	4438	" "	up to ~ 0.1	3
^{16}O $(p,p'\gamma)$	6129	" "	up to ~ 0.1	3
Neutron Capture				
^{1}H $(n,\gamma)^2$H	2223	Solar Flares	up to ~ 1	3,12
^{1}H $(n,\gamma)^2$H	2223	10June74 Transient	1.5×10^{-2}	6
(Redshifted)	1790	" "	3×10^{-2}	6
^{56}Fe $(n,\gamma)^{57}$Fe	5947	" "	1.5×10^{-2}	6
(Redshifted)				
Cyclotron Emission	20-70	Gamma Ray Bursters	up to 3	13
& Absorption in	20-58	X-Ray Pulsators	$1-3 \times 10^{-3}$	14
$\sim 10^{12}$ gauss fields	73-79	Crab Pulsar Transient	4×10^{-3}	15

References: 1. Haymes et al. 1975, Leventhal et al. 1978, 1989, Riegler et al. 1981, Riegler et al. 1985, Leventhal et al. 1989, Matteson et al. 1989; 2. Mahoney 1988, Share et al. 1988; 3. Chupp et al. 1973, Chupp 1984, Yoshimori et al. 1983; 4. Mazets et al. 1979, 1981, Teegarden and Cline 1980; 5. Leventhal et al. 1977, Ayre et al. 1983; 6. Jacobson et al. 1978, Ling et al. 1982; 7. Nolan and Matteson 1983, Ling et al. 1987, Ling and Wheaton 1989; 8. Cook et al. 1988, Mahoney et al. 1988, Matz et al. 1988, Sandie et al. 1988, Matteson et al. 1989, Rester et al. 1988, Teegarden et al. 1989, Tueller et al. 1989; 9. Matz et al. 1989, Tueller et al. 1989; 10. Mahoney et al. 1984, Share et al. 1985, v. Ballmoos et al. 1987; 11. Murphy et al. 1989; 12. Hudson et al. 1980, Prince et al. 1982; 13. Mazets et al. 1981, Dennis et al. 1982, Hueter 1984, Murakami et al. 1988; 14. Trümper et al. 1978, Wheaton et al. 1979, Gruber et al. 1980, Tueller et al. 1984, Maurer et al. 1982; 15. Ling et al. 1979, Strickman et al. 1982, Ayre et al. 1983.

new theoretical work to try to understand its origin. In addition to their importance to nucleoynthesis, observations of ^{26}Al in the present-day interstellar medium have important consequences for the origin of the solar system, since ^{26}Mg anomalies in Al-rich meteorites have been attributed to the decay of ^{26}Al injected into the protosolar nebula by some external nucleosynthetic event (Lee et al. 1977).

Because of its $\sim 10^6$ yr mean life, the diffuse galactic 1.809 MeV emission is most likely produced by the cumulative contribution of a large number of sources. New calculations (Woosley et al. 1990) have shown that the yield of ^{26}Al in Type II supernovae could be larger than the values calculated previously by as much as a factor of two and, furthermore, the yield might be further enhanced by convection, so that such events may be a major source of the observed emission. However, a variety of potential stellar sources of ^{26}Al have also been suggested: Wolf-Rayet stars (Dearborn and Blake 1985; Prantzos and Casse 1986), novae (Wallace and Woosley 1981; Hillebrandt and Thielemann 1982; Clayton 1984), and red giants (Norgaard 1980; Truran 1985). Although each of these sources have characteristic longitude and latitude distributions (Leising and Clayton 1985, Higdon and Fowler 1989), the angular resolution of the HEAO-3 and SMM detectors, 42° and 130°, respectively, could provide only very limited information on the angular distribution of the emission (Mahoney et al. 1985).

More sensitive observations are clearly necessary in order to understand the origin of this emission. In particular measurements of the longitude and latitude distribution with few-degree resolution coupled with measurements of the line profile and energy determined to tenths of a keV resolution as a function of longitude can provide a map in galactic position and velocity space similar to that made with the 21-cm line in radio astronomy. With such measurements the relative contributions to the cumulative ^{26}Al distribution of the population I, Wolf-Rayet stars and Type II supernovae, residing predominantly in spiral arms, and the older disk-spheroid population, red giants and novae will be determined.

Galactic Positrons — The primary source of positrons in the diffuse interstellar medium is expected (Lingenfelter and Ramaty 1989a) to be the escape of positrons produced by the decay of either ^{56}Ni to ^{56}Fe (Clayton 1973a, Ramaty and Lingenfelter 1979) or ^{44}Ti to ^{44}Ca (Woosley 1987) in the ejecta of Type I supernovae, depending on the positron escape probability. Detailed observations of the recently discovered (Share et al. 1988) diffuse 0.511 MeV line emission can thus give important new information on the positron escape fraction and the current galactic rate and spatial distribution of explosive ^{56}Fe and ^{44}Ti nucleosynthesis in Type I supernovae during the last 10^6 yr, since the annihilation mean life of positrons in the interstellar medium is $\sim 10^6$ yr (Bussard, Ramaty, and Drachman 1979). The determination of the galactic ^{56}Fe synthesis rate is essential for an understanding of the evolution of elemental abundances in the galaxy, which can at the present only be indirectly inferred from estimates of the yield of nucleosynthetic ^{56}Fe in extragalactic Type I supernovae, determined by optical and infrared observations (Axelrod 1980; Graham et al. 1986).

With high resolution spectroscopy the phases of the interstellar gas in which these positrons annihilate can be determined from spectral decompositions of the positron annihilation radiation. The annihilation line at 0.511 MeV emitted from cold ($\sim 10^2$ K) H I gas consists of two components, a very narrow feature with a FWHM of ~ 1.5 keV resulting from direct annihilation with atomic electrons and a broader feature with a FWHM of ~6.5 keV (Brown et al. 1984), resulting from annihilation of singlet positronium, a hydrogen-like atom in which the proton is replaced by a positron. The broad width of the positronium line results from the motion of positronium atoms formed in flight by charge exchange with neutral H. In warm (~10^4 K) partially ionized gases the positrons thermalize before they form positronium so that the annihilation line is very narrow for both direct and singlet positronium annihilation. In hot plasmas (~10^6 K), the annihilation line is thermally broadened to a FWHM of ~10 keV. In both the cold and warm media, the annihilation line should be accompanied by characteristic continuum resulting from triplet positronium annihilation, while in the hot plasma this continuum is significantly suppressed. The fraction of the positrons annihilating via positronium is 0.90 in the cold gas (Brown et al. 1986) and essentially 1 in the warm medium (Bussard, Ramaty and Drachman 1979). A decrease in the positronium fraction from 1 to 0.9 leads to a decrease of 30% in the ratio of the triplet continuum flux to the 0.511 MeV line flux, and this difference can be measured. Thus, observations of both the shape of the 0.511 MeV annihilation line and the triplet continuum-to-0.511 MeV line ratio will determine whether the positrons annihilate in the warm phase or in cold clouds.

Galactic ^{60}Fe — Radioactive ^{60}Fe is expected to be produced by explosive helium burning in Type II supernovae (Clayton 1973b, Woosley, Axelrod, and Weaver 1981). Because its decay mean life of 2.2×10^6 yr is much longer than the typical period between galactic Type II supernovae, ~44 yr (Tammann 1982), the narrow line emissions at 0.059, 1.173, and 1.333 MeV are the cumulative contributions of ~5×10^4 sources. Although the nucleosynthetic sites and yield of ^{60}Fe is very uncertain, if only 1% of the ^{60}Ni has ^{60}Fe as its progenitor, then we would expect a galactic flux of 3×10^{-5} photons/cm^2-s-rad which would trace the nucleosynthesis of neutron rich isotopes in the galaxy.

Extragalactic ^{56}Fe — Gamma rays generated by the decay of ^{56}Ni through ^{56}Co to ^{56}Fe, produced in all galaxies throughout all time, should generate broad cosmologically red-shifted features in the cosmic gamma-ray background with relatively sharp edges at the rest energies of the lines which may be as great as 10% of the background (Clayton and Silk 1969, Clayton and Ward 1975). The NAE should be able to resolve these features and thus provide a direct measure of the universal ^{56}Fe synthesis rate as a function of time.

DISCRETE SITES — High resolution observations of gamma ray lines from discrete sites can also be used to study explosive nucleosynthesis in novae and supernovae. There are several gamma-ray lines from ^{44}Ti → ^{44}Sc → ^{44}Ca decay that may still be discovered from the recent Type II Supernova 1987A. Observations of these lines from ^{44}Ti decay can also identify the as-yet-undiscovered locations of the Type I supernovae which occurred in our galaxy during the past several hundred years, allowing them to be investigated in radio and other wavelengths. Measurements of the gamma ray line widths can also distinguish between the detonation and deflagration models of such supernovae. These models can also be tested through observations of the gamma-ray lines from ^{56}Ni → ^{56}Co → ^{56}Fe decay in extragalactic supernovae. Observations of the intensity ratios and widths of these lines can provide an essentially model-independent measure of both the velocity and mass distribution of nucleosynthetic ^{56}Ni in such supernovae. Nucleosynthesis in novae can also be studied directly by observations of the ^{22}Na decay line.

Supernova 1987A — The detection of gamma-ray line emission from the brightest and nearest supernova seen in 383 years has provided important constraints on the nucleosynthesis and mixing in the supernova ejecta. SMM first detected (Matz et al. 1988) gamma-ray line emission from the supernova in August 1987 and the series of balloon experiments launched from Alice Springs have subsequently measured (Cook et al. 1988, Mahoney et al. 1988, Rester et al. 1989, Sandie et al. 1988, Teegarden et al. 1989, Matteson et al. 1990) the time-dependent intensities and profiles of two strongest lines, 0.847 and 1.238 MeV from ^{56}Ni $\rightarrow$ ^{56}Co $\rightarrow$ ^{56}Fe decay. The time at which the gamma-rays were first detected and the subsequent time history of their intensities have allowed us to clearly establish that the supernova light curve is, in fact, powered by the absorption of gamma rays from the decay of ^{56}Co in the expanding ejecta, and that iron in nature is indeed made as radioactive ^{56}Ni. Moreover, the study of this gamma-ray line emission has provided important new information on the dynamics of the supernova, clearly showing that that there is very significant mixing of the ^{56}Co in the supernova ejecta and requiring important changes in the theoretical models. The high resolution measurements of the lines have yielded line centroids and widths that set important additional constraints on the mass-velocity distribution and dynamics of the ejecta.

The radioisotope ^{44}Ti, the progenitor of ^{44}Ca, may also be detectable from SN 1987A. The quantity produced in a Type II supernova, such as SN 1987A, although quite uncertain, is estimated to be of the order of $\sim 10^{-4}$ M$_\odot$ (Woosley, Pinto, and Weaver 1988; Thielemann, Hashimoto, and Nomoto 1989). This implies gamma-ray line fluxes of $\sim 4 \times 10^{-6}$ exp (-t/78 yr) photons/cm^2-s in each of the lines at 0.068, 0.078, and 1.157 MeV from ^{44}Ti and ^{44}Sc (Woosley, Pinto, and Hartmann 1989); and about half of this for positron annihilation at 0.511 MeV, taking into account the expected annihilation via positronium. Because it is synthesized in the deepest layers next to the neutron star, ^{44}Ti would also be ejected with the slowest velocity. The lines would therefore be narrow and could be detectable by the NAE. Observations of these gamma-ray lines would be very exciting not only for the information one would obtain regarding the nucleosynthesis of ^{44}Ca and conditions in the deepest layer to be ejected in the supernova (^{44}Ti is made in a Type II supernova by the "α-rich freeze out" from nuclear statistical equilibrium), but positrons from ^{44}Ti have been suggested as a possible contributor to the Galactic diffuse 0.511 MeV emission. It would be very valuable to know the efficiency with which the pairs are annihilating at very late times in the supernova (albeit a Type II). This would come from comparing the strength of the 0.511 MeV line with that of the 0.068, 0.078 and 1.157 MeV lines. At the times at which NAE would observe, the supernova would be essentially transparent to gamma-rays. The profiles would provide the tightest constraints upon the explosion mechanism and the amount of mass that fell back after the explosion.

Recent Galactic Supernovae — Because the decay of ^{44}Ti $\rightarrow$ ^{44}Sc $\rightarrow$ ^{44}Ca has a relatively long mean life of 78 yr, its three equal-intensity, gamma-ray lines at 0.068, 0.078, and 1.156 MeV serve as tracers of unknown, young ($<$ 400 years old) supernova remnants in our own galaxy. The discovery and location of these supernovae with sensitive observations would enable them to be investigated in radio and other wavelengths. Measurement of the shift of the line center may also allow us to determine the age of the supernova. In addition, observations of these lines will give a quantitative measure of nucleosynthesis of intermediate-mass nuclei by He burning in Type I supernovae, and provide valuable information on supernova hydrodynamics, positron escape, the supernova rate, and the supernova distribution.

Type Ia supernovae, and even more so Type Ib supernovae that may be related to helium detonation, would produce at least as much if not more ^{44}Ti than Type II supernovae (Woosley, Taam, and Weaver 1986). They would do so by a different nucleosynthetic process (explosive helium burning) and in a different region of the star (near the surface rather than in the deepest layers). The velocity of the ^{44}Ti would therefore be much greater in Type I's. One immediate benefit would be the ability to determine the supernova's type from the profiles of its ^{44}Ti emission lines hundreds of years after the explosion. The amount of ^{44}Ti and its velocity profile again provide important constraints on the explosion (detonation and deflagration?) and the nature of the pre-supernova star (capped by a layer of helium or not?).

Detonation models of helium dwarfs for Type Ib have the highest yields of ^{44}Ti, producing as much as 9×10^{-3} $M_\odot$ of ^{44}Ti, while detonation models of accreting C-O white dwarfs for Type Ia produce about 5×10^{-4} $M_\odot$ of ^{44}Ti (Woosley, Taam and Weaver 1986), and deflagration models for Type Ia (Nomoto, Thielemann, and Yokoi 1984) produce only about 8×10^{-5} $M_\odot$ of ^{44}Ti. Since in the first few hundred years only the very outermost regions of the ejecta would have decelerated, large predicted expansion velocities give line broadening of $\sim 20\%$ in the detonation models and $\sim 5\%$ in the deflagration model. With a Type I galactic birthrate of one every 36 yr (Tammann 1982) and even the lowest ^{44}Ti yield of 8×10^{-5} $M_\odot$, there should be at least ~ 5 supernova remnants whose line fluxes exceed the NAE sensitivity of 6×10^{-6} photons/cm^2-s to the 4 keV broadened 78 keV line.

In addition, measurement of the line profile can also give a unique determination of the age of each supernova (Chan and Lingenfelter 1987). This is possible because the gamma rays observed at different doppler-shifted energies, E, at any particular time, t, were emitted at different times, t' = (E/E_o)t, where E_o is the rest energy of the line. These time differences reflect different light travel times across the nebula. Thus we look back at gamma rays emitted at earlier times when we observe the red-shifted half of the line from the more distant half of the nebula and look forward at later times when we observe the blue-shifted half of the line from the nearer half. This leads to a net red-shift of the observed line that increases with time, since the more red-shifted photons from the far side were emitted at earlier times when less of the the ^{44}Ti had decayed and the emission rate was thus higher. Calculations of the expected time-dependent line profile of the 78 keV line for a helium dwarf detonation model show that the peak of the line shifts by ~ 0.5 keV/100 yr over the first 500 years, so that measurement of this red shift can give a direct measure of the age of the supernova remnant.

Future Extragalactic Supernovae — ^{56}Ni created in supernova explosions is the most abundant radioactive isotope generated in stellar nucleosynthesis (Clayton 1973b) and its decay, ^{56}Ni $\rightarrow$ ^{56}Co $\rightarrow$ ^{56}Fe, produces the most intense gamma-ray line emission. Although the bulk of the ^{56}Ni with a mean life of only 8.8 days will have decayed before the supernova ejecta becomes transparent to its gamma-ray line emission, the gamma-ray lines from its much longer lived (111 day) daughter ^{56}Co will remain within half of their peak intensities for $\sim$6 months. The predicted maximum fluxes of the principal ^{56}Co lines at 0.847 and 1.238 MeV are about $\sim 1.5\times10^{-5}$ photons/cm^2-s 100 days after the explosion of a Type Ia supernova located in the Virgo cluster (Gehrels, Leventhal and MacCallum 1987, Woosley and Pinto 1988, Chan and Lingenfelter 1988).

Supernovae somewhat closer than the mean distance to Virgo, should be detected and studied by the OSSE on the GRO at low resolution. To really study the lines, however, and to learn from their time histories and profiles, SN 1987A has taught us the need for high energy resolution (presently afforded only by Germanium detectors) with a sensitivity at least 5 times lower than the peak flux. Typical line widths in Type Ia's are $\sim 5\%$, or ~ 40 keV at 0.847 MeV, but we would like to study the shapes of the lines and learn more than just their widths. At times earlier than 100 days the lines are narrower, because we see a smaller fraction of the cobalt, and greatly blueshifted. The time evolution at high energy resolution would be interesting.

Measurements of the line intensities and profiles would allow the determination of the yields and velocity dispersions of the radioactive ejecta. These, in turn, would provide a critical test of Type I supernova models, and measurements of the line intensity ratios in the early weeks of the supernova expansion will permit a determination of the attentuation at different line energies, and hence give a measurement of the mass distribution in the ejecta, providing a further test of the models.

The NAE threshold for observing the most intense line from ^{56}Ni at 0.158 MeV is 4 to 6×10^{-6} photons/cm^2-s and from ^{56}Co line at 0.847 MeV is 1.4 to 2×10^{-5} photons/cm^2-s depending on the model. Most important, of course, the NAE will be able to resolve the lines and determine the expansion velocity which the latter instrument can not do. With this sensitivity, line emission should be observable from Type I supernovae in the Virgo cluster at a distance of 18 Mpc, where they are estimated to occur at least 3 times a year (Woosley, Axelrod, and Weaver 1981).

Using the peak 847 keV flux recorded from SN 1987A, $\sim 10^{-3}$ photons/cm^2-s, we may infer that cobalt lines, which are also the brightest gamma-ray lines from *Type II* supernovae will be undetectable at a distance greater than 1.5 Mpc, even by an instrument capable of detecting lines as weak at 10^{-6} photons/cm^2-s. Thus, barring the unlikely occurrence of a Type II in our own galaxy, Andromeda, or (again) in the Magellanic clouds, we are unlikely to detect gamma-ray lines from any Type II other than SN 1987A in the near future.

Galactic Novae — Clayton and Hoyle (1974) first suggested that ^{22}Na synthesis in nova could be identified by its 1.275 MeV gamma-ray line emission. Observations of this line can critically constrain models of the thermal history and dynamics of nova outbursts, which are generally interpreted as thermonuclear runaways in the accreted H-rich envelopes of white dwarfs in short-period binary systems (Gallagher and Starrfield 1978). However, detailed predictions of ^{22}Na production in nova explosions are unreliable at present because of uncertainities in the relevant cross sections and the temperature history of ejecta (Truran 1985), especially the role of convection. Moreover, large variations in ^{22}Na yields from novae are expected. For example, nova V693 CrA 1981 has been observed to have large (> 10) overabundances in Na, Al, Mg, and Ne relative to He (Willams et al. 1985). Such observations have been interpreted as evidence of H-rich envelope mixing with the underlying core of a ONeMg white dwarf. It is expected that such novae, which constitute $\sim 1/4$ of all galactic novae, can make significant contributions to interstellar ^{22}Na (Starrfield, Sparks, and Truran 1986; Woosley 1986). Higdon and Fowler (1987) determined an upper limit on the mean nova ^{22}Na yield of 7×10^{-7} M$_\odot$ from a comparison of a Monte Carlo simulation of the galactic nova distribution and upper limits on galactic 1.275 MeV line emission of Mahoney et al. (1982). Even if the ^{22}Na yield is 100 times less (i.e. $>6\times10^{-9}$ M$_\odot$), the NAE with a flux threshold of 6×10^{-6} photons/cm^2-s for the Doppler broadened (FWHM of 8 keV) 1.275 MeV line from nova ejecta, should observe a nearby (<2 kpc) nova about once a year (e.g., Patterson 1984).

Gamma ray line emission at 0.478 MeV may also be detectable (Clayton 1981, Leising 1988) from the decay of short-lived (77 day) ^{7}Be to ^{7}Li, made in nearby (< few kpc) novae (Starrfield et al. 1978). Observations of this line are important not only for novae but also for big bang nucleosynthesis, as they should enable us to determine which is the primary source of ^{7}Li.

INTERSTELLAR MIXING AND ABUNDANCES

High resolution gamma ray spectroscopy can provide new information on the overall abundances and evolution of nucleosynthetic products through unique measurements of the mixing of interstellar material and of the elemental abundances in the interstellar gas and dust.

Interstellar Mixing — The long ($\sim 10^6$ yr) decay and annihilation mean lives of ^{26}Al and positrons in the interstellar plasmas make them excellent tracers (Higdon 1988) for interstellar mixing. High resolution measurements of their gamma-ray lines would also produce the first map of sites of high-velocity plasma motions in the galactic corona and disk. Measurements of such velocity dispersions can constrain the intensity of plasma turbulence, a mechanism which efficiently mixes freshly synthesized stellar ejecta with the ambient gas. ^{26}Al is an excellent tracer (Higdon 1988) for investigating these flows because its mean life is close to the cooling time of the average tenuous plasma (e.g., McKee and Ostriker 1977), and the galaxy is transparent at 1.809 MeV. In contrast, the X-ray emission from the intercloud plasma ($\sim 10^6$ K) can only be observed in the solar vicinity (~ 100 pc) due to photoelectric absorption in nearby HI clouds. Moreover observations of 0.511 MeV line emission from positrons residing in the tenuous intercloud plasma can also be used to investigate mass exchange on a galactic scale between the disk and the corona. A cyclic process of mass exchange, known as the galactic fountain, has been suggested (e.g. Shapiro and Field 1976, Cox 1981) where the disk outflow of tenuous plasma cools, recombines, and condenses to warm H I at about one kpc above the plane, and finally returns to the disk in the form of H I clouds. The galactic fountain scale height of ~ 1 kpc would result in an $\sim 8°$ FWHM latitude distribution, i. e. about three times that of the disk population.

Interstellar Gas and Dust Abundance — Nuclear deexcitation line emission is expected from interactions of interstellar gas and dust with low energy (< 100 MeV) cosmic rays (Ramaty, Kozlovsky and Lingenfelter 1979). The detection of such line emissions can measure the as yet uncertain density of low energy cosmic rays, which could be the primary source of ionization and heating in molecular clouds (e.g. Spitzer 1978). Low energy cosmic ray excitation would produce very narrow (FWHM <5 keV), diffuse galactic line emission in various deexcitation lines, such as that at 6.129 MeV from ^{16}O, with intensities of as much as 4×10^{-6} photons/cm^2-s-rad, based (Higdon 1987) on models of cosmic ray intensities and estimates (Watson 1978) of cloud ionization rates. The very narrow lines result from dust grains where recoiling excited nuclei come to rest before they deexcite (Lingenfelter and Ramaty 1977). Detection of these very narrow lines can provide a direct determination of the composition, size and galactic distribution of interstellar dust grains, as well as the intensity and distribution of low energy cosmic rays.

COMPACT OBJECTS

Another major area in high energy astrophysics which can be uniquely addressed by high resolution gamma ray spectroscopy is that of compact objects. Gamma-ray spectroscopy has already made important contributions to our understanding of neutron stars, such as gamma-ray bursters and X-ray pulsars, as well as the possible accreting black hole at or near the galactic center.

Galactic Center — The galactic center region contains a compact source of 0.511 MeV line emission and gamma-ray continuum extending to at least a few MeV. The line emission is superimposed on distributed annihilation radiation from the galactic plane (discussed above). The strongest evidence for this compact source comes from its time variability (see review by Lingenfelter and Ramaty 1989b). Line emission at 0.511 MeV was observed with Ge detectors in balloon flights in 1977, 1979 and 1988, but the line was not seen with identical or similar detectors in 1981 and 1984. Furthermore, comparison of observations with the Ge spectrometer on HEAO-3 in the fall of 1979 and spring of 1980 also showed significant variability in 6 months (Riegler et al. 1981). The variations in the line flux are probably correlated with variability in the flux of the continuum at energies greater than 0.511 MeV, since the same HEAO-3 observations which showed that the line flux decreased by more than a factor of 3 also showed that the continuum decreased by an order of magnitude (Riegler et al. 1985). Particularly important is the recent detection of the 0.511 MeV line in observations with a balloon-borne Ge detector (Leventhal et al. 1989; Matteson et al. 1990). These show that the compact source, after not being seen for 9 years, has become active again.

The line and continuum varibility suggest (e.g. Lingenfelter and Ramaty 1982) that a single compact object is the source of both the positrons and the continuum. Little, however, is known for certain about this very exciting object which produces electron-positron pairs with a luminosity of 10^{37} erg/s. The ratio of the 0.511 MeV line flux to that greater than 0.511 MeV is $> 5\%$, which constrains the positron production to photon-photon interactions by the > 0.511 MeV photons, since any other process would give a ratio $< \alpha$ or $1/137$, the fine structure constant. If photon-photon interactions are the positron source and the observed flux > 0.511 MeV is isotropic, the the source size must be $< 10^8$ cm (Lingenfelter and Ramaty 1982, 1989), which would imply an esipsodically accreting black hole of mass $< 10^3$ M$_\odot$. Such a source need not be exactly at the Galactic center.

The identification of the source has been severely limited by its poor position determination. The only direct measurement was made with the HEAO-3 detectors by Riegler et al. (1981) in the fall of 1979. This measurement gives a point source centroid position of galactic longitude $l^{II} = 3.9 \pm 4.0°$, assuming $b^{II} = 0°$. This only constrains the source to be within about half a kpc of the Galactic Center, and this region contains a large number of candidate X-ray counterparts. At $\sim 10-100$ keV known hard X-ray sources occur every few degrees, and at energies of a few keV there are 10 known sources within a degree of the galactic nucleus. Recent observations with imaging detectors on Spacelab 2 at 3–30 keV (Skinner et al. 1987) and on a balloon at >35 keV (Cook et al. 1989) show that during the past few years the most intense source at 30 keV and above is 1E1740.7-2942 at $l^{II} \approx -0.9°$ and $b^{II} \approx -0.1°$ has. Future imaging observations must locate the compact annihilation source to ~ 0.1° in order to determine if it is associated with the galactic nucleus, with 1E1740.7-2942, or with any one of the other X-ray sources near the Galactic Center. Based on some similarities in their light curves McClintock and Leventhal (1989) suggested that the X-ray counterpart of the compact annihilation source is the X-ray source GX 1+4, but any reliable identification must await an accurate positional determination.

The energy spectrum of the source must also be measured with high precision over the entire 20 keV to 10 MeV range. The determination of the spectrum of a small patch of sky around the compact object would be particularly useful. Current information on the spectrum of annihilation radiation from the direction of the Galactic Center was obtained with detectors whose fields of view were 15° or larger, and these cannot unambiguously distinguish the contributions of the compact and distributed sources. Because the line centroid energy, line shape and positronium fraction of the two sources could be quite different, the available measurements cannot be easily used to determine the physical conditions near the compact object. When unambiguously determined, the line centroid will provide the distance of the annihilation site from the compact object, the line shape will give information on the temperature, state of ionization and velocities of bulk motion, and the positronium fraction (discussed above) will set limits on the gas, dust and UV radiation densities. It has been suggested (Lingenfelter and Ramaty 1989a) that a sufficiently large UV density near a black hole could photoionize orthopositronium and thereby prevent its annihilation.

We are looking forward to the precise determination of both the position and spectrum of the compact source, to determine its nature and study in depth its physical properties.

Active Galactic Nuclei — Gamma ray emission has been observed from a number of sources which are thought to be accreting black holes, based on their large luminosities and small sizes, as inferred from their rapid time variations. These include the active galactic nuclei, 3C273, NGC 4151, and Cen A, the Galactic Center, and the x-ray binary, Cyg X-1. The gamma ray observations of these sources together with observations at lower frequencies extending their spectra down to the radio range show that all of these sources appear to have peak luminosities at energies around an MeV. They also appear to show great variability at these energies.

Although the nature of these galactic nuclei is still highly speculative, both theory and observation suggest that these objects can also emit detectable fluxes of gamma-ray lines. Models of the positron annihilation radiation from our galactic nucleus suggest (e.g. Matteson 1983, Shtern 1985) that fluxes of $> 10^{-5}$ photons/cm^2-s in 0.511 MeV emission could also be expected from Cen A, NGC 4151 and other active galaxies. HEAO-3 has set (Marscher et al. 1984) limits of only 3×10^{-4} photons/cm^2-s on narrow 0.511 MeV line emission from several active galaxies. As more complete data are obtained on the spectra and variability of active galaxies, observations of both narrow and broad 0.511 MeV lines will play an essential role in our understanding of the nature of their energy generation and radiation processes.

X-Ray Pulsators — Electrons in the ionized plasma accreting towards the surface of magnetic neutron stars, known as X-ray pulsators, can undergo transitions between discrete Landau levels in the intense ($\sim 10^{12}$ gauss) field of the star, giving rise to both gamma-ray emission and absorption lines at the fundamental cyclotron frequency and its harmonics. Observation of such lines gives a direct measurement of the magnetic field intensity, and determination of the line profiles can be interpreted in terms of the spatial distributions of the field and the accreting matter (Meszaros and Nagel 1985a,b). Observations of lines have been reported from four accreting X-ray pulsators and the Crab pulsar (see Table 1). None of these observations, however, had the necessary combination of sensitivity and energy resolution (most have been with scintillation spectrometers) to accurately determine the spectrum as a function of pulse phase. As a result, even in the brightest pulsator, Her X-1, the sinusoidal variation of the line centroid with pulsation phase (Gruber et al. 1980; Voges et al. 1982) has limited the interpretation of the best high resolution measurement (Tueller et al. 1984) and thus it is not clear whether the line is seen in absorption or in emission. Pulse-phase-resolved high resolution measurements $\Delta E \sim 1$ keV, of the line profiles with the NAE should resolve the absorption/emission question and provide a clearer understanding of the emission region.

Gamma-ray line emission produced near the surface of the neutron star can also be used to determine the equation of state of the neutron star (Brecher 1977). From the line profiles the gravitational redshift can be determined and consequently the ratio of the stellar mass to the radius can be derived. Such redshifts alone can place constraints on the models of the equation of state. And if the neutron star mass can be determined by other independent means (e.g. Rappaport and Joss 1983), or if the neutron star pulsation period can be measured (e.g. Detweiler and Lindblom 1985), the equation of state could be essentially uniquely determined from the redshifts. Surface line fluxes as large as a few time 10^{-5} photons/cm^2-s have been predicted (Brecher and Burrows 1980) from such accreting neutron stars. Thus we should be able not only to study the intense magnetic fields of neutron stars but also to determine the neutron star equation of state.

Gamma Ray Bursters — Gamma ray bursts, first observed in 1967 (Klebesadel, Strong and Olson 1973) are intense (as much as 10^{-3} erg/cm^2) flashes of hard X-ray and gamma ray emission lasting anywhere from a few milliseconds to over a hundred seconds. They are thought to occur on or near the surface of highly magnetic neutron stars as a result of either impulsive accretion of matter onto the star, thermonuclear runaway of more slowly accreted matter on the surface, or quakes in the interior caused by sudden phase transitions (e. g. Lingenfelter, Hudson and Worrall 1982, Woosley 1984, Petrosian and Liang 1986). Although individual burst sources have not yet been identified, observations (see Table 1) of spectral features provide compelling evidence that neutron stars are their source. Scintillation spectrometer measurements of unresolved absorption features at 20 to 70 keV are attributed to cyclotron absorption in intense (a few times 10^{12} gauss) magnetic fields associated with neutron stars (Mazets et al. 1981, Dennis et al. 1982, Hueter 1984, Murakami et al. 1988) and emission features around 0.4 MeV are attributed to gravitationally redshifted 0.511 MeV positron annihilation radiation corresponding to surface redshifts (z of $\sim$ 0.3) expected on neutron stars (Mazets et al. 1981, Teegarden and Cline 1980). Although there is clearly a need for much more observational work with scintillation spectrometers, it is already evident that observations with high resolution spectrometers will be necessary to resolve the cyclotron lines and any narrow components, $\Delta E/E \cdot 0.05$, of higher energy lines.

The NAE, during its six months to a year of galactic plane observations of diffuse line emission and discrete sources, should also obtain high-resolution spectra of ~ 80 gamma-ray bursts per year with fluences $> 2 \times 10^{-7}$ ergs cm^{-2}. Each of these will have at least 2×10^3 photons with E $>$ 30 keV, large enough to accurately resolve the absorption and emission features. This estimate assumes that the bursts are of galactic origin and that the isotropically distributed $\sim$ 80 gamma-ray bursts per year with fluences $> 10^{-5}$ ergs cm^{-2} observed by the KONUS experiment lie within a galactic disk scale height of $\sim$ 1 kpc so that the total gamma-ray burst rate $> 2 \times 10^{-7}$ ergs cm^{-2} would be $> 5 \times 10^3$ bursts per year which is equivalent to $\sim$ 400 bursts per year per radian of galactic longitude. The NAE with a field-of-view of $\sim$ 0.2 radians, would thus observe about 80 bursts per year. Moreover, since the NAE detection threshold for gamma ray bursts is roughly an order of magnitude lower than this, such bursts should be observable all the way across the Galaxy. From a measurement of the size frequency distribution we may also be able to determine the absolute luminosity distribution of the bursts.

If gamma-ray bursters also produce nuclear lines (Teegarden and Cline 1980), then high resolution observations will allow them to be used as diagnostics of the burst phenomenon. That copious nuclear line emission can occur in non-solar gamma-ray transients is indicated by the spectrum of the unqiue 10 June 1974 transient (Jacobson 1978, Ling et al. 1982), which consisted primarily of gamma-ray lines (see Table 1) that have been interpreted as due to nuclear processes on a neutron star (Lingenfelter, Higdon and Ramaty 1978).

SOLAR PHYSICS

Intense gamma ray line emission from nuclear interactions of solar-flare accelerated particles with ambient gas in the solar atmosphere was discovered in 1972 (Chupp et al. 1973) and has been observed many times since (see Table 1). Lines have been detected from the most abundant elements and many more are predicted (Ramaty, Kozlovsky and Lingenfelter 1979). Only one observation (Prince et al. 1982) was made with a high resolution spectrometer; the others were made with scintillation spectrometers which did not allow the weaker lines to be revealed, nor the line widths and shapes to be determined. Nevertheless, these results, especially those from the SMM (e.g. Chupp 1984), which also made essential solar flare neutron observations, have provided (1) new insights into the problems of particle acceleration and transport (Forman, Ramaty and Zweibel 1986; Ramaty and Murphy 1987, Murphy et al. 1989), and (2) new methods for determining solar atmospheric abundances (Murphy et al. 1985; Hua and Lingenfelter 1987; Ramaty 1989).

Solar Flare Particle Acceleration — The observations (Chupp et al. 1984) of impulsive gamma-ray emission from many flares has shown that particle acceleration is a common property of impulsive energy release in astrophysical plasmas. The simultaneous observation of lines and continuum from flares indicates that this energy release involves the acceleration of both protons and electrons. The observed gamma-ray time profiles have placed short upper limits on the particle acceleration times (e.g. Ramaty, Dennis and Emslie 1988). Proton can be accelerated to energies as high as a GeV in less than 10 sec, and electrons can be accelerated tens of MeV in less than 2 sec. The exact nature of the mechanism which achieves this is not known.

The acceleration problem is closely connected with the problem of particle transport, which serves as the essential link between the acceleration and gamma-ray production. The gamma rays are most likely produced in the chromospheric and photospheric portions of magnetized loops (e.g. Hua, Ramaty and Lingenfelter 1989). Through the effects of magnetic mirroring, transport in these loops produces particle anisotropies which can be directly studied by observing the shapes of the gamma-ray lines (Murphy et al. 1989). High resolution measurements are needed to study these shapes, which will provide information not only on particle acceleration and transport but also on the structure of the magnetic fields in the chromosphere and photosphere.

Atmospheric Abundances — Observations of gamma-ray lines from solar flares have provided two new techniques for determining abundances in the solar atmosphere. The first technique is based on nuclear deexcitation line spectroscopy. Nuclear deexcitation line fluences are directly proportional to the abundances of elements of the ambient gas in the region in which the accelerated particles interact. Since this region is most likely located in the chromosphere and upper photospheric portions of flare loops, the technique provides information on abundances at these sites. The analysis of the observed spectrum indicates a significant underabundance of C and O and gives a direct determination of the Ne abundance (Murphy et al. 1985; Ramaty and Murphy 1987; Reames, Ramaty and von Rosenvinge 1988). An NAE observation of a large flare would produce hundreds of counts in most lines which would be resolved and stand out well above the flare continuum.

The second technique is based on the analysis of the time dependent flux of the 2.223 MeV line which can determine the abundance of ^3He in the photosphere (Wang and Ramaty 1974; Hua and Lingenfelter 1987). This abundance is relevant to our understanding both of big bang nucleosynthesis and of solar nucleosynthesis and mixing. Analysis of the observed flux has provided the first direct measure of the photospheric ^{3}He/H ratio (Hua and Lingenfelter 1987), which was found to be similar or slightly lower than the corresponding ratio obtained from solar wind observations (Geiss and Bochsler 1986).

High resolution measurements of the gamma-ray spectrum are now required to resolve individual lines in order to study a wide range of elemental abundances with high precision. High resolution observations are also needed to systematically study the 2.223 MeV line, which is intrinsically very narrow (FWHM < 100 eV).

ACKNOWLEDGEMENTS

We thank W. A. Fowler, N. Gehrels, W. A. Mahoney, J. L. Matteson, B. J. Teegarden and J. Tueller for valuable discussions and NASA for partial support of this work under contract NAS5-30338.

REFERENCES

Arnett, W. D. 1977, *Ann. N. Y. Acad. Sci.*, **302**, 90.
Axelrod, T. S. 1980, in *Type I Supernovae*, ed. J. C. Wheeler (Austin: Univ. of Texas Press), p. 80.
Ayre, C. A., et al. 1983, *Mon. Not. Roy. Soc.*, **205**, 285.
Ballmoos, P. v., Diehl, R., and Schonfelder, V. 1987, *Ap. J.*, **318**, 654.

Brecher, K. 1977, *Ap. J.*, **215**, L17.
Brecher, K., and Burrows, A. 1980, *Ap. J.*, **240**, 642.
Brown, B. L., et al. 1984, *Phys. Rev. Letters*, **53**, 2347.
———. 1986, *Phys. Rev.*, **A33**, 2281.
Bussard, R. W., Ramaty, R. and Drachman, R. J. 1979, *Ap. J.*, **228**, 928.
Chan, K. W., and Lingenfelter, R. E. 1987, *20th Int. Cosmic Ray Conf.*, **1**, 164.
———. 1988, in *Nuclear Spectroscopy of Astrophysical Sources*, eds. N. Gehrels and G. Share (New York: Am. Inst. Phys.) p. 110.
Chupp, E. L. 1984, *Ann. Rev. Astr. Ap.*, **22**, 359.
Chupp, E. L., et al. 1973, *Nature*, **241**, 333.
Clayton, D. D. 1973a, *Nature*, **244**, 137.
———. 1973b, in *Gamma-Ray Astrophysics*, eds. F. W. Stecker, and J. I. Trombka (Washington: NASA SP-339), p. 263.
———. 1981, *Ap. J.*, **294**, L97.
———. 1984, *Ap. J.*, **280**, 144.
Clayton, D. D., Colgate, S. A., and Fishman, G. E. 1969, *Ap. J.*, **155**, 75.
Clayton, D. D., and Hoyle, F. 1974, *Ap. J.*, **187**, L101.
Clayton, D. D., and Silk, J. 1969, *Ap. J.*, **158**, L43.
Clayton, D. D., and Ward, R. A. 1975, *Ap. J.*, **198**, 241.
Cook, W. R., et al. 1988, *Ap. J.*, **334**, L87.
———. 1989, in *The Center of the Galaxy*, ed. M. Morris, (Dordrecht: Riedel), 581.
Cox, D. P. 1981, *Ap. J.*, **245**, 534.
Dearborn, D. S. P., and Blake, J. B. 1985, *Ap. J.*, **288**, L21.
Dennis, B. R., et al., 1982, in *Gamma Ray Transients and Related Astrophysical Phenomena*, eds. R. E. Lingenfelter, H. S. Hudson and D. M. Worrall, (New York: Am. Inst. Phys.), p. 153.
Detweiler, S., and Lindblom, L. 1985, *Ap. J.*, **292**, 12.
Forman, M.A., Ramaty, R., and Zweibel, E.G. 1986, in *The Physics of the Sun*, ed. P.A. Sturrock (Dordrecht: Reidel), Vol. II, p. 249.
Gallagher, J. S., and Starrfield, S. 1978, *Ann. Rev. Astr. Ap.*, **16**, 171.
Gehrels, N., Leventhal, M., and MacCallum, C. J. 1987, *Ap. J.*, **322**, 215.
Graham, J. R., et al. 1986, *Mon. Not. R. Astr. Soc.*, **218**, 93.
Gruber, D. E., et al. 1980, *Ap. J.*, **240**, L127.
Haymes, R. C., et al. 1975, *Ap. J.*, **201**, 593.
Higdon, J. C. 1987, *20th Int. Cosmic Ray Conf.*, **1**, 160.
———. 1988, in *Nuclear Spectroscopy of Astrophysical Sources*, eds. N. Gehrels and G. Share (New York: Am. Inst. Phys.) p. 194.
Higdon, J. C., and Fowler, W. A. 1987, *Ap. J.*, **317**, 710.
———. 1989. *Ap. J.*, **339**, 956.
Higdon, J. C., and Lingenfelter, R. E., 1986, *Ap. J.*, **307**, 197.
Hillebrandt, W., and Thielemann F. K. 1982, *Ap. J.*, **255**, 617.
Hua, X.-M. and Lingenfelter, R. E. 1987, *Ap. J.*, **319**, 555.
Hua, X.-M., Ramaty, R., and Lingenfelter, R.E. 1989, *Ap. J.*, **341**, 516.

Hudson, H. S., et al. 1980, *Ap. J.*, **236**, L91.
Hueter, G. J., 1984, in *High Energy Transients in Astrophysics*, ed. S. E. Woosley, (New York: Am. Inst. Phys.), p. 373.
Jacobson, A. S., et al. 1978, in *Gamma-Ray Spectroscopy in Astrophysics*, ed. T. L. Cline and R. Ramaty (Greenbelt: NASA), p. 228.
Klebesadel, R. W., Strong, I. B., and Olson, R. A. 1973, *Ap. J.*, **182**, L85.
Kurfess, J. D., et al. 1983, *Adv. Space Res.*, **3**, 109.
Lee, T., Papanastassiou, D. A., and Wasserburg, G. E. 1977, *Ap. J.*, **211**, L107.
Leising, M. D. 1988, in *Nuclear Spectroscopy of Astrophysical Sources*, eds. N. Gehrels and G. Share (New York: Am. Inst. Phys.) p. 130.
Leising, M. D., and Clayton, D. D. 1985, *Ap. J.*, **294**, 591.
Leventhal, M., MacCallum, C., and Stang, P. D., 1978, *Ap. J.*, **225**, L11.
Leventhal, M., et al. 1977, *Ap. J.*, **216**, 491.
Leventhal, M., et al. 1989, *Nature*, **339**, 36.
Ling, J. C., et al. 1979, *Ap. J.*, **231**, 896.
———. 1982, in *Gamma Ray Transients and Related Astrophysical Phenomena*, eds. R. E. Lingenfelter, H. S. Hudson and D. M. Worrall, (New York: Am. Inst. Phys.), p. 143.
———. 1987, *Ap. J.*, **321**, L117.
Lingenfelter, R. E. 1988, *20th Int. Cosmic Ray Conf.*, **8**, 7.
Lingenfelter, R. E., Higdon, J. C., and Ramaty, R. 1978, in *Gamma-Ray Spectroscopy in Astrophysics*, ed. T. L. Cline and R. Ramaty (Greenbelt: NASA), p. 252.
Lingenfelter, R. E., Hudson, H. S., and Worrall, D. M. eds. 1982, *Gamma Ray Transients and Related Astrophysical Phenomena*, (New York: Am. Inst. Phys.), 500 pp.
Lingenfelter, R. E., and Ramaty, R. 1982, in *Galactic Center*, eds. G. Riegler and R. Blandford (New York: Am. Inst. Phys.) p. 148.
———. 1989a, in *High Resolution Gamma Ray Cosmology*, eds. D. B. Cline and E. Fenyves, *Nuclear Physics B, Proc. Supp.*, **10B**, 67.
———. 1989b, *Ap. J.*, **343**, 686.
Mahoney, W. A., et al. 1982, *Ap. J.*, **262**, 742.
———. 1984, *Ap. J.*, **286**, 578.
———. 1985, *19th Int. Cosmic Ray Conf.*, **1**, 357.
———. 1988, *Ap. J.*, **334**, L81.
Marscher, A. P., et al. 1984, *Ap. J.*, **281**, 566.
Matteson, J. L. 1983, *Adv. Space Res.*, **3**, 135.
Matteson, J. L., et al. 1990, *21st Int. Cosmic Ray Conf.*, **2**, 174.
Matz, S. M., et al. 1987, *IAU Circ. 4419*.
———. 1988, *Nature*, **331**, 416.
Maurer, G. S., et al. 1982, *Ap. J.*, **254**, 271.
Mazets, E. P., et al. 1979, *Nature*, **282**, 587.
———. 1981, *Nature*, **290**, 378.
———. 1982, *Ap. Space Sci.*, **84**, 173.
McClintock, J. E. and Leventhal, M. 1989, *Ap. J.*, **346**, 143.
McKee, C. F., and Ostriker, J. P. 1977, *Ap. J.*, **218**, 148.

Meszaros, P., and Nagel, W. 1985a, *Ap. J.*, **298**, 147.
———. 1985b, *Ap. J.*, **299**, 138.
Murakami, T., et al. 1988, *Nature*, **335**, 234.
Murphy, R. J., et al. 1985, *19th Int. Cosmic Ray Conf.*, **4**, 249.
Murphy, R. J., Hua, X.-M., Kozlovsky, B., and Ramaty, R. 1990, *Ap. J.*, in press.
Nolan, P. L., and Matteson, J. L. 1983, *Ap. J.*, **265**, 389.
Nomoto, K., Thielemann, H. K., and Yokoi, K. 1984, *Ap. J.*, **286**, 644.
Norgaard, H., 1980, *Ap. J.*, **236**, 895.
Patterson, J. 1984, *Ap. J. Suppl.*, **54**, 443.
Petrosian, V., and Liang, E. P., eds. 1984 *Gamma-Ray Bursts*, (New York: Am. Inst. Phys.) 206pp.
Prantzos, N., and Casse, M. 1986, *Ap. J.*, **307**, 324.
Prince, T., et al. 1982, *Ap. J.*, **255**, L81.
Ramaty, R. 1989, in *Cosmic Abundances of Matter*, C. J. Waddington (ed.), (AIP: New York), p. 91.
Ramaty, R., Dennis, B.R., and Emslie, A.G. 1988, *Solar Physics*, **118**, 17.
Ramaty, R., Kozlovsky, B., and Lingenfelter, R. E. 1979, *Ap. J. Supp.*, **40**, 487.
Ramaty, R., and Lingenfelter, R. E. 1977, *Ap. J.*, **213**, L5.
———. 1979, *Nature*, **278**, 127.
Ramaty, R., and Murphy, R.J. 1987, *Space Science Rev.*, **45**, 213.
Rappaport, S. A., and Joss, P. C. 1983, in *Accretion Driven Stellar X-ray Sources*. eds. W. H. G. Lewin and E. P. J. van Heuvel (Cambridge: Cambridge University Press) p. 1.
Rester, A. C., et al. 1989, *Ap. J.* **342**, L71.
Riegler, G. R., et al., 1981, *Ap. J.*, **248**, L13.
———. et al. 1985, *Ap. J.*, **294**, L13.
Sandie, W. G., et al. 1988, *Ap. J.*, **334**, L91.
Shapiro, P. R., and Field, G. B. 1976, *Ap. J.*, **205**, 762.
Share, G. H., et al. 1985, *Ap. J.*, **292**, L61.
———. 1988, *Ap. J.*, **326**, 717.
Shtern, B. E. 1985, *Sov. Astron.*, **29**, 306.
Skinner, G. K., et al. 1987, *Nature*, **330**, 544.
Starrfield, S., Sparks, W. M., and Truran, J. W. 1986, *Ap. J.*, **305**, L5.
Starrfield, S., et al. 1978, *Ap. J.*, **222**, 600.
Strickman, M. S., et al. 1982, *Ap. J.*, **253**, L23.
Tammann, G. A. 1982, in *Supernova: A Survey of Current Research*, eds. M. J. Rees and R. J. Stoneham (Dordrecht: Reidel), p. 371.
Teegarden, B. J., and Cline, T. L., 1980, *Ap. J.*, **236**, L67.
Teegarden, B. J., et al. 1989, *Nature*, **339**, 122.
Thieleman, F. K., Hashimoto, M., and Nomoto, K. 1989, preprint.
Trumper, J., et al. 1978, *Ap. J.*, **219**, L105.
Truran, J. W. 1985, in *Nucleosynthesis: Challanges and New Developments*, eds. W. D. Arnett and J. W. Truran (Chicago: Univ. of Chicago Press), p. 292.
Tueller, J., et al. 1984, *Ap. J.*, **279**, 177.

Voges, W., et al. 1982, in *Accreting Neutron Stars*, eds. W. Brinkmann and J. Trumper (Munich: MPI), p. 125.
Wallace, R. K., and Woosley, S. E. 1981, *Ap. J. Suppl.*, **45**, 389.
Wheaton, W. A., et al. 1979, *Nature*, **282**, 240.
Willams, R. E., et al. 1985, *Mon. Not. R. Astr. Soc.*, **212**, 753.
Woosley, S. E., ed. 1984, *High Energy Transients in Astrophysics*, (New York: Am. Inst. Phys.) 714pp.
———. 1986, in *Saas Fee Lecture Notes on Nucleosynthesis and Chemical Evolution*, ed. B. Hauck and A. Maader, Swiss Acad. Sci. p.1.
———. 1987, personal communication.
Woosley, S. E., Axelrod, T. S., and Weaver, T. A. 1981, *Comments Nucl. Part. Phys.*, **9**, 185.
Woosley, S. E., Hartman, D., Hoffman, R., and Haxton, 1990, *Ap. J.* in press.
Woosley, S. E., and Pinto, P. A. 1988, in *Nuclear Spectroscopy of Astrophysical Sources*, eds. N. Gehrels and G. Share (New York: Am. Inst. Phys.) p. 98.
Woosley, S. E., Pinto, P. A., and Hartman, D. 1989, *Ap. J.*, **346**, 395.
Woosley, S. E., Pinto, P. A., and Weaver, T. A. 1988, *Proc. Astron. Soc. Aust.*, **7(4)**, 355.
Woosley, S. E., Taam, R. E., and Weaver, T. A. 1986, *Ap. J.*, **301**, 601.
Yoshimori, M., et al. 1983, *Solar Phys.*, **86**, 375.
Zweibel, E.G., and Haber, D. 1983, *Ap. J.*, **264**, 648.

FUTURE X-RAY OBSERVATIONS OF 'NORMAL' GALAXIES

G. Fabbiano

Harvard-Smithsonian Center for Astrophysics

ABSTRACT

This paper starts with a brief review of the status of our knowledge of the X-ray properties of normal galaxies after the *Einstein Observatory* survey. It then summarizes some of the currently open issues on this subject, and addresses how substantial increases in sensitivity, angular resolution, bandwidth, spectral capabilities, and the capability to do timing observations are needed to further our knowledge of the X-ray properties of galaxies. It concludes by proposing a large throughput X-ray lunar observatory as a goal for the high-energy astrophysical community, but this should be preceded by a series of increasingly ambitious missions.

I. Brief summary of *Einstein* results

The study of the X-ray emission of normal galaxies is a very recent part of astronomy. This work has been made possible by the sensitive X-ray imaging observations of the *Einstein* satellite, launched by NASA in November 1978. Before then, with the exclusion of the bright X-ray sources associated with Seyfert nuclei, only four galaxies were known to emit X rays: the Milky Way, M31 (Andromeda), and the Magellanic Clouds. The *Einstein* satellite observed over 200 galaxies during its two and a half year life-span. Some were detected with enough detail to allow a study of their X-ray morphology, spectra, and individual sources, and to make comparisons with optical, infrared, and radio data. For all, fluxes and/or upper limits can be used to estimate the average X-ray properties.

An up-to-date review of the X-ray properties of normal galaxies can be found in Fabbiano (1989). Here, I give a very brief summary of some of these results.

Spiral and Irregular Galaxies - The (0.5-3.5 keV) luminosity of 'normal' spiral galaxies (i.e. galaxies where the X-ray emission is not dominated by a nuclear source) is in the range of a few 10^{38} to a few $10^{40} erg\ s^{-1}$, and the ratio of X-ray to optical emission is constant, suggesting that X-ray sources are a constant fraction of the stellar population. This emission is likely to be dominated by binary X-ray sources and SNRs, although Cataclysmic Variables and stars are also likely to contribute some (e.g. Watson 1989). The relative contributions of different kinds of X-ray sources appears to vary in different galaxies (see Fabbiano 1989). Diffuse components, possibly due to hot gaseous emission are also seen in some spirals. Some of these could be due to a hot phase of the ISM, but with very low X-ray luminosities, typically smaller than a few $10^{39} erg\ s^{-1}$. Others are convincingly identified with hot gaseous outflows from starburst nuclei. These can be as luminous as $10^{39} - 10^{40} erg\ s^{-1}$.

Elliptical and S0 Galaxies - These galaxies can be brighter in X-rays than spiral galaxies, reaching luminosities of a few $10^{42} erg\ s^{-1}$ or even higher for group and cluster dominant galaxies. Their X-ray luminosity is not proportional to the optical luminosity, but follows a steeper relation ($L_X \propto L_B^{1.7-2}$ Trinchieri and Fabbiano 1985; Donnelly, Faber and O'Connell 1989) although with a large amount of scatter. Two components are likely to contribute to the X-ray emission: a baseline component of low-mass X-ray binaries, similar to those seen in the bulge of M31, and a gaseous hot component, which clearly dominates the emission of X-ray bright galaxies.

II. Open Issues and Unexplored Territory

Although our knowledge of the X-ray properties of normal galaxies has made a tremendous leap from what it was before the *Einstein* mission, there are many questions opened by the work done so far that cannot be answered without new, more sensitive, observations. In the following I summarize some of the most pressing issues.

Detailed studies of individual spiral galaxies will give us precious information for understanding the formation and evolution of X-ray sources in a variety of environments. In particular, we need to explore the nature of the very luminous 'super-Eddington' sources revealed by the *Einstein* images of nearby galaxies. Moreover, external galaxies offer us the ideal situation for studying the astrophysical properties of classes of sources (e.g. SNRs, QPOs) by removing the distance bias intrinsic to all galactic studies. Sensitive observations in the soft X-ray band will allow us to detect or set stringent limits on hot gaseous components in these galaxies, and so constrain models of the ISM (e.g. Cox and McCammon 1986). The observations of mini-AGN in spirals give us a tool for studying the whole range of the AGN phenomenon.

Starburst galaxies are important because they may offer us a nearby laboratory for the study of the processes connected with galaxy formation. Understanding the properties of their nuclear outflows is important for shedding light on the interaction between galaxies and the surrounding gaseous medium. In particular, we can explore the importance of these outflows for the enrichment of the intergalactic medium and the formation of the intracluster medium. Moreover, a detailed knowledge of the properties of the X-ray emission of these galaxies is essential for investigating their relevance for the explanation of the X-ray background (e.g. Fabbiano 1989; R. Griffiths, this volume).

Many issues need to be addressed before we can claim an even partial understanding of the X-ray properties of early-type galaxies (E and S0). In particular we need to establish how much of their emission is due to a population of binary X-ray sources, and how much is thermal emission of a hot gaseous component. Understanding the nature of the emission in different galaxies, and measuring in detail its spectral characteristics are needed to attempt sensitive mass measurements of these systems. A comparative study of the different phases of the interstellar medium can give us insights on the evolution of these systems and on the presence or absence of nuclear activity.

Finally, it is important to investigate the sample properties of galaxies, and see, for example, how the X-ray luminosity and spectral properties depend on morphology, luminosity class, and other galaxian characteristics. In particular, we know very little on how different environments influence the X-ray emission of galaxies, and we know nothing on the behaviour of the X-ray properties as a function of the look-back time, although theoretical predictions suggest intense X-ray emission at the epoch of galaxy formation (e.g. Bookbinder et al 1980).

III. Requirements for Future Observatories

Detecting Galaxies

As shown in Section I, normal galaxies are relatively faint X-ray sources. Therefore, if we want to study them in any more depth than allowed by the exploratory *Einstein* survey, and if we want to be able to detect them in large numbers and out to large redshifts to pursue sensible statistical studies of their sample properties and evolution, we need to have large collecting areas of imaging telescopes. Fig. 1 shows a plot of the X-ray flux as a function of galaxy distance, for galaxies with X-ray luminosities in the range of those detected with *Einstein*.

The corresponding B magnitudes are also given. These assume the average X-ray to optical ratio for spiral galaxies from the *Einstein* survey. For elliptical galaxies the corresponding optical fluxes could be larger by a factor of ~ 100 or more. Superimposed on this graph are curves that give the limiting sensitivity for a 5σ detection in a 5000 s observation for the *Einstein* IPC, the AXAF CCD experiment (ACIS), and the planned larger throughput missions XMM and LAMAR. Although longer observing times are of course possible, I am not considering them here, since I assume that the observations of galaxies will only be one of the topics that will be pursued with future observatories, and that therefore there will be constraints on the available observing time. Since I wish to emphasize here the capability of detecting large samples of galaxies in a moderate observing time, I am not showing curves for the high resolution experiments on *Einstein* and AXAF.

Fig. 2 gives the typical angular sizes for a typical galaxy diameter at different distances, and these range from something of the order of a degree for a few nearby objects to a few arcseconds (2-3) for galaxies at high z.

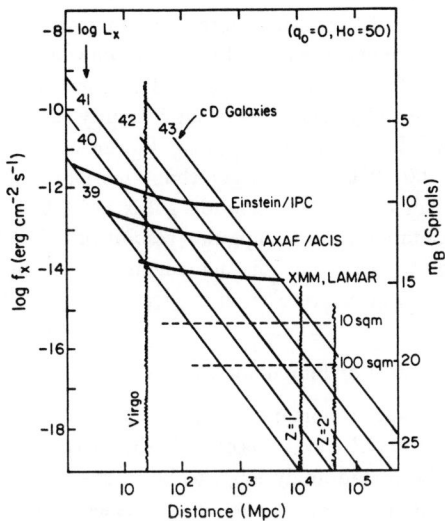

Figure 1: Galaxy detection limits with different instruments (see text).

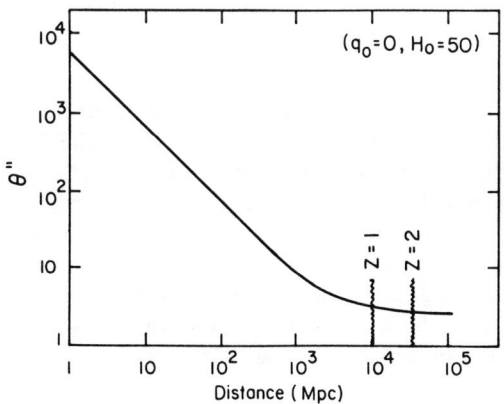

Figure 2: Angular sizes corresponding to 30 kpc at different distances.

With *Einstein* we could detect the least luminous galaxies only in our immediate surroundings, and we could only study the higher luminosity galaxies in the Virgo cluster. XMM and LAMAR will allow the detection in reasonably short observing times of the entire luminosity range of galaxies in the Virgo cluster. This will produce excellent science. However, if we want to study galaxies at larger distances, and therefore sample a variety of cluster environments, we need to go deeper. This is especially true if we want to study in some detail galaxies at the epoch of formation. In the absence of X-ray luminosity evolution in galaxies, larger collecting areas, 100-1000 times larger than those presently planned, and longer observing times will be needed to study galaxies at high z in X-rays. It is also important to remember that an angular resolution of $\sim 1''$ is needed to resolve spatially galaxies at high z, and distinguish them from point-like sources, such as QSOs (fig. 2). This angular resolution is over 10 times that planned for XMM and LAMAR.

High Resolution Studies of Individual Galaxies

A different and not less important aspect of the study of galaxies is that based on detailed high resolution observations of individual objects. Here again the observing times should not be forbiddingly long, to be in the optimal situation of exploring the properties of the X-ray emission components as a function of a reasonably large range of galaxian properties.

Fig. 3 shows a plot of the angular resolution needed to resolve structures of a given linear size at different distances from the Solar System, ranging from 100 pc to cosmological distances. This graph shows that the high throughput missions XMM and LAMAR will not be able for example to resolve supernova remnants in external galaxies, while the higher resolution *Einstein* HRI could resolve them in the LMC, and the AXAF HRC will resolve them as far as M31. However, to be able to resolve SNRs in Virgo galaxies, an angular resolution of $\sim 10^{-2}$ arcsec will be needed. This same angular resolution will allow the study of a few kiloparsec scale structures in high z galaxies, and of planetary scale structures within 100 pc from the Sun.

However, there is no point in studying galaxies at very high resolution without a very large collecting area. Fig. 4 shows the limiting sensitivity for detecting individual sources of a given luminosity in galaxies from the LMC to the Virgo cluster and beyond. I assume 5σ detections in 10^4 sec observations. The AXAF HRC will be able to detect down to the luminosity of the brightest stars in the LMC, but will detect only the brightest binary X-ray sources in the Virgo cluster galaxies. A much larger collecting area will be needed to explore the luminosity function of the X-ray sources in Virgo galaxies down to the luminosities accessible with the *Einstein* data in Local Group galaxies.

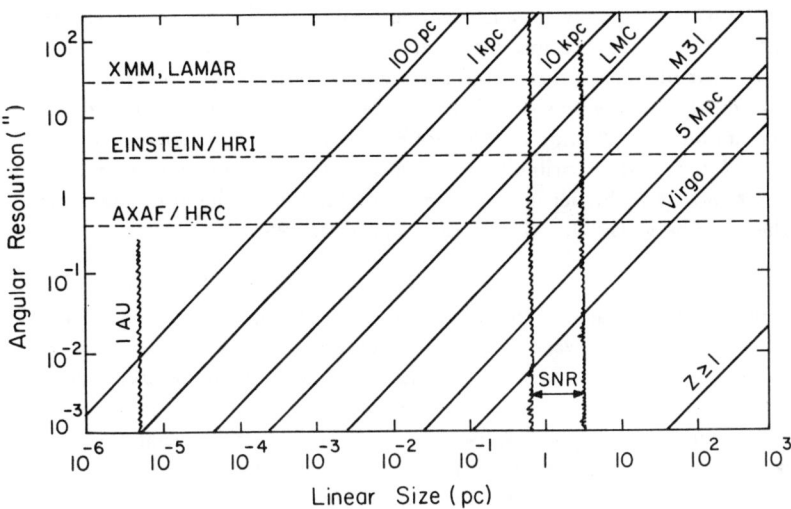

Figure 3: Angular resolution needed to resolve structures of given linear sizes at different distances.

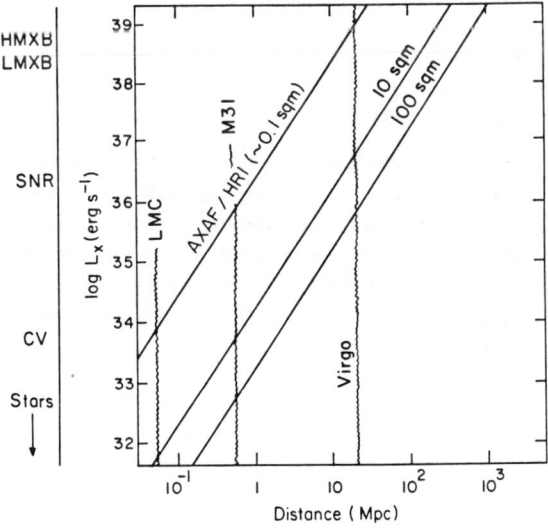

Figure 4: Limiting sensitivities for the detection of point sources in galaxies.

Bandwidth

The X-ray band is extremely broad, covering easily two decades of the spectrum (equivalent to combining the mid-infrared, optical and ultraviolet bands). This breadth has great diagnostic power and the ideal X-ray telescope to study galaxies should cover most of this bandwidth (from $\sim$0.1 keV to 10-20 keV), because different galaxian components contribute to different energy bands. The stellar component, old SNRs and hot ISM contribute to energies $\leq$ 1keV; young SNRs and X-ray binaries emit mostly at higher energies; and small active nuclei are likely to emit both in the soft and in the hard bands (e.g. Fabbiano 1988).

Spectral Studies

Spectral information is essential to understand the physical mechanisms giving rise to the X-ray emission. We want to be able to gather spectral data on galaxies as a whole (e.g. to discriminate between gaseous emission or the integrated output of binary X-ray sources as the predominant source of X-ray emission in X-ray faint early-type galaxies); we want also to be able to study separately specific structures (e.g. bulges vs. spiral arms), or to obtain temperature profiles of elliptical galaxies to measure their masses; finally we want to be able to study individual sources in more nearby galaxies. Therefore, the first requirement for spectral studies of galaxies is the ability to obtain spatially resolved spectral information.

There are two approaches one can take to spectral studies. The first, similar to optical photometry, is to obtain low resolution spectra ($\lambda/\Delta\lambda$ $\sim$a few) or colors over a larger bandwidth. In this case, the incoming photons will be divided in a relatively small number of spectral bins. The second approach, spectroscopy, is to obtain mid-high resolution spectra ($\lambda/\Delta\lambda$ $\sim$100-1000) over different bandwidths. Low resolution spectra can be used to characterize the global spectral properties of different galaxies or populations of X-ray sources; high resolution spectra will be invaluable for exploring the physical state of gaseous components in galaxies (e.g. Canizares, Markert and Donahue 1988).

Even without exploring in detail the characteristics of spectral detectors, it is clear that very high throughput missions (i.e. large collecting areas) will be needed for spectral studies if we do not want to limit ourselves to relatively nearby objects. This can be seen by looking at fig. 1 and considering that the number of incoming photons will have to be divided in a number of separate spectral channels. This is equivalent (in a first approximation) to moving the sensitivity threshold curves upwards by equivalent amounts.

Time Variability Studies

We know from X-ray observation of sources in the Milky Way that X-ray sources vary in time. This time variability gives us precious insights on the nature of sources (see D. Lamb's contribution), especially if joined with spectral analysis. As in the case of spectral studies, large collecting areas are needed to study source variability in external galaxies, even more so if one wishes to explore the spectral behavior in different intensity states. Moreover, it will be desirable to have the flexibility to devote the telescope, or even better a part of it, to long and repeated observations of nearby galaxies to obtain light curves of variable sources on the common characteristic timescales that are longer than a single observation.

IV. The 'Ideal' X-ray Telescope for the Study of Galaxies

Based on the previous discussion I will try to sketch here the characteristics of the ideal X-ray telescope for studying galaxies. I consider this a goal to which the high energy community and NASA should aim, but it is clear to me that the way to reach this goal is by intermediate incremental steps (see e.g. P. Gorenstein's contribution). These intermediate missions will allow us to keep studying galaxies in X-rays, and to test new developments in X-ray optics and instrumentation that will be needed to achieve our final proposed goal.

The characteristics of this ideal X-ray telescope are given below:

- Collecting area > 100 sqm.
- Large bandwidth (0.1 -20 keV).
- Angular resolution of $\sim 1''$. High resolution studies of individual galaxies will require higher angular resolution (up to 10^{-3} arcsec, if there is enough collecting area).
- Spectral capabilities for both X-ray photometry and X-ray spectroscopy.
- Capability to devote part of the area to variability studies of nearby galaxies.

I think that the only possible way of thinking of this X-ray facility is that of an array of X-ray telescopes to be put on the Moon.

This work was supported by NASA contract NAS8-30751.

REFERENCES

Bookbinder, J., Cowie, L. L., Krolik, J. H., Ostriker, J. P., and Rees, M. 1980, *Ap. J.*, 237, 647.
Canizares, C. R., Markert, T. H., and Donahue, M. E. 1988, in *Cooling Flows in Clusters and Galaxies*, ed. A. Fabian (Dordrecht: Kluwer), p. 63.
Cox, D. P., and McCammon, D. 1986, *Ap. J.*, 304, 657.
Donnelly, R. H., Faber, S. M., and O'Connell, R. M. 1989, *Ap. J.*, in press.
Fabbiano, G. 1988, *Ap. J.*, 325, 544.

Fabbiano, G. 1989, *Ann. Rev. Astr. Ap.*, 27, 87.
Trinchieri, G., and Fabbiano, G. 1985, *Ap. J.*, 296, 447.
Watson, M. G. 1989, in *Windows on Galaxies*, ed. G. Fabbiano, J. Gallagher and A. Renzini, (Dordrecht: Kluwer), in press.

DISCUSSION

W. Cash - Long exposures $\geq 10^6$ sec can give the needed sensitivity without major increases in instrument size. While this would greatly reduce the sample size, over a period of years a substantial sample could be collected.

G. Fabbiano - I certainly think that while aiming to the goal outlined in my talk we should not stop studying galaxies in X-rays with whatever facility is available. Moreover, I think that an incremental approach should be followed towards establishing a very large area lunar X-ray telescope, consisting of intermediate size missions. These will both allow us to keep doing science, as you suggest, and to test the new developments in X-ray optics and instrumentation that will be needed for the lunar mission. However, I believe that it is very important to keep our goal in mind. A future generation telescope will not be dedicated solely to the study of galaxies, so unless observations can be done in a reasonable amount of time, the science return will be frustratingly low. Moreover, variability and spectral/variability studies *require* the capability of collecting a large number of photons in short time scales, which cannot be achieved with longer integration times and smaller telescopes. Finally, we must remember that X-ray studies are now beginning to be part of astronomy, and as such we should aim to match the sensitivities achievable or planned in the optical and radio range in a comparable observing time, to insure a truly multi-wavelength approach to the study of astrophysical objects.

C. Canizares - I want to thank you for your inspirational talk. A comment on the spectroscopy is that things are both better and worse than you showed. The XMM reflection gratings have difficulties with extended sources, especially the very diffuse E galaxies for example. The AXAF spectrometers could do better, but that depends on what the final instrument complement really is (i.e., will it include the XRS, the BCS, both or neither?)

R. Burg - Isn't it true that you can trade large field of view optics for collecting area?

G. Fabbiano - This is not true in general. Large field of view optics can help if your aim is to collect a X-ray flux limited sample of galaxies and recognize them as such (i.e. extended objects) or if you want to study in detail a nearby extended galaxy without requiring multiple exposures to cover the observing area with comparable angular resolution. However, large field of view optics alone are not going to help in the study of large samples, if your selection criterion is other than X-ray, as it is likely to be in the case of galaxies, because you will still need to point in arbitrarily different directions. Moreover, they are not going to help in the case of very distant objects, where the main thing is going to be the ability to collect enough photons, and similarly in spectral and timing studies. However, large field of view optics are very desirable and should definitely be considered in designing future X-ray telescopes.

S. Murray - Large area, high angular resolution optics will require large area, high spatial resolution detectors, and both the telescope and the focal plane instrumentation need to be considered in developing new observing capabilities. With 100 m^2 telescopes one needs $\sim$ 10 m^2 detectors to fill $\sim$ 1 deg field of view, unless the packing efficiency of a telescope is significantly improved over that of *Einstein* and AXAF. For a telescope that has a 10m focal length, (such as AXAF) the focal plane scale is 50 microns / arc second. For large area telescopes with fast optical systems, the focal plane scale will be in the range of 10 to 100 microns / arc second. If you wish to have sub-arc second resolution you need 1 to 10 micron detector resolution, and since we want the detector to oversample the desired angular resolution, the 1 micron range is most appropriate. The good news is that these relatively fast telescopes would only require relatively small detectors to cover a 1 degree FOV (40 to 400 mm with the approriate spatial resolution would do it). The bad news is that such detectors really don't exist - HRC comes close and new small pixel CCD's may become available, but not in the overall size needed. Micron resolution is not easy. To overcome this the optical design is driven to higher f numbers which means longer focal length (that's okay if you want to get to higher energy anyway) and now the detector size to cover 1 degree FOV goes up to the few meter range, but the pixel size for arc second resolution is 100 to 1000 microns (again at or below 100 microns to oversample the telescope). With the longer focal length, it is possible to envision milli-arc second resolution goals (along the lines of Cash's approach or just really good fabrication...) and the detector resolution required becomes a micron or less. Such large areas and spatial resolution in detectors are a technical challenge.

G. Fabbiano - Building a very large throughput lunar telescope *is* going to be challenging. For this reason it is important to think in terms of a multistep approach of intermediate missions, including improvements in the optics and in the detectors.

X-ray Observations and the Properties of Active Galactic Nuclei

Richard F. Mushotzky
Laboratory for High Energy Astrophysics
NASA/Goddard Space Flight Center

Active galactic nuclei (e.g. quasars, Seyfert galaxies, radio galaxies, BL Lac objects etc) emit over an extremely broad wavelength range from 10^6-10^{25} Hz. To first order they have roughly equal energy per unit logarithmic bandwidth from the far IR to ~ 1Mev. However, there seem to be 3 local peaks in the energy distribution 1) in the far IR (~10^{13} Hz) perhaps due to the reprocessing of "primary" radiation by dust, 2) at ~ $10^{15.5}$ Hz (the big blue bump) perhaps due to thermal radiation from an accretion disk and 3) a "high energy bump from ~1-3000 kev, the origins of which are unknown.

The x-ray band has several unique properties that make a detailed understanding of this radiation crucial for any understanding of AGN. 1)The 2-10 kev x-ray band seems to show the shortest observed timescales for variability in a given object and thus may originate close(est) to the "central object." 2)There are indications that there are x-ray spectral components (Fe line, edge and a 10-35 kev "excess") that may originate in a reprocessing of the "non-thermal" x-ray radiation in cold matter very near the central engine. 3) the 1-100 kev radiation seems to have a simple "non-thermal" form not seen in other wavelength bands and thus may represent the "initial un-reprocessed" radiation.

Even though x-rays are a major component of the total energy budget of AGN (0.1-0.3 of the bolometric luminosity) we do not have a "viable" theory which directly connects the emission seen in the x-ray, UV, IR and optical bands- for example, based on our present observations of time variability, if the UV-optical radiation comes from an accretion disk then it is clear that the 1-10 kev x-rays do not originate from this disk. In addition, the evolution of the x-ray radiation with cosmological epoch seems to have a different timescale than that in the optical or radio band. However, there are strong statistical connections between the radio, optical and high energy radiation which have dependences on the absolute luminosity of the object.

The highly successful x-ray and γ-ray missions of the past 20 years, while discovering the broad outlines of the high energy radiation from AGNs and providing much pioneering data, have barely scratched the surface of the subject. One hopes that the wealth of new data to be obtained in the next 5 years from the new generation of high energy experiments (e.g.Granat, BBXRT, Rosat, GRO, Astro-D, Spectrum X-Γ, SAX and XTE) in addition to other new instrumentation (e.g. HST,the new ground based telescopes, the VLBI array and ISO) will provide a firmer foundation to posit the interesting questions for the experiments of the next millennium. I thus approach this talk with great trepidation for, no matter what I say, most of it is surely going to be incorrect in detail. For example, if I had been giving this talk two years ago there would have been no mention of "direct" observations of accretion disks or of the determination of the mass of central objects via Fe line spectroscopy, subjects that are now central to the study of AGN .

I hope that in 5 years time we will be able to meet again, and be able to, with some more confidence, "predict" what will be interesting for the next generation of missions: XMM, AXAF and Astro-E. However, even then, I am sure that prediction of the interesting science for the next century will be an extremely risky proposition for if we have learned anything from the past, it is that the new data will fundamentally change our thinking on the nature of active galaxies, their evolution and the mechanism of energy generation and spectral formation. Thus new issues and problems will arise that more powerful missions will be required to refine.

However even given the awesome capabilities of AXAF and XMM and our present ignorance there are some fundamental problems that we are able to think concretely about now, even in the absence of a fundamental theory, that will require enhanced capability.

II Direct Observation of the Evolution of Luminosity and Density of AGN

As of 1989, hard (E>2 kev) selection seems to be the least biased way of detecting active galaxies (in particular quasars, Seyfert I's and Bl Lac objects; the surveys have not been sufficiently sensitive to determine if this is also true for Seyfert IIs, radio galaxies or LINERs). A sufficiently sensitive hard x-ray survey can directly observe the entire luminosity function, from 10^{42}-10^{47} ergs/sec, over a large range in redshift and, in a relatively model independent way, directly observe the evolution of AGN.

The other commonly used methods for finding AGN (e.g. IR color selection, optical line emission, UV excess, optical colors, radio emission) all have large biases. We do not know (yet) of any objects that are devoid of hard x-ray flux. However at low (E <2 kev) energies the effects of absorption by cold matter, reprocessing and "complex" spectral corrections, are substantial . These effects seem to be at a minimum in the 2-10 kev band . In addition the presence of a well measured "diffuse" flux in this band provides a strong integral constraint to AGN evolution.

The next two generations of high energy missions viz Astro-D, Spectrum X-Γ, AXAF and XMM with their great sensitivity and broad energy coverage will make important contributions in this area but will not sufficient sensitivity to directly observe the evolution of AGN out to z~5 down to low x-ray luminosities. What is needed is a hard x-ray system that can observe objects as faint as the z=0 characteristic luminosity (the position of the low luminosity "break" in the luminosity function) $L(x) \sim 10^{43}$ ergs/sec, at large redshift--this amounts to a sensitivity requirement of 5×10^{-16} ergs/cm^2-sec, in the rest frame 2-10 kev band.

The optical luminosity function of AGN can be modeled by $F(L) \sim L^{-a} \exp(L/L^*(z))$; with $L^*(z)$ increasing by a factor of 15 between z~1 and z~2.8 .If we assume that the ratio of optical to x-ray luminosity, log<F(opt)/F(x) >~0, is not a strong function of redshift (clearly an assumption that will be tested by ROSAT and Astro-D data) and that the characteristic x-ray luminosity is not evolving (not yet clear from the x-ray results) then to probe to $L^*(z=0) \sim 5 \times 10^{43}$ ergs/sec (which corresponds to $L^*(opt) = M_v \sim -22.5$), at z=5 (which corresponds to an object of m~ 28th optical magnitude) requires a sensitivity of $F(x) \sim 5 \times 10^{-17}$ ergs/cm^2-sec this is about a factor of 10 below the AXAF sensitivity limit

However if the x-ray L* also evolves in the same way as in the optical then one the limit is considerably brighter ~ 5×10^{-16} ergs/cm^2-sec , in the rest frame 2-10 kev band, which is within the reach of AXAF for very long ~10 day, exposures. However, AXAF will obtain very small samples at this limit.

An additional requirement is sufficient angular accuracy, < 10", to avoid source confusion and allow optical identification. While the value of the x-ray log N-log S is simply not known at these faint fluxes one can try to scale from optical data which gives values of ~500 AGN/sq degree at 24th mag.- extrapolating a evolutionary model gives ~ 1-3000/sq degree. Using the standard confusion criteria of 40 beams/source requires an angular resolution of ~ 10" such that the AGN is uniquely located. Alternatively to distinguish AGN from galaxies requires $\delta\theta \sim 2$" since there are ~ 500 galaxies /sq arc-min at 28th mag and if it proves difficult or impossible to identify the source from optical data, higher quality x-ray imaging is required.

This experiment should have sufficient spectroscopic capability to obtain redshifts of AGN from x-ray spectroscopy at interesting distances/luminosities in exposure times which are not dissimilar from those required for optical spectroscopy. This results in a "requirement of -1) for continuum measures a collecting area of >10x that of XMM; 2)and for line redshifts E/dE>40, A> 20,000cm^2 at Fe to reach 2×10^{-15} in 50,000 second exposures. This collecting area would allow measurements of redshifts for z~1, Log L(x)~ 43 or z~3 log L(x)~44 objects which should be about $m_v \sim 24$th mag.

These collecting areas are sufficient that a continuum spectrum can be determined to sufficient accuracy so that its shape can be determined (~500 photons for a XMM like bandpass) in ~1 day of exposure time for objects at 2×10^{-15} ergs/cm^2-sec. Scaling to the recent Ginga Fe line results for AGN (150 ev EW lines at 6.4 kev) and assuming that the line is narrow, the expected line flux is ~4×10^{-5} ph/cm2-sec/mC. At the μCrab level (2×10^{-14}) a CCD like system with an area of ~2×10^4 cm^2 would obtain a ~3σ detection of the Fe line in exposures of 2.5×10^5 seconds (a calorimeter would have a ~5σ detection for a narrow line).

In order to survey a sufficient solid angle of the sky in a "reasonable" time a solid anglextime factor of ~1×10^7 cm^2-arcmin^2 is required since measurement of many (>1000) objects is necessary to determine a high quality luminosity function as a function of redshift. Because of the high space density of AGN however, a survey of only 2-3 square degrees at flux limit of 2×10^{-15} is necessary to obtain such a sample. With imaging optics, where one is limited by vignetting only, a field of view of diameter 20' is possible at 10 kev. A 3 square degrees solid angle requires ~40 exposures of 50ks each so our putative AGN survey can be done in less than one month. Of course one would like to survey several other fields to determine other interesting results such as determination of the two-point correlation function and its evolution with redshift, a search for large scale structure at moderate to high z as traced by AGN and a measurement of the time variability characteristics of high redshift objects.

III Direct Measurement of the Mass and Change with time of the Mean Mass of the Central Object

A. If AGN have some characteristic timescale in the x-ray domain which is related to their mass, then determination of this timescale in an ensemble of objects will allow measurement of their relative mass. Scaling to those few objects whose mass is known by other means (e.g. from optical measurements of the stellar velocity field near the central object) would allow an "absolute" calibration. Extension of these results to high redshift and a broad range in luminosity would allow a measure of the evolution of the "mass to light ratio" of AGN and its variance with absolute mass.

Scaling to the best observed stellar black hole, Cyg X-1, and assuming that the characteristic time varies only with the mass gives two options for a characteristic timescale: $3ms/10M_0$ or $1sec/10M_0$ (using either the fastest timescale seen in Cyg X-1 or the knee of the autocorrelation function). These scale to 30×10^3 or 10^7 sec respectively for a 10^8 M_0 object. Using $L_x \sim 0.1 L_{bol}$ and a characteristic Eddington ratio for moderate to low luminosity objects of $L_{bol}/L_{edd} \sim 0.1$ one has a characteristic "fast" timescale of 3×10^3 sec at $L(x) \sim 10^{44}$ - which "seems" to be consistent with observations. To well determine the average counting rate requires ~ 50 photons per time bin and thus a counting rate of 0.016cts/sec for our putative 3×10^3 sec timescale. A mission with 10 times the area of XMM could determine the timescale of variability of a $L_x \sim 10^{43}$ object at a redshift of $z \sim 0.5$ and a $L_x \sim 44$ one at $z \sim >>1$ (depending on the value of q_0). This would allow measurement of the evolution in the M/L ratio at redshifts at $z \leq 1$.

However, it is equally important that the actual timescales of variability of AGN be determined for a large sample. If the "knee" of the autocorrelation is the "important" timescale then it is very important to measure timescales of > 1 day for sources of $L_x > 10^{44}$ ergs/sec. A large sample can be easily obtained by a small "moderate sky" x-ray monitor which can reach fluxes of ~ 0.5mC in 1/2 day.

B. Use of Fe-K Lines to Derive the Mass and Size of the Central Object :
Recent indications are that the Fe K lines in some AGN arise from cold material (perhaps an accretion disk) very close to the central engine. This evidence comes from the very rapid response of the Fe-K line to continuum variations seen in some objects (e.g. a delay time of less than 400 seconds in NGC6814), the very large depth of the Fe-K edge and the existence of a high energy "bump" in the spectra of ~1/2 of the AGN with high quality Ginga spectra . If this is true then measurement of the shape of the Fe line (predicted to be double horned with the high energy peak stronger than the low energy peak and with the centroid of the line shifted by gravitational redshifting) combined with the time variability characteristics of the line shape can give the mass of

"thick disks" far from the central engine which also provide the gas to feed the nucleus and radiation supported thick disks near the black hole. Each of these possibilities produce unique x-ray/spectral temporal signatures. These "objects" may be directly connected to the jets seen in Seyfert galaxies. It is not clear if they are related to the much more powerful jets in radio galaxies and quasars.

To study these three phenomena in more detail requires very broad bandwidth (0.1-300 kev), good, $E/\delta E > 40$, energy resolution in the 0.1-9 kev band (to observe the atomic spectral features), long observations (to study the reverberation phenomena in the line and the continuum and to compare the time variability characteristics of the "different" continuum components), and sufficient sensitivity to study the relations amongst these components in a wide variety of AGN.

V. Origin of the Energy and the Continuum

At present we have no "reliable" theory for either the origin of the energy in the high energy continuum . Most of the proposed physical models for photon creation such as pair-dominated plasmas, thermal compton scattering, synchro-compton scattering, or shock acceleration of protons are "best" tested by looking at time variable spectral shape and/or spectral features at $E \gg 20$ kev. It is not clear if we have any "testable" theory for the origin of the energy either . However if it is due to "relativistic" phenomena (such as tapping the spin of the black hole, shock acceleration of particles or magnetic reconnection) similar data is also required. Since the energy budget of AGN is diverging as $E^{0.3}$ (at least until $E \sim 1$ Mev) it is clear that measurements at the highest energies are very desirable.

While results from GRO, Granat, Ginga, SAX and XTE will probably suggest a "best" theory for low redshift, low luminosity objects these missions are not sensitive enough to test the evolution with cosmic time of the underlying physical conditions. There are strong reasons to believe that the physical mechanism(s) should vary with cosmic time (e.g the spin and mass of the central object, the relative accretion rate and angular momentum of material etc) and luminosity (compactness, ratio of "disk" to non-thermal luminosity) . Not only must such a mission be able to study the fainter, more distant objects (Einstein observatory data indicates that the most luminous quasars at $z \sim 2$ have a 0.3-3.5 kev flux of $\sim 10^{-12}$ ergs/cm^2-sec) but it must also have sufficient sensitivity to study possible spectral changes on short timescales in brighter objects. Missions with sensitivity >10x that of XTE with a similar bandpass are required to start such a study.

While the existence of spectral features at $E > 9$ kev is at present very speculative it seems likely that, if pair dominated plasmas are indeed the origin of the high energy continuum, that detailed study of the 511 kev annihilation feature will prove to be very important. It is possible that GRO will provide the first indications of the existence of such a feature in "normal" AGN (of course our galactic center - a very low luminosity AGN- has an annihilation feature). We await the GRO results to attempt a calculation of the required sensitivity of a follow on mission.

the central object and the size of the Fe line region. Measurements of the change of the shape and the intensity of the line in response to continuum variability, so-called "reverberation mapping", can give both the emissivity distribution of the region responsible for the Fe line and the depth of the gravitational potential. This type of measurement thus has the potential for "proving" the existence of a black hole and determining its mass.

Since it requires at least ~50cts/timebin/Fe line, to measure the line shape, one needs~3000 cm^2 at 6.4 kev for a 1mC source assuming a time bin of 500 seconds- about a factor of two better than XMM . While it is important to study these "fast variability" low luminosity objects at higher redshifts (and thus lower flux) it will be technically very difficult to go to z>0.1 with this method (an L_x ~1x10^{43} ergs/cm^2-sec AGN would have a flux of ~2x10^{-13} ergs/cm^2-sec). Thus to study the evolution of mass will require the use of more luminous objects which should have longer timescales. To study the evolution of moderate luminosity quasars L_x~10^{45} with a Fe line emiting region of size, R(Fe)~20R$_G$ and thus a reverberation time of (10^{4}sec/M$_8$)sec, at z~1 requires a collecting area of ~10,000 cm^2 at E~3.5 kev (roughly ~3 times XMM), with an energy resolution ~2x better then a CCD. To go to z~2 requires ~50,000 cm^2. In order to study these objects requires quasi-continuous monitoring on timescales of ~(3x10^4/M$_8$)seconds so that the variability can be well sampled.

IV Material in/near the Central Object
A. Study of the Accretion Flow

Because much of the accreting material is sufficiently ionized that it cannot be "seen" in the IR-UV band it has been very difficult to determine how the central object in AGN is "fed". However, as has been recently pointed out , x-ray spectroscopy and timing may be one of the few ways of studying the flow of material into and near the central object. Even very low filling factors (10^{-12}) of gas at high densities (n~10^{14}) leaves a strong imprint on the observed EUV-x-ray spectrum and very weak signatures in any other wavelength band . While these effects have not yet been "detected" due to a lack of sufficient spectral resolution and sensitivity in the past if they do indeed exist they should be easily detected by the next generation of experiments.

B Study of Accretion disk:
Recent results (viz sec III B) on the presence of 1)strong Fe edges 2) a flattening of the spectra of some Seyfert galaxies at E>15 kev 3) the ubiquity of ~100 ev EW Fe lines 4) and the presence of "thermal" soft components have led many workers to suggest that these are the signs of reprocessing of x-ray radiation by an accretion disk . Alternative explanations of the soft components as the high energy extension of the UV "bump" (often interpreted as the signature of an accretion disk) are also viable. In either case it is clear that x-ray spectroscopy and time variability studies can determine many of the properties of the so-far unobservable accretion disk.

C. Study of Collimating Structures:
Recent theoretical work, optical spectroscopy and polarimetry and radio imaging have indicated that there may exist very optically thick structures which "collimate" the radiation from the central engine. The nature of these structures is not well understood but 2 possibilities are dense molecular

"thick disks" far from the central engine which also provide the gas to feed the nucleus and radiation supported thick disks near the black hole. Each of these possibilities produce unique x-ray/spectral temporal signatures. These "objects" may be directly connected to the jets seen in Seyfert galaxies. It is not clear if they are related to the much more powerful jets in radio galaxies and quasars.

To study these three phenomena in more detail requires very broad bandwidth (0.1-300 kev), good, $E/\delta E > 40$, energy resolution in the 0.1-9 kev band (to observe the atomic spectral features), long observations (to study the reverberation phenomena in the line and the continuum and to compare the time variability characteristics of the "different" continuum components), and sufficient sensitivity to study the relations amongst these components in a wide variety of AGN.

V. Origin of the Energy and the Continuum

At present we have no "reliable" theory for either the origin of the energy in the high energy continuum . Most of the proposed physical models for photon creation such as pair-dominated plasmas, thermal compton scattering, synchro-compton scattering, or shock acceleration of protons are "best" tested by looking at time variable spectral shape and/or spectral features at $E \gg 20$ kev. It is not clear if we have any "testable" theory for the origin of the energy either . However if it is due to "relativistic" phenomena (such as tapping the spin of the black hole, shock acceleration of particles or magnetic reconnection) similar data is also required. Since the energy budget of AGN is diverging as $E^{0.3}$ (at least until $E \sim 1$ Mev) it is clear that measurements at the highest energies are very desirable.

While results from GRO, Granat, Ginga, SAX and XTE will probably suggest a "best" theory for low redshift, low luminosity objects these missions are not sensitive enough to test the evolution with cosmic time of the underlying physical conditions. There are strong reasons to believe that the physical mechanism(s) should vary with cosmic time (e.g the spin and mass of the central object, the relative accretion rate and angular momentum of material etc) and luminosity (compactness, ratio of "disk" to non-thermal luminosity) . Not only must such a mission be able to study the fainter, more distant objects (Einstein observatory data indicates that the most luminous quasars at $z \sim 2$ have a 0.3-3.5 kev flux of $\sim 10^{-12}$ ergs/cm^2-sec) but it must also have sufficient sensitivity to study possible spectral changes on short timescales in brighter objects. Missions with sensitivity >10x that of XTE with a similar bandpass are required to start such a study.

While the existence of spectral features at $E > 9$ kev is at present very speculative it seems likely that, if pair dominated plasmas are indeed the origin of the high energy continuum, that detailed study of the 511 kev annihilation feature will prove to be very important. It is possible that GRO will provide the first indications of the existence of such a feature in "normal" AGN (of course our galactic center - a very low luminosity AGN- has an annihilation feature). We await the GRO results to attempt a calculation of the required sensitivity of a follow on mission.

VI X-ray Emission from Jets:

One of the main problems of extragalactic astronomy has been to discern the mechanism of particle acceleration and confinement in jets. This requires the determination of the environment of the jet and spatially resolved spectra of jets. Because synchrotron loss times are much shorter in the x-ray band, x-ray imaging will reveal the sites of particle acceleration. Measurement of inverse-Compton scattered x-rays combined with radio maps will allow measurement of the magnetic field and particle densities along the jet. Direct measurement of the polarization of the x-ray signal will allow direct determination of the origin of the photons. (X-ray observations of the SS433 jets suggest that there may be x-ray spectral features associated with the jets that provide information on the bulk velocity of the particles.)

The best data we have to data on jets has come from the VLA. Scaling from these maps shows that a spatial resolution of $\delta\theta < 5"$ is necessary, combined with broad spectral coverage to determine the nature of the radiation (e.g thermal due to shocked gas, synchrotron or compton) combined with "good" energy resolution. Until Rosat and AXAF results are obtained it is not clear what area, spectral resolution or angular resolution are really required. It seems clear that polarimetry is highly desired if the radiation turns out to be due to synchrotron radiation. It is also thought that the x-ray radiation from Bl Lac objects is due to a jet in our line of sight and thus may be beamed synchrotron radiation. Time and energy resolved polarization studies of these objects is highly desirable.

VII Study of the Intervening Material Along the Line of Sight

A. The Broad Line Clouds

Standard and "non-standard" models of the "clouds" emitting the broad optical and UV lines in AGN require relatively thick clouds $N_H \sim 10^{22}$-10^{24} atms/cm^2 in order to reproduce the anomalous Balmer decrements and Balmer to Lyman line ratios. These clouds are extremely optically thick in optical and UV resonance lines but not in x-ray resonance absorption lines. High resolution x-ray spectroscopy of absorption edges and resonance absorption lines will allow direct determination of the abundance ratios and absolute column densities. These lines should be intrinsically broad. If we assume widths ~1000km/sec and optical depth >1 over the whole line then a resolution of ~300 is required. To detect such features in AGN requires sensitivities of ~3×10^{-6} ph/bin/10^{-11}ergs/cm^2-sec. Thus to obtain results in exposures of ~50ks requires effective areas of >400cm^2 for sources of flux 10^{-11}ergs/cm^2-sec.-thus the AXAF calorimeter (and perhaps the high energy grating) can obtain results for many Seyfert I galaxies. However to go to fainter, high redshift systems of flux 0.1mC requires the area of XMM at Fe (~1500cm^2) with a high throughput system with a resolution of >300.

B. Broad Absorption Line Quasars:

these enigmatic systems have very broad absorption lines ($\delta V > 10,000$ km/sec) in some of the resonance absorption lines. The systems are very difficult to understand due to the high ionization of these features and the inability of optical spectroscopy to measure their true optical depth. X-ray spectroscopy would be enormously useful. Moderate

(E/δE~60) to high (E/dE~1000) resolution is required. The only putative detection of such a feature in the x-ray has been a OVIII Lyα absorption feature seen in several Bl Lac objects. The "classical" BAL quasars are very dim in the x-ray band(in general they were not detected by the Einstein observatory, perhaps due to strong absorption expected in these systems) and it is unclear what collecting area is required to study them.

C. The Ly-α forest clouds and the Mg II and C IV absorption features:
These absorption features are seen towards many high redshift quasars and are probably due to intervening "galaxies" and intergalactic gas . All of these systems are very narrow with broadening of less than 300km/sec. In particular the Ly-a forest clouds have widths of ~10-20 km/sec. X-ray spectroscopy can determine the total column density of these features and the relative abundance of the elements. If some of these objects are proto-galaxies this may allow direct determination of the evolution of elements in the ISM . This requires a system with a resolution >1000 and sufficient sensitivity to detect O, Mg and perhaps Si and S absorption in the continuum of dim high z quasars.

VIII Conclusion
While many of the requirements for the study of AGN in the 21st century are very uncertain it seems as if there are several general statements one can make. To obtain the data necessary to directly study evolution, determine redshifts, study accretion disks and determine characteristic timescales of variability requires δE~100 ev, a collecting area at E~6 kev of >3000 cm^2 , a broad energy range of 0.1-10 kev, good (δθ<10") spatial resolution and the capability for long observations (e.g. a non-low earth orbit). To further the study of the origin of the energy and the continuum and the accretion disk requires a bandpass of 2-300 kev and a sensitivity more than 10x that of XTE. To further the study of accreting material and material along the line of sight (broad line clouds, intervening galaxies etc) requires large area combined with "high" E/δE>300 spectral resolution. To further pursue the study of jets and collimating structures requires high spatial resolution combined with high sensitivity.

However, the most important requirement for the study of AGN in the 21st century is the continued existence of numerous small missions in the post Rosat/Astro-D era so that the technology to develop the "Great Observatories" of the next century can be tested, new scientific ideas tested out (imagine if optical astronomy only one 10m telescope available in the next 25 years) and specialized research areas followed up.

THE ALL-SKY X-RAY FOREGROUND: OUTSTANDING ISSUES

Elihu Boldt

Laboratory for High Energy Astrophysics
NASA / Goddard Space Flight Center
Greenbelt, MD 20771

Introduction

Although there seem to be structures in the universe on scales up to about a tenth of the event horizon (Tully 1989) a basic tenet of modern cosmology is that the universe on larger scales is isotropic and homogeneous. Under the assumption that the CXB (cosmic X-ray background) is dominated by X-rays emitted at such high redshifts (i.e., z>>0.1, HR>>30,000km/s) the only global anisotropy that would then be expected is the weak dipole variation of apparent surface brightness arising from the solar system's motion with respect to the proper frame of this radiation (Compton and Getting 1935). The global anisotropy of the foreground, however, involves more considerations. In particular, the coherent 600km/s velocity of the local group of galaxies with respect to the proper frame of the microwave background is at least an order of magnitude larger than the individual velocities of recession for constituent galaxies expected from the Hubble expansion; this indicates an appreciable gravitational acceleration from an anisotropic mass distribution outside the local group. Observations of galaxies in the IR and optical (Lynden-Bell et al. 1989) suggest that this acceleration arises mainly from a foreground of anisotropically distributed mass within z=0.013 (i.e., HR<4000km/s). Since the X-ray luminosity of bright extragalactic X-ray sources provide a good mass measure of the radiating objects involved and can be observed relatively free of galactic obscuration effects, such sources are likely candidates for serving as reliable tracers of the total underlying mass (i.e., dark as well as visible) responsible for the acceleration of the local group. In this connection, we note that the local gravitational dipole implied by the fifty X-ray brightest clusters of galaxies at z>0.013 considered by Lahav et al. (1989) is relatively small compared with that inferred from the only three clusters at lower redshifts. Since the local space density of AGNs is about two orders of magnitude greater than rich clusters, however, such compact sources have the potential of providing a vastly improved statistical sample for tracing mass in the low-redshift region of particular interest. Furthermore, recent HEAO-1 A2 dipole analysis of the X-ray flux from resolved AGNs indicates that they are indeed strong tracers of this matter (Miyaji and Boldt 1989). The implications of this for the very pronounced large-scale foreground anisotropies to be measured via AGNs observed in more sensitive all-sky surveys are discussed here.

Dipole Moments

We know that the extragalactic X-ray sky is dominated by a remarkably isotropic unresolved CXB and suspect that there exists a proper frame in which it shows precise global isotropy. If this frame is anchored to that of the microwave background, then the solar velocity $v_\odot$=380km/s relative to it yields an apparent dipole anisotropy in the observed surface brightness (B), with an apex having galactic coordinates l=267°, b=+50° (Lynden-Bell et al. 1988). The dipole moment (D) of this distribution over the whole sky is defined as

$$D \equiv \int_{4\pi} B \cos(\theta) \, d\Omega \quad (1)$$

where θ is the angle relative to the preferred axis. The monopole (M) is:

$$M \equiv \int_{4\pi} B \, d\Omega. \quad (2)$$

For a power-law energy spectrum of index (α) the dipole/monopole ratio (D/M) is

$$D/M = (v_\odot/c)[1+(\alpha/3)]. \quad (3)$$

For α=0.4, the value appropriate to the 2-10keV CXB, D/M =1.4×10^{-3} (i.e., a small number). Although this is compatible with the CXB global anisotropy estimated with HEAO-1 A2 (Shafer 1983), the precision of that determination is severely limited by foreground fluctuations due to unresolved sources and possible extended regions of weakly enhanced surface brightness (Jahoda and Mushotzky 1989).

In sharp contrast to the small value of D/M for the unresolved CXB as a whole, the D/M ratio derived from the all-sky extragalactic foreground of discrete sources (at RH<4000km/s) is very large. For examining this foreground we define

$$D \equiv \Sigma_i S_i \cos(\theta_i) \quad (4)$$

and

$$M \equiv \Sigma_i S_i \quad (5)$$

where S_i is the flux from each individual source and θ_i the angle with respect to a direction ($l=268°$, $b=+27°$) defined by the peculiar velocity of the local group (Lahav et al 1989). In particular, for galaxies (mainly spirals) observed with IRAS, $D/M \approx 0.2$ (Lahav et al. 1988). For optically observed galaxies (ellipticals as well as spirals), $D/M \approx 0.4$ (Lynden-Bell et al. 1989). With sources observed in X-rays $D/M \approx 0.5$ for AGNs (Miyaji and Boldt 1989) and $D/M \geq 0.5$ for the statistically weaker sample of clusters (Lahav et al 1989).

According to linear perturbation theory (Peebles 1980), the peculiar velocity v_{LG} of the local group (LG) is given by

$$v_{LG} \approx g_{LG} [2/(3H)](\rho/\rho_c)^{-0.4} \quad (6)$$

where H is the Hubble constant, ρ is the mean total mass density, ρ_c is the critical closure density and g_{LG} is the net gravitational acceleration vector at the local group. The magnitude of g_{LG} is given by

$$|g_{LG}| = \Sigma_i g_i \cos(\theta_i) = 4\pi G(\rho/b)(D/M)R \quad (7)$$

where the sum indicated is over gravity contributions (g_i) arising from all mass elements within a radial distance (R), G is the gravitational constant and $b \equiv$ (luminous contrast)/(mass contrast) is a "bias" factor characterizing radiative tracers of mass. In particular, using the values of D/M observed in various electromagnetic bands we infer from equations (6) and (7) that $b(X\text{-ray}) \approx b(\text{optical}) \approx 2 \times [b(IR)]$, where $b(\text{optical}) \approx 2.7(\rho/\rho_c)^{0.6}$; see Lynden-Bell et al.(1989).

Future Prospects

Defining $h \equiv H/(50 \text{km s}^{-1}\text{Mpc}^{-1})$ in assuming a present-epoch density of about $10^{-6}h^3\text{Mpc}^{-3}$ for rich ($R \geq 1$) clusters (Bahcall 1988) the number expected within the critical foreground region (HR<4000km/s) corresponds to only a couple of objects or so. This is clearly too few for their effective utilization in tracing the anisotropic distribution of matter responsible for the peculiar motion of the local group. On the other hand, for an AGN space density of about $10^{-4}h^3\text{Mpc}^{-3}$ for $L_x > 10^{42}h^{-2}\text{ergs/s}$ (Persic et al. 1989) the number expected in this foreground region would be a few hundred. And the number expected over twice the critical depth (i.e., within HR$\approx$8000km/s) would be more than a thousand, ample for studying the dipole growth with respect to redshift in the interesting region where preliminary results based on a limited sample of AGNs (Miyaji and Boldt 1989) indicates the approach of a plateau (i.e., saturation). The spectral homogeneity (2-10keV) of AGNs and the good correlation of their X-ray luminosity with the mass of the underlying supermassive blackhole (Wandel and Mushotzky 1986) renders these sources potentially powerful and interesting tracers of all gravitational mass. A suitable all-sky X-ray foreground survey of AGNs could be used to answer the question: "How do supermassive blackholes trace the overall underlying mass distribution responsible for the peculiar velocity of the local group?"

The sensitivity (2-10keV) required for the all-sky survey of foreground AGNs described above would be 3.3×10^{-13} ergs cm^{-2} s^{-1}. Although the ROSAT all-sky survey is expected to be at this sensitivity level below 2keV, it does not have sufficient response within the higher energy band needed for an unbiased AGN survey (i.e., one minimizing the effects of galactic absorption and spectral variations). Deepening the all-sky survey to HR≈30,000km/s (i.e., z≈0.1) would require a sensitivity of about 2×10^{-14} ergs cm^{-2}s^{-1} (2-10keV), comparable to that of the high sensitivity small-field survey (1-3keV) of the Einstein Observatory.

Acknowledgements

I thank Ofer Lahav and Takamitsu Miyaji for stimulating discussions.

References

Bahcall, N. 1988, Ann. Rev. Astron. Astrophys., **26**, 631
Compton, A. and Getting, I. 1935, Phys. Rev., **47**, 817
Jahoda, K. and Mushotzky, M. 1989, Ap. J., **346**, 638
Lahav, O., Rowan-Robinson, M. and Lynden-Bell, D. 1988, M.N.R.A.S., **234**, 677
Lahav, O., Edge, A., Fabian, A. and Putney, A. 1989, M.N.R.A.S., **238**, 881
Lynden-Bell, D. et al. 1988, Ap. J., **326**, 19
Lynden-Bell, D., Lahav, O. and Burstein, D. 1989, M.N.R.A.S., **241**, 325
Miyaji, T. and Boldt, E. 1989, Ap. J. (Letters), submitted
Peebles, P. J. E. 1980, The Large Scale Structure of the Universe (Princeton: Princeton University Press)
Persic, M. et al. 1989, Ap. J., **344**, 125
Shafer, R. 1983, PhD thesis, University of Maryland, NASA TM 85029
Tully, B. 1989, in Large Scale Structure and Motions in the Universe, M. Mezetti et al. (eds), Kluwer Academic Publishers, p.41
Wandel, A. and Mushotzky, R. 1986, Ap. J., **306**, L61

DISCUSSION

Alan Bunner

What angular resolution and effective area would be needed for this anisotropy survey?

Boldt

An all-sky scan with a mission like AXAF would be a good first step.

Steven Kahn

I'm worrying about the source ID problem. You have to know that your detected sources are AGN. How many AGN do you expect out to 32,000 km s^{-1}?

Boldt

The number of AGN with $L > 10^{42}$ ergs s^{-1} up to 4000 km s^{-1} is about 80. Up to 32,000 km s^{-1} it would be 512 ($=8^3$) times 80 (i.e., a large number).

Giuseppina Fabbiano

Why do you want to use the X-ray band? What is special about the X-ray band with respect to, e.g., the IR?

Boldt

Since the X-ray luminosity of an AGN is a remarkably fixed fraction of the Eddington luminosity, it is a linear measure of the black-hole mass. It would be interesting to know how supermassive black holes trace the total underlying gravitational mass.

Martin Elvis

To put some numbers on this: we did some calculations for the Space Station mission LAMAR. Using the slewing data only ($\sim 10\%$ of the whole) produces a survey to ROSAT depth, but in the 2 – 10 keV band, in 3 – 5 years.

Boldt

We would have to worry about the uniformity of sensitivity over the whole sky for such an experiment.

Richard Burg

Does evolution of sources play any role?

Boldt

We are just considering redshifts less than 0.1. As such, I don't think AGN evolution plays a role in this question.

Hugh Hudson

Would a suitable survey have to cover the whole sky? Margaret Geller just proposed a relatively deep survey on a slice, instead, for different purposes. Would this require a different instrument?

Boldt

The vector sum of all intensities is proportional to the net gravity vector only if there is an *all-sky* survey at uniform sensitivity.

SCIENTIFIC REQUIREMENTS FOR STUDYING THE COMPOSITION OF THE X-RAY BACKGROUND

R. E. Griffiths

Space Telescope Science Institute

ABSTRACT

A major goal of X-ray astronomy for the 21st century is to make significant contributions to observational cosmology, via the detection of objects at fluxes of 10^{-19}ergs cm^{-2} s^{-1}, such as the star-forming galaxies at moderate redshift which make up the x-ray background. In order to meet this sensitivity requirement, which is at least two to three orders of magnitude deeper than AXAF, the challenge for the remainder of the present century is to develop the economical and practicable technology of very large mirror arrays.

INTRODUCTION

We need to be able to develop facilities which can detect the star-forming galaxies responsible for the x-ray background, i.e. galaxies which have already (1988) been observed in deep ground-based optical surveys. Otherwise, X-ray astronomy will fall behind other branches of astrophysics and will become the domain of an increasingly powerless interest group, lacking the support of the general astronomical community. We need to be able to detect these galaxies at moderate to high redshift in the 21st century, so that x-ray astronomy will be able to contribute to the study of cosmology, i.e the evolution of star-forming activity and the formation of active galactic nuclei. Appropriate technology needs to be developed in the remainder of this century to enable x-ray astronomers to participate in the probing of the epoch of galaxy formation. The alternative is for x-ray astronomy to be left on the sidelines in what promises to become the greatest epoch of research in astronomy.

DEEP SURVEYS AND THE XRB

Although one of the primary goals of the Einstein Observatory was to resolve the problem of the X-ray Background (XRB), that telescope was capable of detecting directly only 25% of the sources responsible for the XRB, but also allowed indirect evidence that AGN lead to a contribution of up to 50%. The discrepancy between the AGN spectra and that of the XRB has been discussed by many authors, and there remain only two current popular hypotheses for the bulk of the XRB between 3 and 10 kev, viz. a hard component from AGN, such as that which might be expected from AGN dominated by reflection spectra (Fabian 1990), and the contribution from x-ray binaries in star-forming galaxies (Griffiths and Padovani 1990). Given the fluctuation constraint of Hamilton and Helfand (1987), an origin in galaxies would seem more likely, but a combination of these sources is probably present.

The Einstein observatory was capable of detecting in the deep surveys those AGN which have redshifts of unity or less (Griffiths et al 1988) Optically luminous UV-excess quasars were largely undetected in these surveys, and their detection awaits the X-ray missions of the remainder of this decade. The deep surveys also contain a minority of objects which are at relatively low redshift $z\sim 0.4$, have narrow emission lines, and are probably star-forming galaxies. The detection of these objects in very small numbers is consistent with their domination of the number counts at fainter fluxes, rising to equal the x-ray selected AGN at about $10-15 - 10-16 \text{ergs cm}^{-2} \text{ s}^{-1}$ and then dominating the discrete source counts to about $10-19 \text{ergs cm}^{-2} \text{ s}^{-1}$, at which point they account for the remainder of the XRB.

Fig. 1. The log(N>S) - log(S) for extragalactic X-ray sources

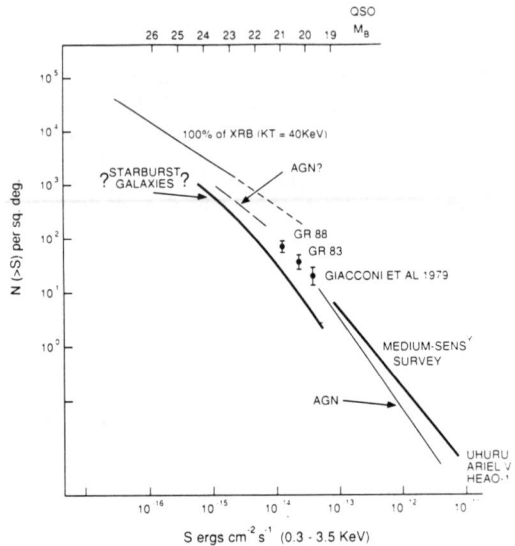

The thick solid line in Figure 1 labelled Medium Sensitivity Survey is from Gioia *et al.*1990, and connects the early sky-survey results from UHURU, Ariel V and HEAO-1 with the original Einstein Deep Survey point of Giacconi *et al.*1979. The AGN component of the Medium Sensitivity Survey is also shown, with a slope of -1.7. The original work in the Pavo field is labelled GR83 (Griffiths *et al.*1983), dominated by AGN in the redshift range 0.4 to 1.2. The re-analyzed Pavo images resulted in the data point labelled GR88 (Griffiths *et al.*1988), dominated by sources detected with the High Resolution Imager and identified with AGN and candidate AGN of optical magnitude 20-21. The dashed curve with slope -1.2 (AGN?) is consistent with the deep survey counts and with the fluctuation analysis of Hamilton and Helfand (1987), as well as with the equivalent slope of optical quasar counts (the upper abscissa shows the average blue magnitude of AGN corresponding to the X-ray fluxes on the lower abscissa).

The solid line at upper left (following Giacconi and Zamorani 1987) shows the number of sources needed at any flux limit to explain the whole of the XRB at 2 keV, on the assumption of extrapolation of the 3-20 keV spectrum to lower energies (kT = 40 keV). If, as indicated by the HEAO A-2 experiment, and by the overall IPC counts (Wu *et al.*1989), the XRB spectrum turns up below 3 keV, then the upper left line is a lower limit (indicated by arrows) and should be raised by about 30%.

STAR-FORMING GALAXIES AND THE XRB

It is relatively easy to show that star-forming galaxies at moderate redshift can account for an appreciable fraction of the XRB: the argument is based on the observed correlation between the x-ray and 60μ fluxes of a sample of starburst and interacting galaxies, which also holds for IRAS galaxies and even for spirals. The measured luminosity function of IRAS galaxies, together with models for its evolution, then lead to estimates of the contribution of the IRAS galaxies to the XRB (Griffiths and Padovani 1990). For these calculations, a hard x-ray spectrum is assumed, as is expected from the massive x-ray binaries which must dominate the x-ray emission from the star-forming galaxies at moderate to high redshift.

Fig. 2 shows the fractional contribution of *IRAS* galaxies to the XRB at 2 keV, as a function of redshift, for the cases of no evolution (dotted line), $\tau = 0.4$ (solid), and $\tau = 0.2$ (dashed) assuming $\alpha_x = 0.3$ and $q_o = 0$ (from Griffiths and Padovani 1990). The value $\tau \sim 0.4$ has been derived from the infrared evolutionary time scale of Franceschini *et al.* (1988), converted to an X-ray time scale; this value is also consistent with the luminosity evolution models of Hacking, Condon, and Houck (1987). The value $\tau \sim 0.2$ is typical of the X-ray evolution of AGN (Maccacaro, Gioia, and Stocke 1984).

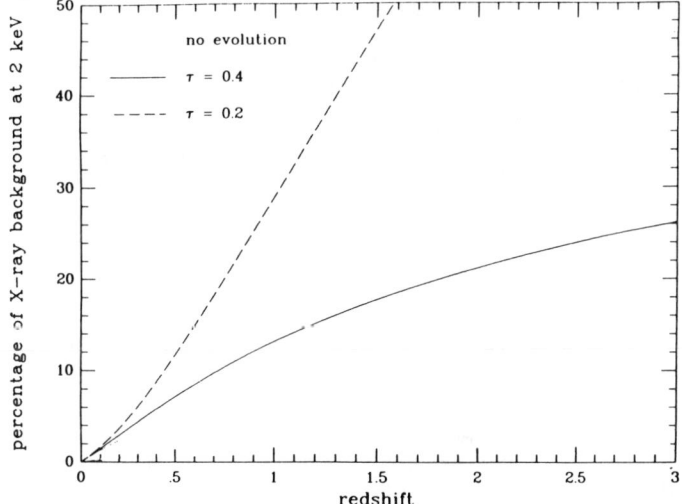

The above estimates on the contribution of star-forming galaxies to the XRB have not included any effects of reduced metallicity at moderate to high redshift, and are therefore conservative. If the average metallicity in star-forming regions at $z \sim 0.7 - 1.0$ is 10^{-2} to 10^{-3}, then the average X-ray luminosity would be higher by a factor of ~ 10, leading to an XRB dominated by star-forming galaxies at moderate redshifts.

BLUE, EMISSION-LINE GALAXIES IN DEEP OPTICAL SURVEYS

Broadhurst, Ellis and Shanks (1988) and others have observed excess blue emission in field galaxies at redshifts as low as 0.3 - 0.4, indicating the evolution of star-formation in galaxies at moderate redshifts. The number of MXRB in these galaxies is expected to follow the size of the HII regions, or the number of O-stars. Such blue galaxies may be expected to have integrated X-ray emission of about 10^{40-42} ergs s^{-1}, and to number perhaps 20–30% of all galaxies at those redshifts.

The colours and counts of the galaxies in the deep images of Tyson (1988) are an indication of star-forming activity, perhaps resulting from merges or interactions, at somewhat higher redshifts. X-ray production at an epoch of galaxy merging is expected to be common and copious.

One immediate conclusion of such an effort is that future X-ray observatories will need to have relatively high angular resolution, $i.e. \leq 3''$ in order to identify individual star-forming galaxies at the limit of the present optical surveys, $i.e.$ at 27th - 29th mag. Without such resolution, sources will be confused and individual spectra will be merged.

INSTRUMENTAL REQUIREMENTS

The main requirement for pursuing the cosmological implications of star-forming galaxies in the 21st century is that of sensitivity, ultimately reaching 10^{-19} ergs cm^{-2} s^{-1}, combined with arcsecond angular resolution.

Star-forming galaxies are expected to have sizes of order 1-2 arcsecs, and to be highly crowded: arcsecond resolution is therefore a minimum requirement. Morphological studies comparable with those which will be performed with the HST would be of great interest, but would lead to spatial resolution requirements which might be prohibitive in view of the large collecting areas required. The energy resolution required is comparable with that already being achieved (1989) from bolometric and Josephson junction detectors.

SCIENTIFIC REWARD IN THE CONTEXT OF C21st SCIENCE

X-ray astronomy potentially has a unique role in the 21st century: to contribute to our understanding of the evolution of galaxies and star-formaton, as well as to the importance of mergers and interactions of galaxies at moderate to high redshift. Global properties of such galaxies will give us information on the formation of the metals, and the enrichment of intergalactic space.

It will also tell us about the formation and evolution of AGN to high redshift, and on the relationship between starburst activity in the nuclei of galaxies, and the formation of massive black holes.

REFERENCES

Broadhurst, T. J., Ellis, R. E., and Shanks, T., 1988, *M.N.R.A.S.*, **235**, 827.
Fabian, A. C., 1990, in proc. ESLAB Symposiun on "AGN and the X-ray Background", in press.
Franceschini, A., Danese, L., De Zotti, G., and Xu, C., 1988, *M.N.R.A.S.*, **233**, 175.
Giacconi, R., et al., 1979, *Ap. J.*, , **234**, L1.
Giacconi, R. and Zamorani, G., 1987, *Ap. J.*, **313**, 20.
Gioia, I. M., Maccacaro, T., Schild, R. E., Wolter, A., Stocke, J. T., Morris, S. L., and Henry, J. P., 1990, *Ap. J. Supp.*, in press.
Griffiths, R. E., et al.1983, *Ap. J.*, **269**, 375.

Griffiths, R. E., and Padovani, P., 1990, *Ap. J., in press*
Griffiths, R. E., Tuohy, I. R., Brissenden, R. J. V., Ward, M. J., Murray, S. S., and Burg, R., 1988, in "The Post-Recombination Universe", NATO ASI Series C, vol. 240, eds. N. Kaiser and A. N. Lasenby, p. 91.
Hacking, P., Condon, J. J., and Houck, J. R. 1987, *Ap. J. (Letters)*, **316**, L15.
Hamilton, T. T. and Helfand, D. J., 1987, *Ap. J.*, **318**, 93.
Maccacaro, T., Gioia, I. M., and Stocke, J. T. 1984, *Ap. J.*, **283**, 486.
Tyson, J. A., 1988, *A. J.*, **96**, 1.
Wu, X., Hamilton, T., Helfand, D. J., and Wang, Q., 1989, preprint.

DISCUSSION

Richard Mushotzky:

You're referring to telescope sizes of the order of the size of the Keck telescope. It seems to me that we should not preclude smaller missions, and we should adopt a modular approach in the construction of future x-ray observatories.

Griffiths:

I agree that to detect the same galaxies in X-rays as have already been detected in the optical we need telescopes with collecting areas comparable with those already in operation on the ground.

Martin Elvis :

The point, I think, of Richard Griffiths' talk, and that of Pepi Fabbiano's earlier today, is to lay out ultimate goals toward which we can work. There will certainly be tremendous missions (such as the one described by Margaret Geller) which can be carried out on the way to this. It is certainly correct that we should not get trapped into planning only the ultimate mission for 20 years. However, it does make sense to plan for intermediate missions with the long-term goals in mind.

Griffiths:

We should also bear in mind the comments made by Webster Cash earlier, viz. that it would not be unreasonable for deep surveys to take up some fraction of the total projected facility lifetime. The Einstein Observatory, with a total lifetime of only a few years, spent several weeks performing deep surveys. With a facility expected to remain in operation for decades, it would be reasonable to spend months on each of a few deep surveys. Such projects would presumably be organized along the lines of the Key Projects about to be undertaken with the HST and with ESO. Such large projects in terms of observing time would then minimize the needed collecting area to a mission of more realistic size.

III. Technology for New Instrumentation

Jonathan E. Grindlay and William C. Priedhorsky, Chairmen

X-RAY IMAGING

Richard C. Catura

Lockheed Palo Alto Research Laboratory

Dept. 91-30, Bldg. 256

3251 Hanover St., Palo Alto, CA 94304

1 INTRODUCTION

In trying to envision the future of x-ray imaging after AXAF, it is interesting to compare the performance improvements of this mission over that of the Einstein Observatory. As shown in round numbers in Table 1, there are large factors across the board in the improved performance of AXAF. Considering the enormous contributions that Einstein has made to astronomy it is clear that AXAF will indeed be a very powerful observatory. It seems clear that it is unlikely we will again make such large factors of improve-

TABLE 1 COMPARISON OF EINSTEIN AND AXAF PERFORMANCES

	Einstein	AXAF	Factor
Angular Resolution	4 arc sec	0.5 arc sec	8
Half Power Radius at 4 keV	9 arc sec	0.5 arc sec	20
Effective Area at 1 keV	400 cm^2	1600 cm^2	4
Spectral Resolution	50	1000	20
Spectral Range	0.2-3.5 keV	0.1-10 keV	6
Sensitivity in 10^5 sec	5x 10^{-14}	5x 10^{-16}	100

ment in all categories in future missions. Thus it seems, at least in the early 21st century, that smaller missions will be favored which are designed to solve specific scientific problems. These problem oriented missions can be kept modest in scope by designs that improve only those instrument parameters required to achieve the selected scientific objectives. This paper will briefly discuss some of the techniques in x-ray imaging that may be used to enhance selected instrumentation capabilities. This serves as an introduction to the more detailed discussions in other papers presented at this workshop. Since it is not addressed elsewhere in the workshop, the scientific objectives, instrumentation and capabilities of a normal incidence multilayer telescope operating at soft x-ray energies will be considered in greater detail. Finally, some of the technologies that need further development to support the needs of x-ray imaging in the early 21st century will be discussed.

2 HIGH THROUGHPUT IMAGING

Any objective, such as high resolution spectroscopy or timing, where it is necessary to sort the data into many bins, requires detection and analysis of many photons. In terms of x-ray imaging this means large collecting area or high throughput. The concept of high throughput imaging, where high angular resolution is traded for large effective area, has been around for many years and has been reviewed recently by Gorenstein (1988). It involves a number of co-aligned grazing incidence telescopes where, for reason of economy, their angular resolution is limited to about an arc minute. The larger background from both the moderate angular resolution and the many detectors involved limit the sensitivity of the system. Thus, sensitivity and source confusion have been compromised for the ability to detect many photons. The strength of a high throughput mission is in spectroscopy and timing observations of sources down to the confusion limit of the telescopes. Three fabrication techniques for high throughput optics are presently viable; thin foil optics, bent glass optics and replicated optics. The first two methods are mature, but developing technologies for achieving sub-arc minute angular resolution in the case of thin foil optics and for weight reduction in bent glass optics is well worthwhile. Replication, either by electroforming or epoxy replication, are under development and continued improvement in the performance of these optics is certain to occur. Papers by Ulmer and Gorenstein on high throughput optics are included in this workshop.

3 HIGH ANGULAR RESOLUTION IMAGING

There are a number of suggestions for improving the angular resolution of x-ray optics beyond that of AXAF. Until one achieves microarc sec resolution, the scientific objectives chiefly involve stellar x-ray observations. Presently, studies of stellar x-ray activity as a function of spectral type are confused by inability to resolve the source of emission in multi-star systems. AXAF will only be able to resolve the visual binaries. Miliarc second resolution would resolve close binaries with separations of 10 solar radii out to 50 pc and separations greater than 1 AU out to 1 kpc. Resolution of a microarc sec would resolve stellar coronae, x-ray binaries and the cores of some AGN.

Normal incidence multilayer telescopes operating at soft x-ray energies should be able to provide angular resolutions of a least 0.1 arc sec (Harvey et. al., 1989). Resolution of an arc sec have already been achieved by Walker et. al. (1988) and Golub (1989) in imaging the solar corona. This type of telescope is able to achieve this resolution over an appreciable field by using classical normal incidence optical designs.

Webster Cash (Cash 1987) has suggested a way to achieve angular resolution approaching a miliarc sec by making use of the fact that in grazing incidence telescopes the effects of scattering and figure errors are far more serious in the plane of reflection than normal to this plane. Thus, if a grazing incidence telescope is stopped down so only a small azimuthal segment of the mirrors is used it is possible to obtain much higher resolution in the dimension perpendicular to the plane of reflections. This is illustrated in Fig. 1, where a long but very narrow image results from rays reflected by the small shaded areas of the telescope. In principal, if the azimuthal segments are kept small, one can improve the resolution in one dimension by a factor of the reciprocal of the grazing angle. If the focal length is made very long and only segments of cylindrical mirrors are used, it is possible to obtain useful effective areas with such a system. Also, using crossed segments one may achieve high resolution in two dimensions and by multiple exposures with such a system and by performing azimuthal image reconstruction a complete high resolution image may be recovered. Such a telescope is described by Cash later in these proceedings.

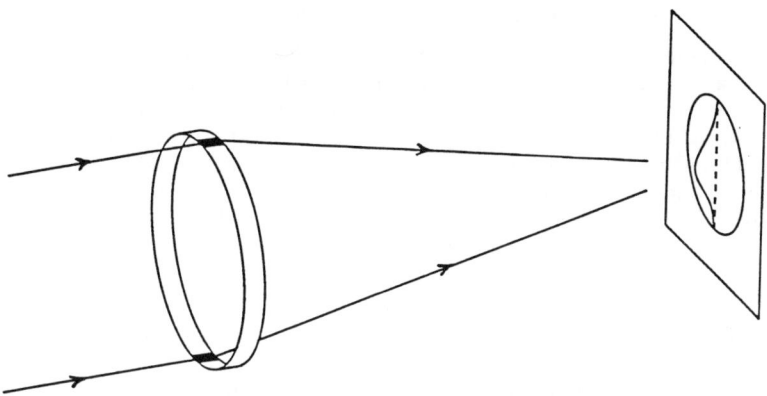

FIG. 1 SCHEMATIC ILLUSTRATING ASSYMETRICAL IMAGE
FROM A SMALL SEGMENT OF A GRAZING INCIDENCE TELESCOPE

Other techniques for achieving high angular resolution that involve x-ray interference effects will be discussed later in this workshop. Novick describes a method for obtaining angular resolution in the 10 miliarc sec range by detecting the Fresnel diffraction pattern in the x-ray shadow of a distant occulting edge. His proposal makes use of the slow scan rate and long baseline provided by a lunar base. Martin discusses a concept for obtaining microarc sec resolution with a lunar based x-ray interferometer. The interferometer uses multilayer reflectors and requires very precise optical surfaces and knowledge and control of optical pathlengths to a few angstroms. While this seems out of reach in the near term, the required technology may emerge later on in the 21st century.

4 HIGH ENERGY REFLECTIVE IMAGING

So far, 10 keV has been the high energy limit on the useful range of focussing x-ray optics used in astronomy. This is because the indices of refraction of all materials rapidly approach unity at higher energies. Thus, the grazing angles required for efficient reflection become very small at energies above 10 keV and in order to achieve useful effective areas either very highly nested mirror systems or very long focal lengths are required. While non-focussing techniques may be used to image the sky in higher energy x-rays, much better sensitivity can be achieved with reflective optics. The possibility of using reflective optics up to energies of 40 keV was discussed by Elvis et al. (1988). They proposed a highly nested system of telescopes of Kirkpatrick-Baez design that could provide several hundred cm^2 effective area. Weisskopf (These Proceedings) discusses a high energy telescope involving a long focal length with a mosaic of commercially available highly polished and metal coated silicon wafers as reflectors. It may be possible to achieve angular resolution of about 5 arc sec with such a design. Although one will likely be able to obtain larger effective areas with modulation collimators, the increase in signal to noise that is possible with a focussing system may ultimately provide higher sensitivity.

5 OTHER OPTICAL DESIGNS

It is possible to somewhat improve the off-axis angular resolution of a grazing incidence system by proper choice of the surface contours for the two mirrors. These optical designs trade on-axis resolution for improved resolution farther out in the field. Polynomial surfaces were investigated by Cash et al. (1979) and they concluded there was little improvement over simply shifting the focal plane in a Wolter I telescope slightly. More recently Nariai (1987) considered the use of hyperboloidal surfaces for both mirrors in a grazing incidence system to improve off-axis resolution. This design has been used in the Soft X-ray Telescope on the Solar-A mission (Bruner et al.,1988) and results in improved angular resolution over a de-focused Wolter I design as shown in Fig. 2. While such improvements are not large, they may be useful in survey applications.

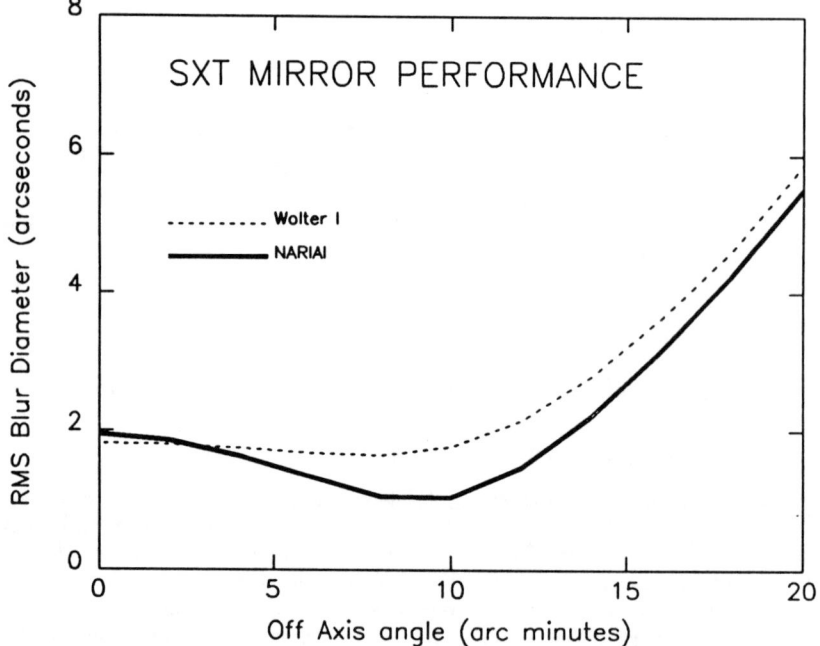

FIG. 2 CALCULATED IMAGE BLUR OF A HYPERBOLOID-HYPERBOLOID (NARIAI) TELESCOPE COMPARED TO A DEFOCUSED WOLTER I

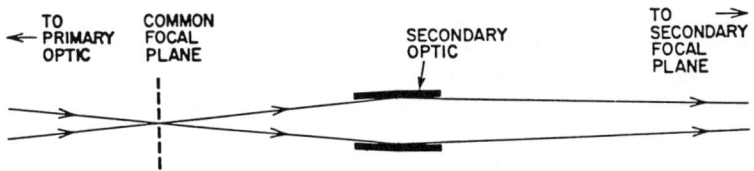

FIG. 3 MAGNIFYING GRAZING INCIDENCE RELAY OPTIC

Improvements in the polishing of optical surfaces have made it possible to achieve near-theoretical x-ray reflectivity with little scattering. This makes it practical to consider the use of grazing incidence relay optics (Chase et al.,1981) requiring multiple reflections for special applications in a telescope system. This is illustrated in Fig. 3 where a secondary optic is shown behind the focal plane and in the diverging beam of a grazing incidence telescope. This may be useful for magnifying an image to better match the telescope plate scale to the resolution of an image sensor or to make the diverging beam parallel to feed a spectrometer. The secondary optic may also be placed before the primary focal plane by using externally reflecting surfaces. Such a system has been fabricated and tested for use in imaging the solar corona by Moses et al.(1986). In fabricating a telescope of miliarc sec angular resolution it will likely be necessary to use magnifying relay optics to match the image sensor resolution to the telescope plate scale.

6 MULTILAYER NORMAL INCIDENCE IMAGING

Multilayer thin film interference coatings, that may be applied to figured optical surfaces by vacuum deposition techniques, provide the ability to reflect soft x-rays at normal incidence. These optics have been used for imaging the solar corona from sounding rockets and the USSR's PHOBOS mission. They will be used for a sky survey on the ALEXIS mini-satellite and for detailed observations of astronomical objects on the SPECTRUM-X-GAMMA mission. All of these optics are rather small in size, and the performance of a large multilayer telescope has not been considered in detail, although Elvis (1981) has discussed a large system in general terms. This paper will present the design of 1 m diameter soft X-ray Multilayer Telescope (XMT), consider some of the scientific objectives one may address with such a system and describe its performance in achieving these objectives.

Multilayer optics are complimentary to grazing incidence optics with each design having advantages and disadvantages. Multilayers operate only in rather narrow bandpasses at energies below about .25 keV and thus fail to use most of the available x-ray photons. However, for applications involving survey observations or where large areas of the sky are to be monitored, the large field of view afforded by the normal incidence design compensates for this inefficiency. The narrow passbands of the multilayers provide good energy discrimination and may be tuned to specific emission lines to obtain images that are sensitive to the temperature or density of the emitting region. Because of normal incidence, multilayer optics provide much more efficient use of their surface area and will thus be much less expensive than grazing incidence optics and weigh less for comparable effective area.

6.1 Scientific Objectives of the XMT

Since interstellar absorption is appreciable for soft x-rays, multilayer telescopes are best suited for study of local objects, although they will likely be able to see out of the galaxy at high latitudes, particularly for observing QSOs that exhibit an excess flux of soft x-rays (Elvis, 1987). There are, however, a number of interesting local phenomena that may be studied by a sensitive soft x-ray telescope able to observe out to about 200 pc in the galaxy. These include study of the the origin and angular distribution of the soft diffuse x-ray flux, study of the density and location of dust clouds in the local interstellar medium and performing surveys to investigate x-ray emissions mechanisms as a function of spectral type, to study stellar flaring and activity cycles and stellar evolution.

There is good evidence that the diffuse x-ray flux below about 1 keV consists of two parts: an extragalactic component which is isotropic except for absorption by interstellar matter and a galactic component produced by emission from a local hot plasma at a temperature of about 10^6 K. The spatial variation of the local component is determined by the distribution of both emitting and absorbing material. Soft x-ray observations may be the only way of verifying the existence and determining the temperature and angular distribution of a local hot plasma.

Stars of nearly all spectral types and luminosity classes have been found to emitt soft x-rays. It is very likely that different physical mechanisms are responsible for the x-ray emission across the HR diagram. As discussed by Rosner, at this workshop, understanding of stellar x-ray emission is incomplete and largely phenomenological. Because of its broad field, a multilayer telescope could easily obtain a large body of data on stellar luminosities and monitor variability and flaring for many stars simultaneously. If several telescopes were used, each tuned to lines formed at different temperatures, an index of stellar x-ray temperatures could be quickly obtained for each star in their field.

Because a multilayer telescope operates at normal incidence, it does not suffer from the incoherent small angle scattering found in grazing incidence systems. Consequently, it provides clean high contrast images that are ideal for studying soft x-ray scattering by interstellar dust grains. Dust halos have been detected in images of a number of galactic sources with the Einstein Observatory (Rolf 1983, Catura 1983, Bode et al. 1985 and Mauche and Gorenstein 1986,1989). Since scattered x-rays must travel greater distances they will be delayed in time from unscattered radiation. Thus, the time dependence of a halo about a variable source contains information on size and locations of dust clouds near the line of sight and on the distance to the x-ray source (Trumper and Schonfelder,1973). The scattering cross section from dust grains varies as the reciprocal of the x-ray energy squared. Thus, soft x-ray measurements may be the only way of investigating local interstellar dust within 100pc of the sun since reddening measurements are rather insensitive within this distance. An important question is whether nearby dust exists or if it has been sputtered away by the hot interstellar gas that produces the local soft diffuse flux. Since a multilayer telescope reflects only in a narrow passband it provides good energy discrimination that is necessary to analyze the scattering data independent of variations in spectra from source to source.

6.2 An XMT Design

Studies of the spatial variation of the soft diffuse x-ray flux require a fast telescope that images a large solid angle onto a small detector area to minimize particle background rates. This fact and the desire to image a large field require a short focal length telescope. A mirror diameter of 1 m was chosen as being reasonable to fabricate in the future. An energy of .13 keV (94 A) was chosen since continuum emission from a 10^6 K plasma peaks near this energy and several emission lines are present in this spectral region. The reflectivity expected for a 100 layer-pair silver-carbon multilayer is shown in Fig. 4. The FWHM of the passband that is reflected is about 2 A or 2.6 eV. Because the multilayer has appreciable reflectivity at long wavelengths, a proportional counter was chosen as the image sensor since it may be made insensitive to this lower energy radiation. Also, the charged particle background may be greatly reduced in such a detector by energy discrimination, since only a narrow energy range need be detected. The telescope angular resolution is limited by the detector near field center. Telescope aberrations limit this to 10 arc min at the edge of the 5 deg radius field. Fig. 5 shows a calculation of the reflectivity of the multilayer as a function of the telescope field angle. Since the incident angle at the edge of a 5 deg radius field is 10 deg, Fig. 5 indicates that there is no difficulty in achieving this size field of view. Other parameters of the telescope are given in Table 2. It is clear that this telescope will be very sensitive to the soft diffuse x-ray flux since its counting rate exceeds the background from particles by about 500.

TABLE 2 PARAMETERS OF AN XMT

Mirror Diameter	1 m	Focal Length	1.4 m
Bandpass Center	0.13 keV	Energy Bite	2.6 eV
Detector	IPC, 25 mm Dia.	Field of View	10 Deg Dia.
Detector Resolution	1 mm pixels	Angular Resolution	2.5 arc min
Geometrical Area	7850 cm^2	Peak Effective Area	740 cm^2
Particle Bkg./Pixel	2x10^{-6}/sec	Diffuse Bkg./Pixel	10^{-3}/sec
Continuum Sensitivity in 10^5 sec	1.5x10^{-4} ph/cm^2 sec keV 2x10^{-7} ph/cm^2 sec A		
Line Sensitivity in 10^5 sec			4x10^{-7} ph/cm^2 sec

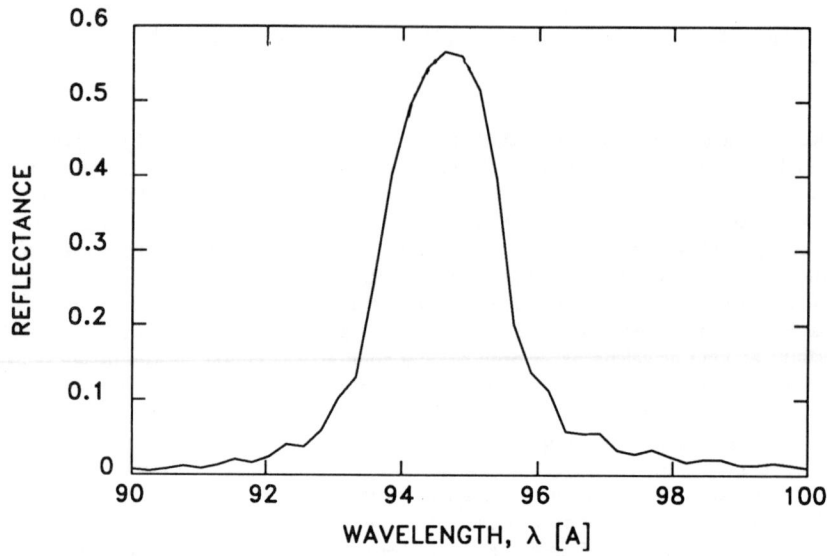

FIG. 4 BANDPASS OF A 100 LAYER-PAIR SILVER CARBON MULTILAYER AT NORMAL INCIDENCE

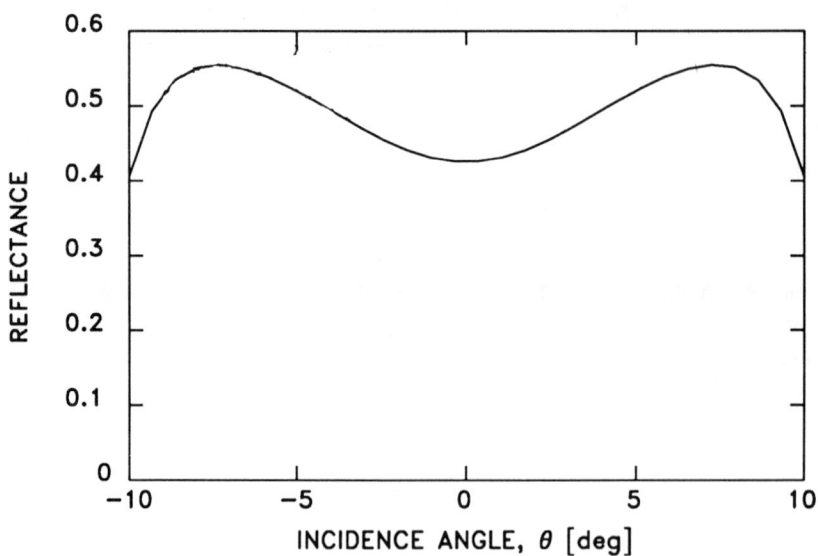

FIG. 5 REFLECTIVITY OF THE MULTILAYER OF FIG. 4 AT 94Å

6.3 XMT Performance

The number of counts acquired from various objects by the XMT in a 10^5 sec observation is given in Table 3. The XMT is ideal for study of the soft diffuse x-ray flux. In a single days observing, one could obtain one percent statistics with an angular resolution of 8 arc min over a 100 square degree area of the sky. The two percent energy discrimination of the XMT would make the data much more readily interpretable than the broad band results from EUVE and the WFC on ROSAT. This would be an even greater advantage if line emission were found in the diffuse flux and the telescope tuned to that energy.

The XMT is able to detect stars at 10 pc down to a luminosity of 10^{27} ergs/sec and 10^{29} ergs/sec at 100 pc for spectra similar to that of Capella. Cooler coronae, whose spectra peak in soft x-rays would be detectable at lower luminosities. Non-flaring stars range in luminosity up to about 10^{31} ergs/sec and much higher in stellar flares. Consequently the XMT would obtain a large body of data on stellar x-ray emission.

TABLE 3 PERFORMANCE OF THE XMT

OBJECT	DISTANCE (pc)	FLUX AT 100 A (ph/cm^2 sec A)	COUNTS/10^5 sec
Diffuse Flux		1000 ph/cm^2 sec sr keV	100/pixel
HZ 43	65	.05	8×10^6
Feige 24	90	.04	7×10^6
Capella	14	.001	2×10^5
Mkn 841		10^{-4}	2×10^4
3C 273		10^{-4}	2×10^4
Particle Bkg.		10^{-3}/cm^2*	0.2/pixel sec keV

* detector area

An optical depth at 0.13 keV is about 100 pc for a hydrogen density of 0.1 cm^{-3}. By using the scattering cross sections of Martin and Rouleau (1989), one expects about 1.5% of the x-rays from a source at this distance to be scattered into a halo about the image core. Thus, the halo will be very well determined for the bright and distant sources HZ 43 and Feige 24. However, most stars will either be too nearby or too faint to observe dust scattering above the strong signal from the diffuse flux, unless they are flaring. For example, Capella is a bright x-ray star about 14 pc away and one expects only 0.2 % of the radiation at 0.13 keV to be scattered by dust. This amounts to only about 400 counts spread over many pixels with a diffuse signal of 100 counts per pixel. However, flaring can increase stellar luminosities by many orders of magnitude and thus provide adequate signals from dust scattering. Another advantage of observing transient emissions is that the time dependence of the halo provides information on both the distance to the source and the distribution of dust along the line of sight. The large field of view of the XMT will include many stars and the probability of observing flare activity is very good. Study of dust scattered x-rays from quiescent stars would be better done at higher energies where the signal is higher and the diffuse flux is lower. This requires developing multilayer fabrication technology further since short period multilayers presently have poor reflectivities.

7 NEEDED TECHNOLOGY DEVELOPMENTS

Some of the efforts and technologies relevant to x-ray imaging that require further development over the next decade to prepare for their use in the 21st century are listed below.

1. Develop superpolishing techniques for large surfaces that will allow the low scatter finishes required for high angular resolution imaging, for efficient reflection of x-ray energies up to 40 keV and for fabricating efficient short period multilayers. Investigate coating materials and their vacuum deposition techniques that optimize high energy reflectivity.

2. Investigate the design and verify the expected performance of one dimensional high angular resolution optics.

3. Improve the angular resolution of thin foil high throughput optics.

4. Reduce the weight of bent glass high throughput optics and further improve their angular resolution.

5. Study the design of an x-ray interferometer to learn its practicality and identify required technology developments.

6. Develop techniques for depositing multilayers on thin plastic films which will make possible large scale soft x-ray telescopes by applying these films to figured substrates in the manner that solar reflection coatings are applied to window glass.

7. Develop the technology for stabilizing and controlling large light weight optical benches for optics of 100 m class focal lengths.

8 ACKNOWLEDGEMENTS

This work was supported by the Lockheed Independent Research Program.

9 REFERENCES

Bruner,M.E., Acton,L.W., Brown,W.A., Stern,R.A., Hirayama,T., Tsuneta,S., Watanabe,T., Ogawara,Y., 1989, AGU Monograph 54 titled "Outstanding Problems in Solar Systems Plasma Physics 1988", J.H. Waite, Jr., J.L.Birch, and R.L. Moore, editors.

Bode, M.F., Priedhorsky, W.C., Norwell, G. A., and Evans, A., 1985, Ap.J. 299, 845.

Catura,R.C.,1983,Ap. J. 275,645.

Cash,W.,Shealy,D.L. and Underwood,J.H.,1979,SPIE Proceedings 184,228.

Cash,W.,1987,Appl. Opt. 26,2915.

Chase,R.C.,Davis,J.M.,Krieger,A.S. and Underwood,J.H.,1981,SPIE Proceedings 316,74.

Elvis,M.,1981,SPIE Proceedings 316,144.

Elvis,M.,1987,Proceedings of the Rutherford-Appleton Lab. Workshop on "Emission Lines in AGN", ed. by P. M. Gondahlekar, RAL-87-109,20.

Elvis,M.,Fabricant,D.G. and Gorenstein,P.,1988,Appl. Opt.27,1481.

Golub,L.,1989,private communication.

Gorenstein,P.,1988,Appl. Opt. 27,1433.

Harvey,J.E.,Zmek,W.P. and Moran,E.C.,1989,SPIE Proceedings 1160,209.

Martin,P.G. and Rouleau,F.,1989,in Proc. of the Berkeley Colloquium on Extreme Ultraviolet Astronomy,ed. by R. Malina (Dortrecht:Reidel)

Mauche,C.W. and Gorenstein,P.,1986,Ap. J. 302,371.

Mauche,C.W. and Gorenstein,P.,1989,Ap. J. 336,843.

Moses,D.,Krieger,A.S. and Davis,J.M.,1986,SPIE Proceedings 691,138.

Nariai,K.,1987,Appl.Opt. 26,4428.

Rolf,D.P.,1983,Nature 302,46.

Trumper,J. and Schonfelder,V.,1973,Astron. and Astrophys. 25,445.

Walker,A.B.C.,Jr., Barbee,T.W.,Jr., Hoover,R.B. and Lindblom,J.K., 1988, Science 241, 1725.

DISCUSSION

William Priedhorsky

One can increase throughput for radiation from 10^6 K plasma by working in the 70 eV Fe IX–XII bands. At such temperatures, 30% of the total energy output from plasma falls in these bands. I second your call for wide fields of view — the ALEXIS experiment has telescopes with 33° diameter fields of view. The continued reflectivity of multilayer mirrors to long wavelengths is indeed a problem, as backgrounds in this region are large. A particular problem is geocoronal flux at 304 Å. Fortunately, this background can be overcome by a "wavetrap" anti-reflective coating which notches out 304 Å reflectivity and reduces it to 10^{-4}. "Wavetrap" mirrors have been successfully demonstrated in our laboratory.

Catura

I agree that multilayer bandpasses can be tuned to specific lines for larger throughput, but for my purposes a worst-case scenario was chosen where only continuum emission was considered.

Claude Canizares

First a comment on optical benches — at MIT we have a Space Engineering Research Center funded by OAST whose main purpose is developing controlled structures with a tolerance of microns over many tens of meters. The drivers for this work are from astrophysics. Second, a question about multilayers: what are the prospects for better control of layer thickness and roughness so that they can operate at shorter wavelengths, say 10 Å or 5 Å?

Catura

I believe that layer thicknesses and substrate surface roughness can be better controlled than at present, and this will help produce efficient multilayers down to about 40 Å 2d spacing. As Bob Rosner points out, optical constants of all materials approach one at shorter wavelengths, and it is a constrast in the optical constants that make the multilayers work. There is very little hope of getting appreciable normal incidence reflectively at 10 Å, but of course they will operate well at larger incident angles that satisfy the Bragg condition.

Richard Mushotsky

(1) Please comment on the use of multilayers to enhance the reflectivity of grazing incidence optics in narrow energy bands. (2) What about the use of a 4-element optical design for X-ray calibration facility?

Catura

(1) In principle, one can deposit multilayers on such optics to provide several hundred cm^2 effective area in narrow bandpasses up to $\sim$ 100 MeV. This requires AXAF quality surface finishes, and it may be more practical to use long focal lenths and flat-sheet approximations to figured optics as Marty Weisskipf describes in his paper at this workshop. This may be less expensive and provide a much broader energy range than multilayer optics. (2) At energies below about 0.5 keV one could conceivably use multilayer normal incidence optics, but a different mirror would be required for each wavelength. If one were to calibrate AXAF using a grazing incidence collimator to illuminate its apertures, the collimator would need to be of quality comparable to the AXAF optics. It is much less expensive and more versatile to use an upgraded facility at MSFC that is presently being fabricated.

Alan Bunner

(1) You claimed that Markarian 841 and 3C273 might yield 10^{-4} photons cm^{-2} s^{-1} Å^{-1} at 100 Å. Is there any evidence for this flux level, or is this only a calculation? (2) Is there any 4-element optical design for an X-ray telescope that offers a substantially larger field of view than a Wolter I?

Catura

(1) In papers by Martin Elvis and Belinda Wilkas, they show spectra of these objects from Einstein IPC observations. The flux levels given are extropolations of these spectra from $\sim .28$ keV down to .13 keV. (2) I think that any relay optics are limited in field of view by the primary mirrors.

Richard Rosner

I have a further comment regarding the ultimate limits on performance of multilayer optics at short wavelengths. As one goes to shorter wavelengths, the *difference* in index of refraction between the reflecting and "spacer" material (which is essential to getting reflections) becomes smaller and smaller. This is one of the principal reasons that more and more layers are needed to build up reasonable reflectivities at short wavelengths. Thus, past the carbon edge, the challenge is to find materials with sufficient difference in index of refraction that the reflection per interface *can* overcome the net absorption of the layers themselves.

Catura

I agree.

AN IMAGING PHOTOEMISSION POLARIMETER FOR SOFT X-RAYS

P. Kaaret, R. Novick, A. Heckler, and P. Shaw
Columbia Astrophysics Laboratory, Columbia University
G.W. Fraser, J.E. Lees, and J.F. Pearson
X-Ray Astronomy Group, University of Leicester

Recent measurements have shown that photoemission from soft X-ray photocathodes depends on the linear polarization of the incident X-rays. An X-ray polarimeter exploiting this effect placed at the focus of a grazing incidence telescope should be more than an order of magnitude more sensitive than any X-ray polarimeter flown to date. In this paper, we review experimental results on the polarization dependence of photoemission in the optical, UV, and soft X-ray bands. We discuss new data on the effect in the soft X-ray band, including measurement of the number of photoelectrons produced by individual X-rays. Finally, we present a concept design of an imaging photoelectric polarimeter and sensitivity calculations for both imaging and non-imaging photoelectric polarimeters.

The typical geometry of experiments on the polarization sensitivity of the photoelectric effect is shown in Figure 1. Photons are incident at a shallow grazing angle on a photocathode. The number of photoelectrons emitted from the surface is measured as the photocathode is rotated about an axis aligned with the incident beam. In all the experiments we will discuss, the incident beam is linearly polarized. We use the term P-polarized to denote radiation polarized normal to the photocathode and S-polarized to denote radiation with the polarization vector in the plane of the photocathode. If the photoemission is polarization sensitive, then the photocurrent will vary as $i = A\cos[2(\phi - \phi_0)] + B$, where ϕ_0 is the angle between the polarization vector and the normal to the surface which produces the largest photocurrent. The coefficients A and B are the magnitudes of the polarization-modulated and constant parts of the photocurrent. To quantifiy what fraction of the photocurrent is modulated by polarization, we define the modulation factor $\mu = A/B = (max - min)/(max + min)$, where max and min are the maximum and minimum photocurrent observed as the photocathode is rotated.

Although the polarization dependence of photoemission for soft X-rays has only recently been discovered, the effect has been observed at other wavelengths since the last century and historically has been referred to as the vectoral photoeffect. In 1894, Elster and Geitel showed that P-polarized optical light produced a significantly higher photocurrent than S-polarized light when incident on a liquid Na/K alloy. Ives (1924) observed the effect for optical light using a large number of different thin alkali metal films deposited on a variety of substrates. McConkey and collaborators (Tomc 1984) observed the effect using UV light incident on gold and lead oxide photocathodes.

The effect was first observed in the soft X-ray band by Oba (1987) and collaborators using CsI, Al_2O_3, and reduced PbO photocathodes and X-ray energies of 0.099, 0.124, 0.165, and 0.248 keV. They reported modulation factors of up to 0.4 for grazing angles of 20°. Independent studies of the X-ray vectoral photoeffect were made by Fraser's group

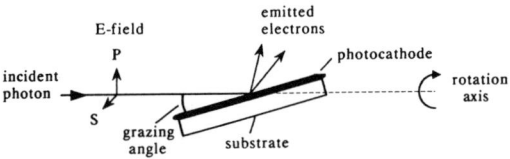

Figure 1. Experiment geometry.

at Leicester (Fraser 1989). The Leicester group measured the photoemission from CsI photocathodes at 0.175, 0.67, 1.647, and 2.7 keV. Modulation factors as large as 0.7 were reported for grazing angles of 5°.

After learning of Fraser's results, Novick and collaborators, at Columbia, began an independent program to verify the existence of the vectoral photoeffect (Heckler 1989). Measurements of the photocurrent versus rotation angle clearly showed the $\cos(2\phi)$ modulation which is the signature of polarization sensitivity, (space prevents us from presenting the data). A modulation factor of 0.12 was obtained at a grazing angle of 10°. The angle between the incident polarization vector and the photocathode surface normal corresponding to the maximum photocurrent is offset by 16° from the pure S-polarized state. This offset was previously noted by Fraser and was given the name the 'phase shift'. The phase shift measured by Columbia at 2.6 keV and by Leicester at 2.7 keV are in agreement.

The phase shift is likely to be of considerable importance in determining the correct theory of the vectoral photoeffect, because it requires a third axis, in addition to the photocathode surface normal and the polarization vector, to break the symmetry of the system. The agreement between the offsets measured by Leicester and Columbia would tend to suggest that the phase shift is a intrinsic property of the photocathode that does not vary each time a new CsI photocathode is deposited.

Measurements were also made, by the Columbia group, of the pulse height for individual X-rays. The pulse height for each event is proportional to the number of electrons photoemitted by the incident X-ray. Comparison of the pulse height distributions for P-polarized and S-polarized X-rays showed a significant excess of low pulse height events for the P-polarized beam and a corresponding excess of high pulse height events for the S-polarized beam. Subsequent measurements performed by the Leicester group clearly resolved single versus double photoelectron events and confirmed the Columbia pulse height measurements. The polarization dependence of the number of electrons emitted suggest that it might be possible to acheive large modulation factors using pulse height data in addition to total photocurrent data.

The polarization sensitivity of the photoelectric effect makes possible a new generation of X-ray polarimeters. Use of CsI-coated microchannel plates (MCP's) is currently planned for the High Resolution Camera for AXAF. It is a straight forward step to a device which will be able to do imaging polarimetry. Polarimetry requires that the photocathode surface be flat in order to maintain a constant angle between the polarization vector of incident X-rays and the surface normal. The rounded pores of conventional MCP's are not suitable.

Imaging Photoemission Polarimeter

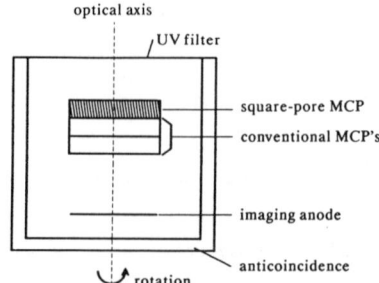

Figure 2. An Imaging Photoelectric Polarimeter.

However, it is possible to obtain MCP's with square pores which are aligned over distances of several millimeters. If CsI is deposited on only one face of each pore then the MCP will act as a flat photocathode.

A schematic drawing of an imaging polarimeter concept is shown in Figure 2. The main component is a square-pore CsI-coated MCP. This MCP is operated at low gain to prevent saturation and allow one to determine the number of electrons produced by the initial X-ray photoemission. Beneath the square-pore MCP is a chevron stack of conventional MCP's for charge multiplication. The charge signal is detected by an imaging (i.e. wedge and strip) anode. An anticoincidence shield is placed around the X-ray detector to reduce the background. The entire device must rotate about the optical axis of the telescope in order to measure the photocurrent modulation due to polarization. The energy band pass of the instrument is determined at the high end by the rapid fall off in the quantum efficiency of CsI above 2 keV and at the low end by the UV filter used. With multiple filters it would be possible to give the instrument two or three color energy resolution.

A soft X-ray polarimeter exploiting the vectoral photoelectric effect is being added to the Stellar X-Ray Polarimeter (SXRP) which will be flown on the Soviet SPECTRUM-X-Gamma mission. The polarimeter will be at the focus of one of the SODART metal-foil grazing-incidence X-ray telescopes. Each of these telescopes has a collecting area of 2000 cm^2 at 1 keV. The angular resolution of the telescopes is expected to be on the order of 2 arcminutes (FWHM). Using the data reported by the Leicester and Columbia groups on the vectoral photoeffect, we have estimated the sensitivity of the SXRP photoelectric polarimeter. Figure 3 shows the minimum detectable polarization at 99% confidence level for a 10^5 second observation of a source with a Crab-like spectrum. The lower solid line is the sensitivity of the SXRP instrument. The dashed line is the sensitivity of the imaging polarimeter, discussed in the previous section, if it were placed at focus of an X-ray telescope with a collecting area of 2000 cm^2 and an angular resolution of 2 arcseconds. The improvement in sensitivity at low source strengths arises from the reduction in background due to imaging. The upper solid line is the minimum detectable polarization for the SXRP Bragg crystal polarimeter. The photoelectric polarimeter offers more than an order of magnitude increase in polarization sensitivity.

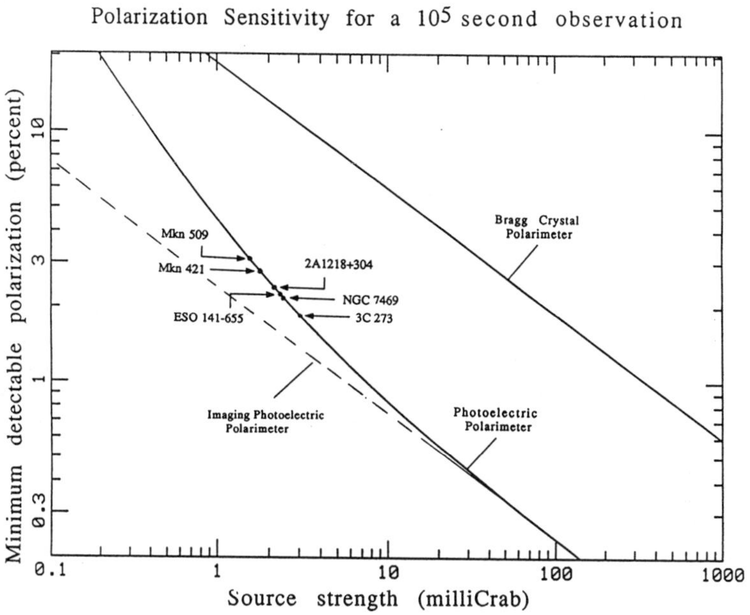

Polarization Sensitivity for a 10^5 second observation

Figure 3. Polarization sensitivity.

Figure 3 also indicates the equivalent source strength and minimum detectable polarization of a few particularly interesting quasars and active galactic nuclei. A few dozen such sources are bright enough to be studied with a polarization sensitivity of a few percent in 10^5 second observations with the SXRP. The photoelectric polarimeter will make possible sensitive measurements of the polarization of a few hundred objects ranging from supernovae remnants to active galactic nuclei to black hole candidates. This new level of sensitivity will make X-ray polarimetry a useful tool for the study of many types of astrophysical objects.

This work was supported by NASA grant NAG 5-618. This is contribution 404 of the Columbia Astrophysics Laboratory.

REFERENCES

Elster, J. and Geitel, H. 1894, *Ann.Phys.*, **52**, 433.
Fraser, G.W., Lees, J.E., and Pearson, J.F. 1989, *Nucl. Inst. and Meth.*, in press.
Heckler, A., Blaer, A., Kaaret, P., and Novick, R. 1989, *Proc. SPIE*, **1160**, 580.
Ives, H., 1924, *Ap. J.*, **60**, 209.
Oba, K., 1987, personal communication to Fraser.
Tomc, J., Zetner, P., Westerveld, W.B., and McConkey, J.W. 1984, *Appl. Opt.*, **23**, 656.

DISCUSSION

Gerald Share

Has the effect been observed in earlier space experiments? Has it impacted any of the observations?

Kaaret

The effect is unlikely to be noticed in earlier experiments, because the detector must be rotated to observe the polarization signature. The polarization dependence of the quantum efficiency could have affected photometric measurements if polarized sources were observed. In this connection it is important that the AXAF HRC be calibrated with a polarized source, since it will employ a CsI photocathode.

Richard Mushotzky

What is the sensitivity of the device for astronomical sources in Spectrum X?

Kaaret

We expect to be able to measure the polarization of a number of AGN's at the 1% to 3% level in 10^5 second observations.

Steve Murray

What is dependence of modulation and stability on CsI deposition and history?

Kaaret

We are planning a number of experiments to test the dependence of the effect on the state of the surface. However, in previous experiments we have left the CsI exposed to varying amounts of atmosphere and moisture without obvious effects on the polarization properties.

THE PROMISE OF REPLICATION FOR X-RAY ASTRONOMY

M. P. Ulmer

Department of Physics and Astronomy
Northwestern University

I. INTRODUCTION

X-ray mirror replication uses electroforming techniques to produce multiple, high quality copies from a single master. This method can also be used to produce light weight, thin walled mirrors. As a result of the capabilities of replication technology, several European missions have chosen this method to fabricate their required X-ray mirrors. An angular resolution of 20″ has been obtained for mirrors up to 12 cm in diameter. XMM test mirrors have been made that are about 70 cm in diameter and have walls only 1 mm thick. The surface roughness of the XMM test mirrors is low enough to reflect 7 keV X rays with a half energy width of better than 1′. For details about recent work, the reader is referred to Ulmer (1989), Jensen et al. (1989), and Citterio et al. (1988), and references therein.

The promise shown by this work and the choice of replication as the technique for several major European space missions should be enough to encourage NASA to consider replication as a technology for the future. But another reason for seriously considering replication is the recent development of new composite materials, which may eventually allow us to produce copies that are as good as their masters in terms of both figure and surface roughness. Below I outline just a *few* of the concepts that might be considered viable if replication were to be used as the technology for producing high quality mirrors.

II. HIGH QUALITY OPTICS

Since replication allows the fabrication of a thin walled system while still producing a high quality optic, it is a method to consider even if only a single mirror is being made. The limiting angular resolution of mirrors fabricated by this technique has not yet been determined since arc-second quality masters have not been made until recently. Matsui et al. (1988) and Citterio et al. (1988) have shown that 20″ resolution is possible, and the XMM project has a goal of 10″ per single mirror. The XMM system is made up of a deeply nested system of very thin walled mirrors (1 mm or less), and therefore better figure quality may be achievable merely by making thicker walls. Based on this, I estimate that 3″ resolution mirrors are possible, and making mirrors that are as good as the best masters is not out of the question.

III. HIGH THROUGHPUT SYSTEM

Although the LAMAR project with a Kirkpatrick-Baez (K-B) design has been selected for the Space Station, I disagree with that selection for several reasons. First, since there is only one reflection per plane, the K-B design requires twice the focal length of a Wolter I system having the same diameter (de Korte 1987). Second, the system uses bent float glass, for which the best angular resolution is $\sim 30''$. Finally, given the tardiness with which Space Station systems are actually likely to fly, it seems much better to use Space Station funding to work towards advanced technology rather than putting so much effort into a system which will have the same area as XMM (at 7 keV), but using a design based on a technology that was available in the 1970's; NASA, however, disagrees.

What are the alternatives to going forward with a K-B LAMAR system? One alternative is to simply join with the Europeans in designing a post-XMM mission with up to 10 times the XMM area. This would take advantage of the exquisite masters that will be produced for the XMM project. Another possibility for U.S. scientists is to develop our own replication technology and count on using the masters from the XMM mission. Yet another alternative is to design a system that is optimized to fit within available U.S. expendable launch vehicles and will not necessarily use the masters developed for the XMM project.

Whatever design is chosen, it may be best to acknowledge that the next large project will be so large as to *require* international collaboration and to begin discussions with ESA on what this next large mission should be. Whatever the mission, as long as super-angular resolution is not the goal, replication will likely be the technique selected to fabricate the mirrors.

IV. A HIGH ENERGY SYSTEM

Several groups are developing designs that would work up to 40 keV, including SAO, Northwestern, and Milan. A possible design is given below. I have chosen 4 meters as the focal length, since ASTRO-D is planned to have a 4 meter focal length, and this demonstrates that relatively modest launch vehicles can be used to place 4 meter focal length systems in orbit. Also, I have judged that a reasonable goal for such a system is a collecting area of over 1000 cm^2.

The design consists of a 10' grazing angle system, which should be optimized for ~ 40 keV, and have a 9 cm outer diameter mirror. Assume the footprint allocated for mirrors to be a square 81 cm on a side. Then if we assume a covering factor of the mirrors of 50% and a 50% net efficiency, the effective area of this system is over 1600 cm^2. If we assume that a point source has a size of about 2 mm on a side, which corresponds to a modest 1.7' HEW, we have a net gain over a non-focusing system in collecting area to detector area of about 500. Assuming that we could use the pixels that are not collecting source photons to collect simultaneous background rates, we should be able to improve the sensitivity over a conventional non-focusing system even further. Correction for background variations is important since time variations in the background are a vexing problem in this energy range. Of course, there is the drawback that the field of view is only 10', but the vast increase in sensitivity makes it worthwhile. Also note that source positioning should be at least as good as 10'', as typical positional accuracy can be determined to 1/10 the size of the resolution element.

While a detailed design has not yet been developed, it is expected that each mirror module would be nested 10 deep and would have a total mass of about 2 kg if the mirrors are made out of nickel. The total mass of a 81 cm × 81 cm system should then be about 300 to 500 kg, including detectors. Since the design is modular, the net area and weight can easily be scaled to an available launch vehicle.

V. A LARGE DEPLOYABLE ARRAY WITH NORMAL INCIDENCE OPTICS

The infrared program has already been studying the fabrication of a large deployable array. Replication may be able to play a role in this by allowing us to manufacture only 5 or so masters, and then to replicate the entire 30 to 100 segments that are needed to make a $\sim$ 30 meter diameter telescope. The segments should also be much lighter than conventional glass segments. If the segments could be over-coated with multi-layers, it would then be possible to use a large deployable array as a normal incidence telescope for X rays. The enormous collecting area available would more than compensate for the narrow band width of the system. This is by far the most ambitious and "far out" of the designs proposed here, but we should be discussing things that *are* out of reach today to motivate us to develop the required technology.

VI. SUMMARY AND CONCLUDING REMARKS

The sample designs I have given above are by no means exhaustive. Almost any project that uses Wolter I optics can, in principle, benefit from replication technology. Wolter I optics made via replication are cheaper and lighter and can be more deeply nested than any others currently proposed. The promise is there. The question is, will NASA be willing to make the necessary investments?

I thank S. M. Matz and W. R. Purcell for providing comments on this manuscript.

REFERENCES

Citterio, O., *et al.* 1988, Proc. SPIE, **830**, 139.
de Korte, P. A. J. 1988, Proc. SPIE, **830**, 172.
Jensen, P. L., Ellwood, J. M., and Peacock, A. J. 1989, Proc. SPIE, **1160**, 525.
Matsui, Y., Ulmer, M. P., and Takacs, P. Z. 1988, Proc. SPIE, **830**, 63.
Ulmer, M. P. 1989, Proc. SPIE, **1160**, 426.

M. P. ULMER: Northwestern University, Department of Physics and Astronomy, 2145 Sheridan Road, Evanston, IL 60208

DISCUSSION

Martin Weisskopf

Have you measured the surface properties of the device that gave you 20″ resolution (roughness, figure, etc.)?

Ulmer

We believe the resolution is limited by the figure accuracy. The HEW is larger and is dependent on surface roughness.

HIGH ENERGY, HIGH RESOLUTION X-RAY OPTICS

Martin C. Weisskopf, Marshall Joy
NASA/Marshall Space Flight Center

Steven Kahn
University of California, Berkeley

The Einstein Observatory aptly demonstrated the advantages of focussing X-ray optics by improving point-source detection efficiency by a factor of 1000 over previous techniques. The Einstein optics were what we now call "classical" high resolution Wolter-1 optics: multiple paraboloid-hyperboloid pairs which were carefully (and commercially) figured, polished, and aligned. In this paper we present a line of reasoning that argues for a different approach for future development of X-ray optics in order to meet both the scientific and the socio-economic needs of X-ray astronomy.

We first consider the scientific goals. There are a number of X-ray observatories, in this country and abroad, which will use focussing X-ray optics for astrophysical studies up to about 10 keV; these include ROSAT, ASTRO-D, Spectrum-X, AXAF, and XMM. The data from these missions will enable great progress to be made in the "thermal" portion of the X-ray spectrum. However, the "hard" portion of the X-ray band (E > 10 KeV) will remain largely unexploited, despite its scientific promise. Hard X-ray observations provide the most direct and unambiguous means of diagnosing nonthermal phenomena in cosmic sources, which are expected in a rich variety of astrophysical settings. Nonthermal processes tend to generate power law spectra which extend up to very high energies. Although nonthermal contributions are expected for the soft and medium energy X-ray bands as well, they are often masked by very intense thermal components which dominate the emission. This is quite apparent in the X-ray spectra of clusters of galaxies, supernova remnants, and solar flares, to name a few diverse examples. In all of these cases, the nonthermal component of the spectrum is only visible at energies greater than ~ 10 keV. Since nonthermal processes are among the most exotic of all astrophysical phenomena, they are of substantial interest for future research.

Interestingly enough, the need for focussing optics is nowhere more apparent than in the hard X-ray band. Nonthermal power law spectra fall steeply with energy, so most sources are faint in this spectral range. In addition, the particle contribution to the background is rising with energy, so non-focussing experiments become severely background limited; thus, vast improvements in hard X-ray detection sensitivity are required in order for this field to develop fruitfully. For the foreseeable future, this implies the use of either coded apertures or grazing incidence optics. Clearly, this is not an either-or question, since coded apertures can operate at very high energies where grazing incidence methods fail. However, in the energy ranges where both methods are feasible, we argue that focussing optics can offer a significant advantage in signal-to-noise *and* in greater scientific return from the use of microelectronic focal plane technology.

The Optics Concept

We now turn to the basic design of hard X-ray optics. We believe that a compelling case can be made for simplicity: classical methods of figuring and polishing high resolution X-ray mirrors, when applied to the production of high energy optics, are both too expensive and too far removed from the control of the scientific community. In the context of these Proceedings and especially in preparation for the next National Academy Survey, this last point is of particular importance. For example, the difference in cost between the AXAF optics (constructed by an industrial contractor) versus the BBXRT optics (which were developed by an active group of research scientists) is far greater than the difference in actual performance. Furthermore, in the case of BBXRT all available effort was made to optimize the performance within the technological and fiscal constraints. This is rarely (if ever) the case in an industrial procurement for state-of-the-art technology. We need not belabor these points further, but it is clear that they are partly responsible for the ever-increasing costs of space missions and the reduction in flight opportunities, which in turn limits scientific productivity.

In order to operate efficiently at high energies, grazing incidence optics must incorporate very small angles of incidence; this, in turn, leads to very long focal lengths and large radii of curvature if reasonable effective areas are to be obtained from optics of finite thickness. Based on the above requirements of simplicity and large curvature, we base our hard X-ray optical design on the simplest of all such surfaces: a flat plate. Assuming for the moment that thin, flat, and highly polished surfaces are available, we examine the theoretical properties of a possible telescope design.

We begin by considering a single reflection at the front end of a flat-plate Kirkpatrick-Baez telescope. In the small angle approximation the projected width of the flat plate is

$$\delta = \frac{ld}{2F},$$

where l is the length of the plate, d is the distance of the plate from the optical axis, and F is the focal length. The angular resolution, $\Delta\theta$, is then

$$\Delta\theta = 10^5 \frac{ld}{F^2} \text{ (arcseconds)}.$$

If we have a number of nested surfaces of thickness T with the outermost set at d=D/2 then it is easy to show that the spacing is given by (see Weisskopf 1973, *Applied Optics*, **12**, 1436):

$$d_n = (\frac{D}{2} - T)\left[(1+c)X^{n-1} - c\right]$$

where n=1 is the outermost surface and

$$c = \frac{T}{(D/2 - T)(1 - X)}, \text{ and } X = 1 - \frac{L}{2F}.$$

The fraction of the total area that is available for (single) reflection is

$$f = f' - \frac{2nT}{D}, \text{ where } f' = 1 - \left[\frac{2(d_{n+1} + T)}{D}\right].$$

The variable f' is the fraction of the total available area filled with reflecting surfaces (the packing fraction). The number of reflecting surfaces on each side of the origin for the first reflection is given by

$$\frac{1}{\log X} \cdot \log\left[\frac{1}{1+c}\left(c + \frac{D[(1-f') - 2T/D]}{2[(D/2) - T]}\right)\right],$$

and the total available geometric area for two reflections is $(fD)^2$.

We begin our design consideration with Figure 1, where we have plotted the square of the theoretical reflectivity (the efficiency of a double reflection) from a gold surface for a sequence of grazing angles from 1 to 10 arc-minutes in 1 arc-minute steps. Based on this figure, and for the purpose of this discussion, we selected a maximum grazing angle of 7 arc-minutes which allows the telescope to operate efficiently up to 40 keV with a reasonable field of view (~7 arcmin). This choice then establishes the ratio d/F. The remaining parameters of interest are the angular resolution and the effective area. These are not independent and, especially the effective area, are effected by the thickness of the reflecting elements. Space does not permit an exhaustive treatment of all the functional interdependencies. Figure 2 shows the variation of several relevant parameters for a 60 meter focal length telescope constructed from 2 millimeter thick plates. The angular resolution shown in the figure is the area-weighted mean resolution, assuming that all plates reflect with equal efficiency.

The finite thickness of the flat-plate reflecting elements leads to the interesting property that there is a an optimum signal to noise ratio which occurs at a finite angular resolution! The "signal to noise" curve presented in Figure 2 is determined by the ratio of the effective geometric area to the plate scale (F·<Δθ>).

Reality

The actual performance of telescopes of this design concept and their feasibility of construction is clearly an open question. However, there are at least two issues that can be immediately put to rest. The first concerns the actual effective area; that is, the geometrical effective area reduced by the efficiency of two reflections. As shown in figure 1, this double-reflection efficiency is above 80% for the grazing angles of interest, even at energies up to 40 keV. Another consideration is the quality of the reflecting surface with regards to scattering. This includes spatial variations of the reflecting surface on submicron scales (usually referred to as surface roughness) as well as the medium- and large-scale flatness of the of the reflecting elements. Scattering will remove flux from the central core of the image and serves not only to reduce the signal to noise for the detection of point sources, but also degrades the capability for studying extended objects of low surface brightness. The consequence of these effects can be seen in figure 3 which shows the fraction of flux after two reflections contained within a cone of a given half angle for 40 keV x-rays for several grazing angles (L. Van Speybroeck, private communication). Even in the worst case, the bulk of the scattered flux (>80%) is contained within a half cone angle of about 1" -- much smaller than the aberrations expected to result from large scale surface errors. The calculations shown in figure 3 are based on AXAF quality surfaces and include three contributions: a 3 angstrom amplitude with a 18 micron correlation length (microroughness); a 25 angstrom amplitude with an 18 mm correlation length (mid frequency); and a 0.275" gaussian error which for AXAF accounts for both figure and alignment errors. Our assumption is that if surfaces of this quality can be fabricated for the AXAF surfaces of revolution then surely they can be obtained on flats. In fact, on a smaller scale, such materials already exist in profusion: the silicon wafers produced for the semiconductor industry. Figure 4 shows a comparison of the power density spectrum of the AXAF surfaces compared to measurements taken by Peter Takacs at Brookhaven on a commercially procured eight inch diameter silicon wafer. This arbitrarily selected sample is already close to AXAF standards and can be obtained in large quantities at a relatively low cost. Based on this result, we would argue that obtaining the required surface characteristics at short spatial wavelengths (<5mm) is not an obstacle to telescopes of this design. The real technical challenge concerns the large-scale flatness of the material and the stability and accuracy of the alignment of multiple plates.

Of course, the simplicity and advantages of the proposed long focal length telescope will be nullified if a rigid, 60 meter long spacecraft is required for operation. However, X-ray astronomy has long exploited single photon counting to simplify telescope design specifications: the relative orientation of the telescope-detector system is monitored using fiducial lights, so the rigidity of the telescope truss is not a crucial design driver. In addition, large deployable structures have recently been developed which would allow the long focal length telescope to be launched in a conventional vehicle; active control experiments such as CASES are also underway with the goal of stabilizing deployable structures at the few arcsecond level.

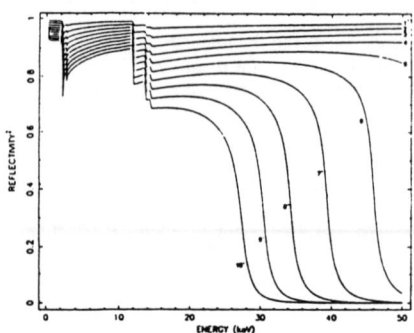

Figure 1. Efficiency for two reflections from a gold surface versus energy for a number of grazing angles.

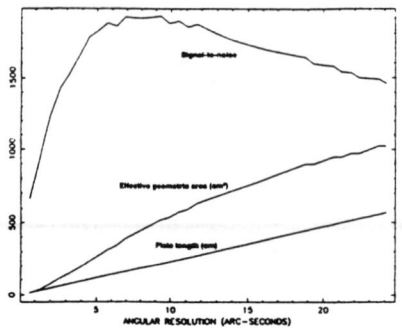

Figure 2. Signal-to-noise (see text), area, and plate length versus aberration-limited angular resolution.

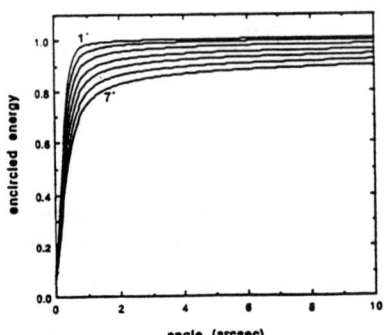

Figure 3. Encircled energy after two reflections at 40 keV as a function of half-cone angle.

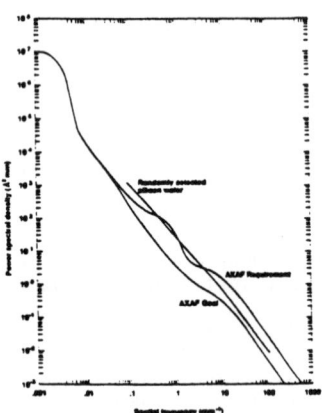

Figure 4. Surface power spectral density as a function of spatial frequency.

DISCUSSION

Charles Hailey

How severe are depth of focusing effects in such a gas counter at large absorption depths and high energies?

Weisskopf

They depend on the particular applications. They are simply determined and one has to be careful not to design a system, i.e., a coded-aperture/detector, where such effects become a problem. Clearly one cannot simply increase efficiency by arbitrarily increasing the depth of the absorption/drift region.

Richard Mushotzky

(1) How much lower is the background in these devices than in standard phoswiches?
(2) How much more sensitive are these detectors than XTE-type phoswiches?

Weisskopf

The typical background in a phoswich is $\sim 10^{-4}$ events cm^{-2} s^{-1} keV^{-1}. Our first look at our flight data, for flourescence pairs, shows a background that is $\sim 10^{-5}$ at the xenon K-edge (where the pair efficiency is highest), rising to about 2×10^{-4} at 60 keV and then falling again. These results include no correction for the krypton contamination (~ 100 picocuries) nor any sophisticated event selection criteria; thus they must be viewed as an upper limit. As far as "sensitivity" is concerned, a detailed comparison must be posed in terms of a specific astrophysical question, taking into account spatial resolution, energy resolution, etc.

SPACE ASTROPHYSICS WITH LARGE STRUCTURES: CASES and P/OF

H.S. Hudson (UCSD) and J.M. Davis (NASA/MSFC)

ABSTRACT

Space instruments for remote sensing, of the types used for astrophysics and solar-terrestrial physics among many disciplines, will grow to larger physical sizes in the future. The zero-g space environment does not inherently restrict such growth, because relatively lightweight structures can be used. Active servo control of the structures can greatly increase their size for a given mass. The Pinhole/Occulter Facility, a candidate Space Station attached payload, offers an example: it will achieve 0.2 arc s resolution by use of a 50-m baseline for coded-aperture telecopes for hard X-ray and γ-ray imagers. The **CASES** experiment (NASA Office of Applications and Space Technology) — deployable on the Space Shuttle as early as 1994 — will provide an engineering and scientific demonstration of active structural control in this context.

I. INTRODUCTION

As space astrophysics and other observing disciplines mature, the premiere objectives in each field demand physically larger instruments. In astrophysics, the Great Observatories program (Gamma-Ray Observatory, Hubble Space Telescope, Advanced X-ray Astrophysics Facility, Space Infrared Telescope Facility) will be placed in orbit within the decade of the '90's. Astrophysics and other disciplines have even more ambitious plans for large future instrumentation. The motivation for large size usually is that it confers advantages in sensitivity and/or angular resolution.

This development poses a challenge for NASA managers and systems engineers. For very large instrumentation, what is the result of a tradeoff between a sophisticated, lightweight structure, and (say) a larger rocket booster? A scientist's typical view of mechanical design of observing instrumentation tends towards the "cast-iron structure" approach, in which brute force ensures the rigidity of the observing platform. This approach may have helped to save the Multiple Mirror Telescope, a six-mirror ground-based telescope on Mt. Hopkins, Arizona; automatic phasing of the array has been problematic since first light, but the strength of its structure has allowed it to be focused and used at least as simple light-bucket imager. Cast-iron engineering probably is inappropriate for space instruments because of the large cost per unit mass of transportation into orbit.

We accept as a premise that large and expensive space experiments will be done most efficiently with full recognition of the systems-level engineering; for the users of the large instruments (the scientists) this means that sophisticated engineering can enable far more capable instruments with less-demanding requirements on the transportation system. This paper describes the application of large structures to obtaining excellent angular resolution for X–ray and γ–ray observations — < 2 arc s for the **CASES** experiment, the technology demonstration, and < 0.2 arc s for the **P/OF** facility. This paper reviews these experiments briefly.

II. THE **CASES** EXPERIMENT

The **CASES** experiment (Controls, Astrophysics, and Structures Experiment in Space) is the flight verification of a NASA engineering program for the development of knowledge regarding the behavior and control of large, flexible structures in a zero–g environment. The engineering aspects of the **CASES** program have heritage in the SAFE program, which successfully deployed a 32–m Astromast structure from the Space Shuttle (the STS 41–D mission). In addition a comprehensive program of sensor, actuator, and algorithm development is centered on laboratory experiments in the one–g environment. Small–scale experiments in zero–g are also being carried out in the Shuttle mid–deck lockers.

These experiments are being conducted by NASA's Office of Aeronautical and Space Technology (OAST), to be distinguished from the Office of Space Science and Applications — the usual home of astrophysics, solar–terrestrial physics, and other observing disciplines. The OAST will fund the entire **CASES** program up to the point of scientific data analysis. This includes all of the scientific instrumentation (position–sensitive hard X–ray counters and coded–aperture masks), which also directly contributes to the engineering verification.

a) *The* **CASES** *Instrumentation*

The **CASES** structure consists of a 32–m boom mounted rigidly in the Orbiter and controlled with a series of sensors and actuators. The sensors include two–axis star trackers, permitting a reference to the fixed stars, and cold–gas thrusters at the boom tip. These thrusters point the boom by orienting the entire Orbiter at the desired target. The engineering experiments include both open–loop and closed–loop excitation and control of the boom structure. The boom is of the Astromast technology, as was the SAFE boom, based on continuous fiberglass longerons that coil up in a deployment canister. The structure is lightweight (estimated at 60 kg, including the canister) and surprisingly rigid. There is good engineering confidence that the pointing can be maintained to the order of 10 arc s absolute, well within the field of view of the X–ray imagers described below.

The **CASES** science instrumentation represents the minimum required to achieve excellent science with coded–aperture arrays: a set of eight position–sensitive Xe proportional counters (already prototyped and flown on a balloon; see Ramsey et al., 1989), and a set of coded–aperture mask elements at the tip of the boom. The counters use K–coincidence gating to achieve low background, and have a total effective area of some 5,800 cm^2. The coded apertures have a sequence of four angular resolutions to give a fairly uniform weighting of the contrast sensitivity (the (u,v)-plane coverage, in Fourier terminology), as indicated in Table 1, over fields of view of order 0.5° (FWHM).

Table 1. **CASES** angular resolution

Aperture	Resolution
10 mm	64 arc sec
3	19.2
1	6.4
0.25	1.8*

*note: The highest-resolution elements require a second aperture mask directly in front of the counters.

b) *The* **CASES** *Science Goals*

The **CASES** experiment plan includes dedicated observing time for solar and non-solar observations, in addition to which the purely engineering portion of the experiments will be done during solar viewing. X-ray observations of the Sun during the structural controls experiments will provide a valuable bonus for the engineering measurements: the X-ray throughput of the coded-aperture array will characterize the effect of high-frequency jittering motions of the whole structure, at frequencies well above the ~ 10 Hz available from the star trackers.

During the dedicated X-ray astronomy observations, **CASES** will study a singularly important source, the Galactic Center, and obtain the first high-resolution observations in hard X-radiation. The significance of this observation is the diagnostic power it gives for the recognition of determination of the nature of the object there — whether a black hole or not — an observation significant for our understanding of the energy sources in quasars and active galactic nuclei. **CASES** will also be able to observe the Crab Nebula and its pulsar, again uniquely important scientific targets as well as a natural calibration source because of the pulsar's regular X-ray pulsations.

CASES will also provide solar observations far beyond the capabilities of any hard X-ray instrumentation prior to the Pinhole/Occulter Facility. The specific target will be "microflares," small non-thermal events discovered with non-imaging detectors in 1981 (Lin *et al.*, 1981). Other phenomena may also be available for observation, given the unprecedented sensitivity of the **CASES** X-ray detectors. The effective area of these counters — some $10 \times$ that of the largest previous *non-imaging* instruments — will provide the first high-contrast observations of solar hard X-rays. For example, the scattering albedo from the solar photosphere should be easily detectable for the first time with this instrument.

III. CONCLUSION: P/OF AND OTHER LARGE EXPERIMENTS

The Pinhole/Occulter Facility (**P/OF**) combines the large–structure technology with sophisticated observing instruments in its "focal" plane, to provide high–resolution X–ray, γ–ray, and coronal observations of the Sun, and for general high–resolution X–ray and γ–ray astronomy (Hudson and Davis, 1989). The limiting angular resolution should be better than 0.2 arc sec, as limited by diffraction at the 50–m structural length under study for the Space Station. The observing instrumentation for P/OF is to be selected by a future NASA Announcement of Opportunity, with candidate instruments including proportional counters, scintillation counters, and solid–state detectors in various configurations.

The Pinhole/Occulter Facility only represents the first application of large–structures technology to space experiments in these disciplines. We expect that other disciplines — optical and infrared astronomy, for example — will also require physically large instruments. Furthermore, Earth–observation programs (for example, EOS) will require similarly large remote–sensing instruments pointed downwards. It is thus clear that the technology verified with the **CASES** program will have abundant applications in enabling experiments and facilities of great sophistication over the next decades of space research. The **CASES** program will allow us to make a first use of such an engineering capability for worthy astrophysics observational objectives, and in so doing to develop instrumentation (large–area position–sensitive proportional counters) of general utility in other domains.

ACKNOWLEDGEMENTS

The authors wish to thank the members of the **CASES** Science Working Group for defining the scientific program. and J. Dabbs, J. Sharkey, and H. Waites of NASA Marshall Space Flight Center for helping to make this pathfinding technology available for high–energy astrophysics.

REFERENCES

Hudson, H.S., and Davis, J.M., 1989, *EUV, X-ray, and Gamma-Ray Instrumentation for Astronomy and Atomic Physics*, SPIE Proc. **1159**, 318.

Lin, R.P., Schwartz, R.A., Kane, S.R., Pelling, R.M., and Hurley, K.C., 1984, *Astrophys. J.* **283**, 421.

Ramsey, B.F., Weisskopf, M.C., and Joy, M.K., 1989, *EUV, X-ray, and Gamma-Ray Instrumentation for Astronomy and Atomic Physics*, SPIE Proc. **1159**, 246.

Skinner, G.K., and 10 co–authors, 1987, *Nature* **330**, 544.

DISCUSSION

Neil Gehrels

How high an energy will the P/OF occulter work to?

Hudson

The P/OF detectors and mask element swould have to be chosen by competitive proposal, since P/OF is to be a general observing facility. How thick do you want? As I recall, about 2 cm of tungsten will absorb almost any neutral radiation except for neutrinos. One other consideration is the tradeoff between angular resolution and field of view, which would be limited by a thick occulter. This might restrict "zoom" operation, for example.

Martin Elvis

Do you have estimates of how large will be the jiggling of the two ends of the boom with respect to one another, since this determines the size of detector one needs?

Hudson

It depends upon many factors for a given application. For CASES our requirement is 2 arc min (rms) or about 2 cm. The engineers don't seem to have any problem with this, even with the light weight Astromast boom that has been chosen.

OFF-PLANE IMAGING FOR MILLI-ARCSECOND X-RAY OBSERVATORIES

Webster Cash

Center for Astrophysics and Space Astronomy
University of Colorado
Boulder, CO 80309-0391

ABSTRACT. Off-plane imaging provides a practical technique for achieving very high resolution imaging with grazing incidence optics. Although the effect is only in one dimension, it can be utilized to vastly increase the resolution of both spectroscopic and imaging systems. In this talk we present a concept that can lead to milli-arcsecond imaging in the x-rays.

1. Goals for the 21st Century

It is my belief that if x-ray astronomy is to flourish in the next century, there are three things that, at a minimum, must be accomplished.

1) Do Spectroscopy Right
Spectroscopy is the mother lode of x-ray astronomy. It will give us a fundamentally altered and much more detailed view of the processes in x-ray sources. The goal is to perform spectroscopy of x-ray sources with resolution of up to 10,000 ($\lambda/\delta\lambda$) and collecting area of up to 3,000cm^2. The spectral resolution would then allow us to resolve all the way to the doppler width of most of the lines. The collecting area would be enough to support observations of all known sources. We know how to build such an instrument, and it is my hope that it will be an effort started in the 20th century.

2) Perform Milliarcsecond Imaging
A three order of magnitude improvement in imaging quality would undoubtedly provide fundamental discoveries. With .001" resolution we could match the exquisite resolution now enjoyed only by radio astronomers. This resolution would allow us to start imaging the x-ray emission from stars, and would allow us to observe galaxies with the same kind of resolution with which we now observe the Milky Way. In this paper I present one concept that could lead to such an observatory in the early years of the 21st Century.

3) Perform Microarcsecond Imaging
The goal of microarcsecond imaging is to image accretion disks. I believe this will be accomplished at some point in the next century. However, no viable design has yet

been identified.

2. Off-Plane Imaging

In a recent paper (Cash 1987), I discussed the use of off-plane response of grazing incidence optics to make a major improvement in the quality of our x-ray imaging. Grazing incidence optics have the property that the resolution in the plane of reflection will be $1/\theta$ times worse than the resolution in the off-plane direction. (θ is the graze in radians.) With a graze angle of 0.5deg, this corresponds to a factor of 100 potential improvement in one dimension.

One application of this effect has been in the design of x-ray spectrographs. Spectrographs image in one direction and disperse in another, so that by properly arranging reflection gratings, the hidden structure of the telescope image may be used to give significant improvement in the spectral resolution. In principle this technique should give improvements of up to 100 in resolution. In practice we have been able to achieve almost a factor of ten in resolution improvement. This technique is presented in Cash (1990).

In Cash (1987) I discussed how two dimensional images could be created from one dimensional data. This is a standard technique used, for example, in CAT scans, and has been applied to x-ray astronomy. (eg Stevens and Garmire, 1973; Rappaport et al, 1973). In the remainder of this paper we present some simulations of a strawman design that would support imaging down to one milliarcsecond.

3. MAX, the Milli Arcsecond X-ray Observatory

To create a milliarcsecond optic we start by using a conventional paraboloid/hyperboloid design with half degree graze angles. It must be fabricated to 0.1" tolerances so that the $1/\theta$ factor will carry us to .001". It must be large to provide a long focal length and hence a reasonable plate scale. Size is also needed to avoid the diffraction limit. Notice that, even at 4Å, an aperture must be greater than 8cm to avoid diffraction limiting at .001".

Next, we must take less than .01 radians from the circumference of the annular aperture. Otherwise the contributions from the range of azimuthal angles will degrade the image. This means that the radius of the telescope design must be at least 8 meters! The focal length will be over 100meters!! Still, one milliarcsecond will be only 0.5 micron wide on the focal plane.

To demonstrate the manner in which such an instrument might function, we performed raytracing that included simulation of the mirror tolerances. First we placed nine sources randomly around a square arcsecond of sky and created figure 1, showing the characteristic line appearance of each source. It is clear that such an instrument would be very powerful even without image reconstruction. One rather astounding result from the raytracing was that the field of view is huge. The .001"

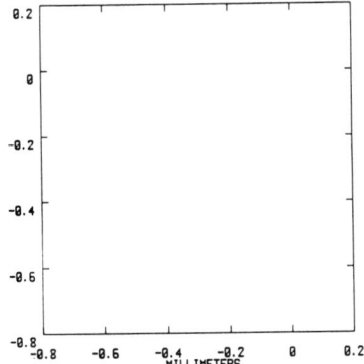

Figure 1: Raytrace of MAX, showing nine sources distributed randomly around a one arcsecond square of sky. The image of each source is 0.1" high and .001" wide.

resolution does not degrade to .002" until 10 *arcminutes* off axis. That is a field of view that is one million resolution elements wide in the one dimension. Similarly, it was interesting to note that, due to the extreme slowness of the beam, the depth of focus for the detector was a full millimeter.

The second raytracing experiment is shown in Figures 2-4. In 2 we show the object. It is a simple version of an AGN core containing a point source, an extended source, and a jet. The extent is just a few milliarcseconds. In 3 we show the spot distribution of the image at the focal plane of MAX, projected onto the x-axis. In 4 we performed the same raytrace, but with the telescope rotated 90 degrees with respect to the target. It is clear from comparison of 3 and 4 that the basic information about the shape of the source has been retained. This demonstrates that milliarcsecond images could, in principle, be obtained with today's technology.

4. MAX - Some Practicalities

Fabrication of the Optics: The mirror surface to support MAX must be a factor of ten smoother and better controlled than the mirrors on AXAF. This sounds daunting at first, but consider that the curvature required is tiny. The deviations from flat are on the order of 100μ. Thus it appears attractive to polish a *flat* mirror and then bend it. This is similar to the approach being taken for fabrication of the Keck Telescope.

Plate Scale: Even at 100 meters from the mirror, a .001" image is only 0.5 microns wide. Currently we do not have detectors that can support this, but the problem can be solved in either of two ways. First, we can wait for technology to catch up. Currently the best detectors resolve about 5 microns, so it is not an outrageously

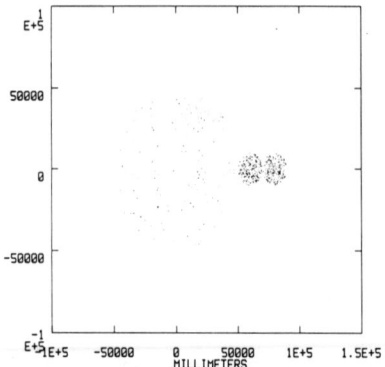

Figure 2: Sample object to be raytraced. Placed at a distance of 10^{12}mm to simulate a large distance, it is meant to resemble the core of an AGN with jet.

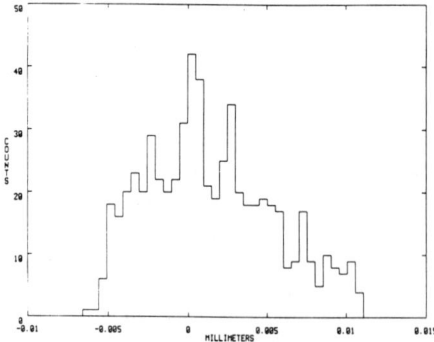

Figure 3: After raytracing the ray distribution was projected onto the x-axis. It is clear that the asymmetric shape of the object is present.

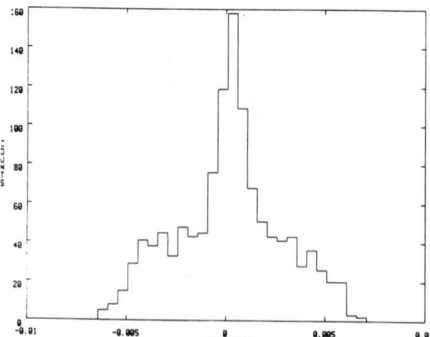

Figure 4: Same as the previous image except that the telescope has been rotated 90 degrees with respect to the target to create a projection onto the y-axis. This distribution is symmetric.

large improvement required. Second, we could use a relay optic to magnify the image as in Chase, Krieger and Underwood (1982).

Small Collecting Area: A single mirror of the type traced would provide only 16cm^2 of collecting area, and most objects would require observations from several angles. Several mirrors, sampling different rotation angles would solve this problem. Nonetheless, it is unlikely that this instrument would ever achieve large collecting area, and would compensate with lengthy observations.

Stability: It is obvious that such a telescope would provide major engineering challenges. It is probably unreasonable to stablize a lengthy beam to .001", and so photon counting coupled with monitoring of the facility would provide the information to reconstruct the image. A laser/cornercube system could provide information on the internal structural drifts of the instrument, while a visible light stellar interferometer could provide a pointing error signal.

This work has been supported by NASA grant NAG5-96.

References

Cash, W, *Applied Optics*, *26*, 2915 (1987)
Cash, W, *Applied Optics*, submitted (1990)
Chase, R. C., Krieger, A. S., Underwodd, J. H., *Applied Optics*, *21*, 4446 (1982)
Rappaport, S., Cash, W., Doxsey, R., Moore, G., and Borken, R., *Ap.J. (Letters)*, *186*, L115 (1973)
Stevens, J. C., and Garmire, G. P., *Ap.J. (Letters)*, *180*, L19 (1973).

DISCUSSION

Alan Bunner

What might be the total weight of a MAX observatory that has sufficient effective area to match the needs of milliarcsecond resolution? What is the radius of curvature of the best focal surface?

Cash

The observatory would contain three basic elements: the optics, the detectors, and the structure. I have absolutely no idea how much the structure would weigh, as this is very sensitive to stability questions which have not yet been tackled. The detector requirements would be comparable to current detector requirements, i.e., in the 50 kg range. The optics would each have 0.3 m^2 of surface area. Twenty would have 6 m^2 of area, comparable to a 4 meter-class mirror. Since weight drops faster than linear with aperture, I guess that the total weight of the mirrors ought to be less than that in the primary of the HST. The raytracing shows no degradation at the milliarcsecond level across a 20 arcminute, *flat* field.

Richard Mushotzky

How much area do you need per module?

Cash

Each module has only 20 cm^2. It is clear that multiple (> 10) modules would be required to study sources of the intensity detected by the Einstein Observatory.

Martin Elvis

The wide field-of-view would be very useful for astrometry, since you will often get more than one source in the field. Why can't you stack plates, as in AXAF?

Cash

I believe you can stack plates as in AXAF. In fact, this is highly desirable. Twenty stacked plates would allow many azimuthal angles to be sampled simultaneously and would increase total collecting area from 20 to 400 cm^2.

PROSPECTS FOR HIGH PRESSURE GAS SCINTILLATION CHAMBERS

T.K. Edberg, A. Parsons, B. Sadoulet, S. Weiss, J. Wilkerson

Department of Physics, University of California, Berkeley;
Space Sciences Laboratory, University of California, Berkeley;
and The Center for Particle Astrophysics

G. Smith

Lawrence Berkeley Laboratory

I. INTRODUCTION

We present here the current developmental status and future potential of what we believe to be a very promising hard X-ray and soft γ-ray detector: the high pressure xenon gas scintillation imaging drift chamber. This detector combines excellent spatial resolution, very good energy resolution, good time resolution, and extremely high sensitivity because of excellent background rejection capabilities; and it offers possibilities for relatively easy extrapolation to large area detectors.

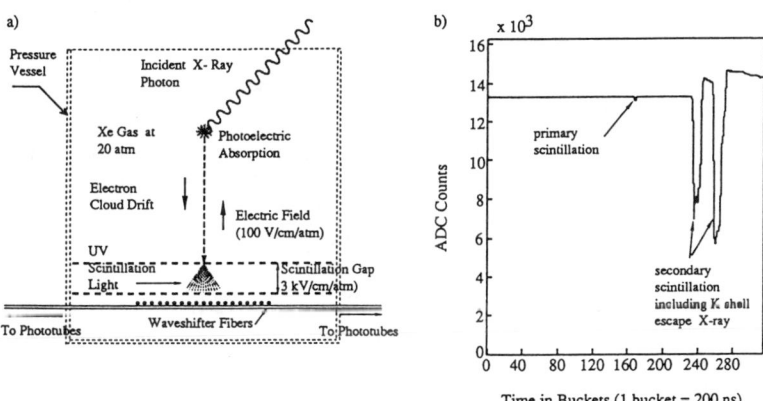

Figure 1. (a) *A two-dimensional schematic diagram outlining the operating principle behind the gas scintillation drift chamber.* (b) *An example of a raw event observed with the prototype chamber. Note the small primary scintillation pulse which precedes the secondary pulses. The two large secondary scintillation pulses represent the detection of both the initial X-ray interaction and the K-shell fluorescence photon.*

The gas scintillation drift chamber, Figure 1a, is based on the combination of gas scintillation and the time projection concept (Sadoulet, *et al.*, 1987, 1988, 1989; Parsons *et al.*, 1988, 1989, 1990). It is based on the detection of UV scintillation light produced in high pressure xenon gas. Electrons produced in an X-ray interaction in the gas will drift in a moderate electric field toward a detection region where a high electric field is maintained (~3 kV/cm/atm). UV scintillation light is emitted in two different processes: a small amount of scintillation light, the "primary scintillation," is produced at the time of the initial X-ray interaction; after drifting through the gas, the extracted electrons also generate an intense flash of scintillation light as they traverse the high field region. This "secondary scintillation" light has an intensity proportional to the energy initially deposited in the chamber. Detection of this secondary scintillation light allows two dimensions of spatial imaging perpendicular to the drift direction, and the comparison of the times of the primary and secondary scintillation gives the drift time of the electrons and, therefore, the position of the initial interaction in the drift direction. Our novel readout method with wave-shifter fibers enclosed in thin quartz tubes running parallel to the scintillation gap allows us to extrapolate the technique to high pressures and large areas. This technique is now fully demonstrated. Figure 1b displays a typical event in such a chamber.

This technique has significant merits in five areas. **High pressure and area:** due to the lack of any sort of sight window, there is in principle no limit to the size nor pressure of a scientific instrument. This allows large effective area. **Background rejection:** in xenon, 76% of the photoelectric absorptions result in a K-shell fluorescence photon which can be used as a tag of a K-shell interaction. The gas pressure can be tailored (~20 atmospheres) so that the mean free path is of the order of 1 cm and the second interaction site is resolvable. The use of this two-site signature to identify photoelectrically absorbed photons provides an extremely effective means of rejecting one- and three-site detector backgrounds due to elastic neutron scattering, β-decays from cosmic ray induced radioactivity, and photons which Compton scatter but do not stop in the detector. These background components, to our knowledge, cannot be rejected by this method in solid or liquid detectors. Combined with large effective areas, the background rejection makes for a very high sensitivity device. **Measurement of the interaction time:** the primary scintillation gives a prompt signal of the occurrence of an interaction which allows us to measure the drift time and to minimize the dead time caused by interactions in the shield. **Energy resolution:** if enough photoelectrons are detected per initial electron, the intrinsic energy resolution given by Fano fluctuations is not degraded and accuracies only 3 times worse than germanium are expected. The ultimate full width half maximum of $14\%/\sqrt{E(keV)}$ can be expected giving 2.0% at 50 keV and 1.3% at 122 keV. Utilization of the constraint of the energy of the K-shell photon can improve these numbers above the K-shell edge. **Position resolution:** with the detection of the primary scintillation and the position sensitivity of the fibers, it is possible to localize point-like interactions in all three dimensions with precision typical of drift chambers (~200 μm FWHM).

II. CURRENT DEVELOPMENTAL STATUS

In order to experimentally test the five merits of the concept, we have built a prototype of a gas scintillation drift chamber (Sadoulet, et al., 1989). We have demonstrated the fiber readout scheme, and verified our light collection efficiency to be 6×10^{-3} photoelectrons per photon incident on the fibers, as we expected. We have operated at pressures up to 20 atm, mapping the light yield using an anger camera geometry. Due to a high voltage breakdown from a flaw unique to the prototype chamber, we were limited in the light yield we could achieve; most of our fiber readout tests were conducted at 15 atm where the light yield was at maximum. We have observed the K-shell signature in the time domain, as evidenced in Figure 1b. The primary scintillation has been observed with the fiber readout; we determined the yield to be one photon per 76±12 eV deposited for a Xe/He (90/10) mixture. Pure xenon gives similar estimates.

We have measured the FWHM energy resolution at 122 keV to be 5.8%, within a factor of five of the ultimate limit of 1.3%. We believe this relatively low resolution is due to the combined effects of the increased statistical error incurred by operating at lowered secondary scintillation light yield (from the limited electric field necessitated by the high voltage breakdowns), coupled with poor pulse height resolution of the phototubes we are currently using. Using our current phototubes at full scintillation yield should give a resolution of 3% at 122 keV; better phototubes should allow us to approach the ultimate resolution limit.

In order to measure the spatial resolution, we use a collimated 60 keV gamma source pointing parallel to the drift field to measure the spread in the reconstructed position at the plane of the fibers closest to the scintillation gap. Using a simple weighted mean to measure the position of the interaction yields a 1.6 mm FWHM, completely dominated by the spread in the source beam. (Future tests will use a better collimator.) From this we can place a conservative upper limit of 400 μm rms position resolution in the fiber plane for conversions occurring one centimeter from the gap. Position resolution in the drift direction should be similar to the transverse resolution.

Although we have not yet attained the ultimate performance of the chamber, the previously described tests partially demonstrate each of the expected properties of gas scintillation drift chambers with wave-shifter readout. These results show that such devices are very powerful and well worth further development.

III. FUTURE POTENTIAL

In the near future, we (along with collaborators Bob Lin, Kevin Hurley, Tom Prince, and Rick Cook) have been approved under NASA's SR&T program to develop a balloon-borne instrument to fly in late 1992 or early 1993. The instrument, called SIGHT (for Scintillation Imaging Gas-filled Hard X-ray Telescope), is based on a 1260 cm^2 active area xenon gas scintillation drift chamber operated at 20 atmospheres, coupled to a coded aperture mask. With a depth of 2 g/cm^2 of xenon, this telescope will offer observations in the 35-300 keV range. Along with the good energy resolution expected, the telescope should have an angular resolving power of 1.3' (and < 10" strong source localization, assuming an improved aspect system).

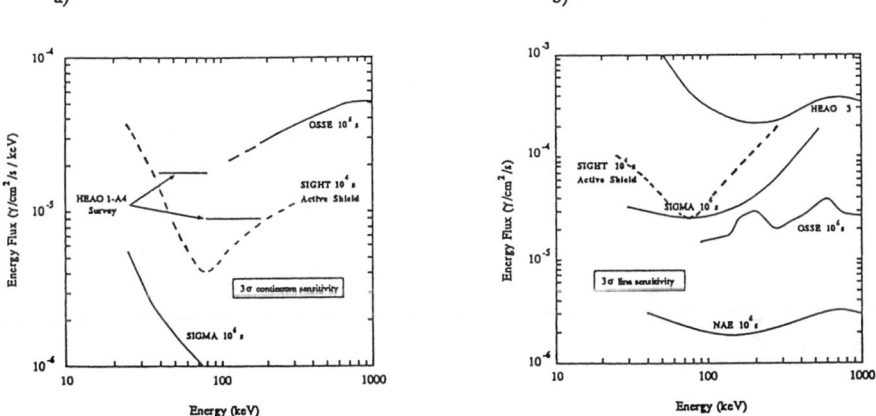

Figure 2. SIGHT 3σ sensitivity with a 5 cm thick BGO shield (for 10^4 seconds integration time, dashed line) as a function of energy: (a) continuum, (b) narrow line. Various satellite sensitivities (for 10^6 seconds integration time) are shown for comparison.

Because the intrinsic background is so low, a thick active shield is required to reduce the leakage background in order to achieve the full potential of this detector. The resulting low background leads to a very sensitive instrument. Taking into account the effective area (400 cm^2 at 100 keV), and assuming a 5 cm thick BGO shield, we calculate the sensitivity of SIGHT for a 10^4 second integration time shown in Figure 2. For comparison, the sensitivity of various satellite experiments are also depicted, assuming 100 times longer integrations.

As for prospects in the more distant future, it is important to realize that there is nothing in this technology to prevent extrapolation to larger detectors and satellite environments. (In practice, active area will be limited to ~1m² by event overlap and deadtime from charged tracks and shield vetoes.) An obvious first application is to extrapolate the SIGHT instrument to larger area and lower background. For example, a detector of 80 cm active diameter, with a gas depth of 4 g/cm² giving an effective area of the order of 8 times that of SIGHT. Properly shielded with 10 cm of BGO, an active 1°x1° collimator and with a larger ratio of gas to inert material, the background can be about a factor of 5-10 lower than for SIGHT. For 10^6 s observations, the 3σ narrow line sensitivity in the hard X-ray range is estimated to be about 4×10^{-7} γ/cm²/s. This type of instrument will allow deep pointed observations.

The imaging properties of the gas scintillation drift chamber may also be applied to a somewhat unorthodox concept. The idea is based on a spherical gas scintillation drift chamber of 1 meter diameter surrounded by a hemispherical coded aperture mask made in the form of approximately half a geodesic dome with ≈200 facets. These can be considered as 200 independent telescopes, each of which has an 11° field of view constantly monitoring nearly the entire half sky away from the earth. An active BGO shield vetoes the hard X-rays and gamma rays coming up from the atmosphere and a plastic scintillator at the surface of the drift chamber rejects cosmic rays. Filled with a xenon-based mixture at 40 atmospheres, this device would provide significant stopping power (14 g/cm²) which would make it sensitive from 20 keV up to a few MeV (using the instrument as a Compton telescope, see below), excellent time tagging capabilities, 200 μm FWHM spatial resolution at low energy leading to high angular accuracy (arc minute mapping, arcsecond localization) and an energy resolution comparable to that of SIGHT. The power of such an instrument is quite impressive for the continuous monitoring of objects with strong transient behavior such as X-ray pulsars, AGN and γ-ray bursts.

A potentially important application of the gas scintillation drift chamber is the possibility of imaging the trajectories of high enough energy electrons. Monte Carlo simulations of 500 keV electron trajectories in pure xenon at 40 atm, taking into account dE/dx and Moliere scattering, show that half the tracks are imagable to better than 20° at that pressure. Using a lower Z mixture (e.g. 95% Ar, 5% Xe which scintillates as well) and/or lower pressure, it would be possible to have longer and straighter trajectories. This opens the possibility of measuring the direction of a Compton-scattered electron, and of making a powerful Compton telescope. The entire (potentially complex) history of a Compton event could be imaged within the detector, which serves as a completely active target. The direction measurement of the scattered electron and photon allows the full reconstruction of the real source position for each event, as well as providing polarization information. The error box for each event will be restricted to an arc (~20°) of an annulus (~15' wide) on the sky, in contrast to the full annulus in a conventional Compton telescope. The angular accuracies are typical values based on estimates of the reconstruction accuracy of the angle of the electron trajectory (arc) and position and energy resolution (annulus). This would reduce source confusion and background drastically.

REFERENCES

Parsons, A., Sadoulet, B., Weiss, S., Smith, D., Hurley, K., Lin, R.P., and Smith, G. 1988, *Proceedings of the Workshop on Nuclear Spectroscopy of Astrophysical Sources*, AIP Conf. Proceedings 170, 478.

Parsons, A., Sadoulet, B., Weiss, S., Edberg, T.K., Wilkerson, J., Smith, G., Lin, R.P., and Hurley, K. 1989, *IEEE Trans. on Nucl. Sc.*, **NS 36**, No. 1, 931.

Parsons, A., Edberg, T.K., Sadoulet, B., Weiss, S., Wilkerson, J., Hurley, K., Lin, R.P., and Smith, G. 1990, *IEEE Trans. on Nucl. Sc.*, in press.

Sadoulet, B., Lin, R.P., and Weiss, S. 1987, *IEEE Trans. on Nucl. Sc.*, **NS-34**, No. 1, 52.

Sadoulet, B., Weiss, S., Parsons, A., Lin, R.P., and Smith, G. 1988, *IEEE Trans. on Nucl. Sc.*, **NS-35**, No. 1, 543.

Sadoulet, B., Edberg, T.K., Weiss, S., Parsons, A., Wilkerson, J., Hurley, K., Lin, R.P., and Smith, G. 1989, *SPIE Proceedings* **1159**, 45.

DISCUSSION

Richard Catura

Does activation of the Xe from on-orbit radiation pose a background problem?

Edberg

For a balloon flight, activation is not a problem. We estimate the activation background to be small, less than the leakage background, even without imposing the K-shell signature requirement; use of the K-shell signature should suppress the activation background to a negligible level. We have not performed the corresponding analysis for a satellite application.

James Kurfess

Are you able to independently measure the energy loss of two simultaneous interactions, so that the K-X-ray suppression technique can require a $\sim$ 30 keV energy loss?

Edberg

Yes, as long as the two interaction sites are far enough apart to be resolvable. We estimate that this requires separation of about 1 mm in the drift direction or about 7 mm in the orthogonal directions, which should occur about 70% of the time in 20 atmospheres of xenon.

Gerald Share

(1) How are you viewing the primary scintillation light? (2) Have you modelled the multiple coulomb scattering in xenon to come up with your 20° Compton arc?

Edberg

(1) In the SIGHT instrument, the detection efficiency for the primary scintillation using the waveshifter fiber readout is a little too low for reliable detection. (We expect about one observed photoelectron at 40 keV, about 8 at 300 keV.) We will therefore install proportional tubes (based on TMAE gas for high quantum efficiency) inside the pressure volume to detect the primary scintillation with virtually 100% efficiency. We have almost completed assembly of a prototype proportional tube and expect to begin testing it within a week or so. (2) Yes. The Monte Carlo simulation was run for pure xenon including the effects of dE/dr and multiple coulomb scattering (and Moliere scattering).

X-RAY SPECTROSCOPIC INSTRUMENTATION FOR THE 21ST CENTURY

Steven M. Kahn

Departments of Physics and Astronomy
and Space Sciences Laboratory
University of California, Berkeley

Abstract: The physical principles and salient characteristics of a variety of technologies currently under development for use in astronomical high resolution X-ray spectroscopic instrumentation are compared and contrasted. Included for discussion are dispersive spectrometers based on crystal diffractors, transmission gratings, and reflection gratings, and non-dispersive cryogenic devices such as microcalorimeters and superconducting tunnel junction detectors. Substantial developments have occurred in most of these areas in the last few years. It is argued that a realistic evaluation of these alternatives cannot be performed out of context of the particular missions envisioned, since the relevant trade-offs usually depend more on the implementation of the technology than on its fundamental characteristics. Nevertheless, in light of the present-day developments, reflection grating devices seem best suited for very high resolution work in the soft X-ray band (< 1 keV), whereas the cryogenic detectors seem to be especially promising at higher energies.

1. Introduction.

The task that has been assigned to me, to review the technology that will be important for X-ray spectroscopy in the 21st Century, presents a number of difficulties that are worth airing "publicly" at the beginning of this discussion. First, "technological forecasting" is notoriously unreliable, particularly in fields that always function at the state-of-the-art. Considering the monumental advances over the last 20 years that have occurred in such relevant areas as microelectronics, solid-state devices, and precision optics, to name just a few, it seems quite presumptuous to suggest that we can even imagine what the X-ray spectrometers of the 21st Century will look like, much less to compare and evaluate them. I believe that I can best serve the purposes of this Workshop by confining my remarks to spectroscopic instrumentation which is under *current* development. Even this presents a challenge in light of the extreme diversity of the relevant technology. I can therefore present only a rudimentary overview of the fundamental considerations in each case. There are several other papers in these Proceedings that supply more of the necessary detail.

A second issue one faces in attempting to compare the various instrument concepts is that the most important considerations often turn out to be associatied with the *implementation* of the technology, not with its fundamental characteristics or properties. In the design of "real" space experiments, nearly all of one's time and most of one's creative input goes into trying to "fit" the instrument into the allowable envelope (mass, size, power, field of view, etc.) afforded by the launch opportunity. In the end, all that matters is the capability of the experiment to make scientific measurements at the required precision and sensitivity. Instruments involving technology with inherently "poorer" capabilities can still outperform more sophisticated approaches if they are implemented in a more effective way. Cost is a major factor here. As a trivial example, detector designs with low quantum efficiency can still achieve very high sensitivity if they can be easily reproduced or made large at relatively low cost. Unfortunately, cost is perhaps the one parameter that can be least reliably predicted for the distant future. Our intuition has very often been wrong on this issue.

Finally, as we look toward the 21st Century, I think it becomes increasingly difficult to define quantitative criteria by which to evaluate candidate instrumentation. For spectrometers, one typically quotes the spectral resolving power and effective area defined in some average sense over some appropriate spectral band. However, for nearly all technologies and implementations that are currently envisioned, both the resolving power and the effective area are strong functions of energy. In the current era, the "average performance" approach is justified because our satellite facilities are so few and far between that we expect each one to have a broad range of capabilities, applicable to a diverse spectrum of astrophysical questions. In the next century, I suspect this will no longer be the case. It is likely that we will want to fly a large number of smaller experiments targeted to answer specific sets of scientific questions. For these "targeted" experiments, spectrally-averaged instrumental parameters are irrelevant. All that matters is the resolving power and sensitivity at the energies of the lines one wants to measure. In this case, specification of the performance criteria requires prior specification of the scientific problem.

In light of these considerations, I believe that it is prudent to continue development of a large suite of candidate instrumentation for astronomical X-ray spectroscopy. Below I review the salient characteristics of crystal spectrometers, diffraction gratings (both transmission and reflection), microcalorimeters, and superconducting tunnel junction detectors. All of these devices can, at least in principle, deliver high resolving power, $E/\Delta E > 10^3$, over some region of the conventional X-ray band, 0.1 - 10 keV. I will not discuss lower resolution instruments like gas counters and photoconductive solid-state devices because I believe they will not meet the resolution requirements of future spectroscopic experiments.

2. Crystal Spectrometers.

X-ray spectrometers based on crystal diffraction have been used extensively in solar astronomy and are still in widespread use in the laboratory plasma community. There were a few rocket experiments utilizing crystals for observations of cosmic sources in the 1970s (e.g. Stark and Culhane 1978), but these produced few results because of limited sensitivity. A crystal spectrometer was flown in the focal plane of the *Einstein* Observatory (Canizares et al. 1977, Giacconi et al. 1979) which, although used sparingly, did produce the highest resolution, nonsolar, astrophysical X-ray spectra that are currently available. A similar experiment is under consideration for the Advanced X-ray Astrophysics Facility (AXAF; Markert et al. 1988). Finally, an objective crystal spectrometer is currently under development for flight on the Soviet Spectrum X, Γ mission.

A crystal diffracts X-ray because of the regular periodicity of the lattice planes formed by the constituent "unit cells". X-rays impinging on the surface of a properly cleaved crystal will only be reflected with high efficiency if the components scattered from subsequent crystal planes constructively interfere. A simple geometrical calculation shows that the X-ray wavelength must then satisfy the Bragg condition:

$$m \lambda = 2 d \sin \theta$$

where θ is the incidence angle made with the crystal planes, m is an integer corresponding to the spectral order, and d is the lattice constant, or the spacing between the crystal planes.

If the angle θ is systematically varied by rocking the crystal relative to the X-ray beam, the intensity of the beam as a function of wavelength can be recorded. Alternatively, if the crystal is illuminated by diffuse radiation, different wavelengths can be focussed in different directions, thereby sampling a finite spectral range simultaneously.

Although crystal spectrometers are often referred to as "dispersive devices", the term is really a misnomer because crystals do not disperse, they selectively reflect, like a narrow-band filter. The key point is that a crystal spectrometer does not multiplex, it does not simultaneously record all wavelengths, from all parts of the beam, at all times. Some designs record the whole spectrum simultaneously by dividing up the beam. Others utilize the whole beam, but divide the spectrum up in time by rocking the crystal. This lack of multiplexing has severe implications in terms of the sensitivity of the spectrometer for surveying a finite spectral band. As compared to other technologies, crystal spectrometers will always have low sensitivity for this purpose since they "throw away" most of the incident photons.

The efficiency of a crystal diffractor can, in most implementations, be shown to be given approximately by:

$$\mathit{Eff} = R_C(\lambda) / (\Delta\theta)$$

where $R_C(\lambda)$ is the integrated reflectivity (defined as the integral of the rocking curve over angle), and $(\Delta\theta)$ is the full angular range covered, either by scanning or by non-collimated illumination. If the spectral range surveyed is narrow and is given by N resolution elements, then this expression can be rewritten in the form:

$$\mathit{Eff} = (\lambda/\Delta\lambda) R_C(\lambda) / N \tan\theta_B$$

where $(\lambda/\Delta\lambda)$ is the resolving power. As it turns out, typical resolving powers are in the range $10^3 - 10^4$, while typical integrated reflectivities are $\sim 10^{-4}$ (Alexandropoulos and Cohen 1974), so, in fact, the efficiency is not especially low if N is kept small, i.e. if the spectrometer is only used to survey individual lines. Hence there is potentially still an important role in X-ray astronomy for crystal spectrometers, but probably not as spectral survey instruments.

The spectral resolving power of a crystal spectrometer can be found by differentiating the Bragg equation:

$$(\lambda/\Delta\lambda) = \tan\theta_B / W$$

where W is the total angular spread. W may include contributions from angular uncertainties (e.g. telescope and camera blur, aberrations), but it must include a contribution from the finite width of the rocking curve itself, W_C. Some crystals are mosaic limited, i.e. they are made up of many smaller crystallites which are themselves misaligned. In these cases, W_C can be rather large, ~ several arc-minutes. Because θ_B is not small typically, the resolving power is still high, ~ $10^3 - 10^4$. For other, more nearly "perfect" crystals, W_C is limited only by absorption and extinction as the X-rays penetrate down to deeper crystal planes. In these cases, resolving powers as high as ~ 10^5 can be achieved. Unfortunately, there is a positive correlation between W_C and R_C, so high resolution is obtained only at the expense of sensitivity. The highest resolutions achievable are at high energies where X-rays are penetrating and can interfere off many crystal planes.

Crystals or multilayers used in Fabry-Perot configurations (Barbee and Underwood 1983) yield resolving powers as high as 10^7, but at very low sensitivity.

There is a rich variety of possible orientations in which a crystal can be implemented into a spectrometer for use in X-ray astronomy. In the classic Rowland circle design (see Figure 1), the crystal is placed behind a slit in the focal plane on a circle which also includes the slit and the detector. If the crystal is bent into an appropriate toroidal shape, all photons of a given Bragg wavelength are focussed by the crystal onto a spot on the detector. If the crystal and detector are scanned along the circle, different wavelengths are sampled. A very different design has been adopted for the instrument which will be flown on Spectrum X, Γ (Schnopper and Byrnak 1985). In this "barn-door" approach illustrated in Figure 2, a flat crystal is placed in front of a telescope concentrator at an angle. For diffuse sources, X-rays at different wavelengths are imaged from different regions of the source simultaneously. As the telescope scans, a complete image can be accumulated in several different lines.

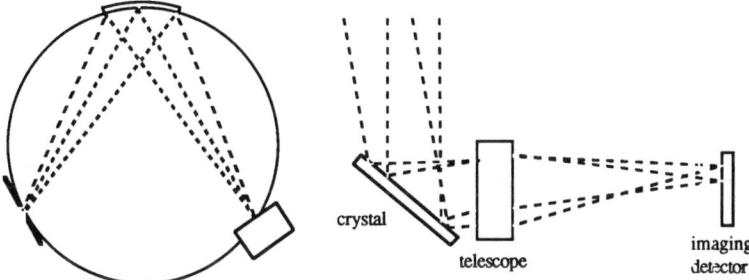

Figure 1: The classic Rowland circle orientation for a crystal spectrometer wherein the crystal, entrance slit, and detector all lie on a common circle.

Figure 2: The barn-door orientation used on the Spectrum X,Γ Mission. Different regions of an extended source are imaged in different wavelengths for each pointing direction.

Crystal technology is at present fairly mature, so there are few technological drivers for the future. Achieving good crystal quality and optical figure for implementations requiring bending of the crystal in one or two dimensions is one area in which some development is still required. In addition, the assemblage of crystals into Fabry-Perot configurations is an exciting frontier area that could lead to very high resolution devices for the future.

3. Transmission Gratings.

A transmission grating spectrometer was also flown on the *Einstein* Observatory (Giacconi et al. 1979, Seward et al. 1982) and provided some of the few moderate resolution X-ray spectra of cosmic sources presently available. Similar experiments were flown on EXOSAT (Taylor et al. 1981) and are currently planned for AXAF (Brinkman et al. 1985, Canizares, Schattenburg, and Smith 1985). The transmission grating consists of a periodic array of metal bars separated by spaces. X-rays passing through the spaces at incident angle, ϕ_i, constructively interfere and are dispersed to an outgoing angle, ϕ_o, given by the dispersion equation:

$$m \lambda = d (\sin \phi_i - \sin \phi_o)$$

where d is the spacing between the centers of the bars, the "grating constant", and m is the spectral order. Photons are accumulated in both positive ($m > 0$) and negative ($m < 0$) spectral orders, as well as in the spectroscopically useless zero order. Imaging of the dispersed photons on a position-sensitive detector thus yields the spectrum.

In most applications, the X-rays are incident along the normal to the grating surface, so $\phi_i = 0$. In that case, the resolution of the system is given simply by:

$$\Delta \lambda = d \cos \phi_o (\Delta \phi) / m$$

where ($\Delta \phi$) is the total angular uncertainty in the system, including telescope blur, detector broadening, and aberrations. For AXAF, and probably for at least some other future X-ray observatories, ($\Delta \phi$) ~ 1 arc-second. With line densities as high as 10^4 1/mm, which can presently be made, first order resolving powers ranging from 250 to 25,000 can be obtained in the 0.1 - 10 keV band. Since the resolution is fixed in wavelength, the resolving power gets much better at lower energies.

The efficiency for a transmission grating spectrometer can be calculated from simple Fraunhofer diffraction theory. If the grating bars are completely opaque to the incident radiation, the efficiency in spectral order m is given by:

$$\text{Eff}_m = [\sin (m \pi a /d) / (m \pi)]^2 ,$$

independent of wavelength, where a is the width of the grating bar and d is the line spacing. This expression peaks at a value of 10% in first order for $a / d = 0.5$. In this configuration, all non-zero even orders vanish. Adding together the two first orders yields a maximum efficiency of 20%.

If, however, the grating bars are partially transparent, the efficiency can be different and may vary with energy. This is shown explicitly in Figure 3 which shows the first order efficiency as a function of energy for a 1000 1/mm gold grating, plotted for several different values of the bar thickness in microns. As can be seen, the efficiency can double in selected wavelength bands. It can be shown (Schattenburg et al. 1988) that peak efficiencies ~ 40% in each of the first spectral orders can be obtained in this way for selected wavelength bands, if the bar thickness is adjusted accordingly. This phase interference effect is used to advantage by both of the transmission grating experiments which will be flown on AXAF.

The original transmission grating experiment flown on a rocket in the 1960s (Vaiana et al. 1968) utilized an objective grating placed in front of an X-ray mirror. In both the *Einstein* and EXOSAT experiments, and in the two AXAF experiments, the grating is instead inserted behind the mirror. Because the beam is converging at this point, aberrations will result if the grating is a flat planar array. These can be corrected using a Rowland torus configuration (Beuermann, Brauninger, and Trumper 1978). The torus, formed by rotating the Rowland circle about an axis perpendicular to both the grating bars and the optical path, also contains the telescope focus and the spectroscopic detector plane.

The grating facets must be normal to the incident rays hitting the torus. Because the facets are of finite size, some residual aberrations do remain.

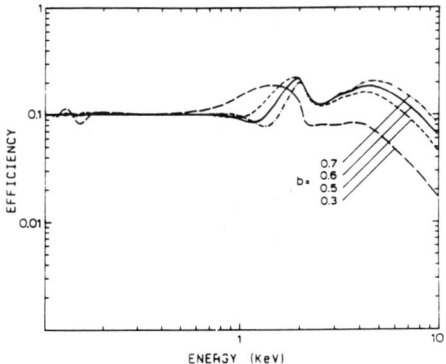

Figure 3: The first order grating efficiency of a gold-bar, 1000 l/mm transmission grating plotted as a function of energy for various values of the bar thickness in microns. (Taken from Brinkman et al. 1985).

Transmission gratings are usually produced holographically, wherein an interference pattern is etched into a photoresist and then transferred by a variety of means to the metallic bars. The MIT group has pioneered the use of X-ray lithography in recent years (Schattenburg et al. 1988), which has enabled the production of very high line density gratings with very thick bars. This is essential for higher energy applications because high dispersion is required as well as large thickness to stop the X-rays. However, the extreme aspect ratio of the bars makes these gratings fragile. As a result, they are supported by a plastic substrate which limits their use at lower energies.

4. Reflection Gratings.

Reflection gratings have not yet been flown on X-ray astronomy experiments, although a rocket payload has been fully developed and is currently awaiting flight (Cash 1988). In addition, a reflection grating design has been recently selected for flight on the European Space Agency's X-Ray Multi-Mirror Mission (XMM; Brinkman et al. 1989). A schematic of a reflection grating is shown in Figure 4. If an impinging ray makes a polar angle θ with the grooves, it will leave at that angle as well. This is the so-called "conical diffraction" condition. The dispersion equation, derived from the interference condition between the grooves, relates the incoming azimuthal angle, α, to the outgoing azimuthal angle, β:

$$m \lambda = d \sin \theta (\cos \beta - \cos \alpha)$$

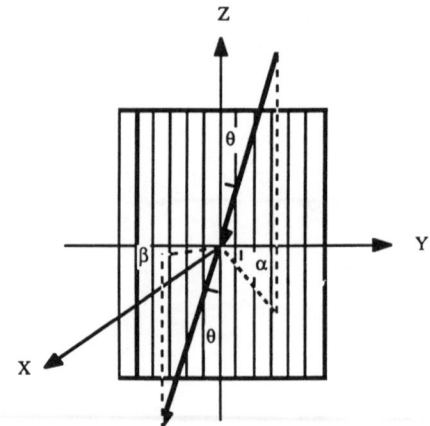

Figure 4: A schematic of a reflection grating showing the orientations of the incoming and outgoing rays.

where d is the groove spacing, and m is the spectral order. $m = 0$ corresponds to pure reflection off the surface. $m > 0$ orders are called "outside orders", and $m < 0$ orders are called "inside orders" in this notation.

Most applications which require high diffraction efficiency invoke "blazed" gratings in which the facets exhibit the sawtooth pattern shown in Figure 5. The tilt angle δ is called the blaze angle. When the incoming and outgoing graze angles with the facets are equal, then each of the facets acts like a tiny mirror, and they all add coherently to give maximum diffraction efficiency. The is called the blaze condition. A simple manipulation of the dispersion equation gives:

$$|m| \lambda_B = 2 d \sin \gamma \sin \delta$$

at blaze, where γ is the graze angle on the facets.

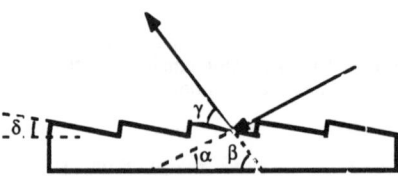

Figure 5: A blazed reflection grating, shown in projection in a plane perpendicular to the grooves.

The efficiency of a reflection grating can be approximately calculated from scalar diffraction theory. At the blaze wavelength, one obtains:

$$Eff = R(\lambda, \gamma)\, \eta$$

where

$$\eta = \sin\alpha / \sin\beta_B.$$

Since η can be chosen close to unity (within a factor of a few in most cases), the efficiency approaches the pure reflection efficiency off the metal surface, which can be very high if γ is chosen appropriately.

There are two kinds of "gratings orientations" which have been advocated in the literature for use in X-ray astronomy (c.f. Kahn, 1990), the "off-plane" mount, and the "in-plane" mount. In the off-plane configuration, the rays come in nearly parallel to the grooves with θ equal to the graze angle γ. The parameter η can then be identically equal to one, so the efficiency is very high. In the more conventional in-plane mount, the light comes in perpendicular to the grooves. For this case:

$$\eta = \sin(\gamma-\delta) / \sin(\gamma+\delta)$$

which can be quite a bit lower.

The two orientations have very different resolution properties, however. For the off-plane mount, in most implementations, the obtainable resolution is very similar to that of a transmission grating with the same line density. Since line densities in excess of 10^4 l/mm can be produced on reflection gratings, respectable resolving powers can be obtained, especially at low energies. For the in-plane case, on the other hand, there is a projection effect, which enhances the resolution even for moderate line densities. Since the light is perpendicular to the grooves and comes in at grazing incidence, the "projected" line spacing is much higher than the ruled line spacing, which leads to high dispersion. A simple calculation gives:

$$(\lambda/\Delta\lambda) = \gamma (1/\eta - 1) / (\Delta\theta)$$

at blaze, where $(\Delta\theta)$ is the angular uncertainty in the system. Note that low η leads to high resolution but low efficiency, and visa versa. With $(\Delta\theta) = 1$ arc-second, $\gamma = 2°$ (applicable out to ~ 2 keV), and $\eta = 0.5$, $(\lambda/\Delta\lambda) = 15,000$. It can reach ~ 65,000 for $\eta \sim 0.1$. The former system would have an efficiency ~ 40% at 1 keV, the latter ~ 8%. The achievable resolution is lower at higher energies because of the γ-dependence.

As in the other cases, reflection gratings can be implemented in a variety of ways. The rocket experiment currently awaiting flight has the gratings placed in a parallel stack in front of the X-ray mirror (Cash 1988). In the XMM design, the gratings are arrayed behind the mirror (Brinkman et al. 1989). To correct for the converging beam aberrations, the gratings are arrayed along a Rowland torus and the groove spacing must be varied slightly along the grating length (Kahn 1990).

X-ray reflection gratings have historically been fabricated by two methods: (1) mechanical ruling directly into a metal coating using a diamond stylus, and (2) holographic exposure followed by ion etching. Both have produced acceptable results (den Boggende et al. 1988).

5. Microcalorimeters.

Microcalorimeters are thermal detectors that "sense" X-rays through the presence of thermal fluctuations in a controlled volume (Moseley, Mather, and McCammon 1984). These devices have received a lot of attention in recent years primarily because of the extensive development efforts undertaken by the Goddard Space Flight Center/University of Wisconsin team which is preparing an experiment of this kind (the X-Ray Spectrometer, XRS) for flight on AXAF (Holt et al. 1988). In its simplest form, the calorimeter consists of an absorber, a thermometer, and a thermal link to a heat bath at a controlled temperature. When an X-ray is absorbed, a temperature pulse is generated which decays exponentially with a time constant given by the ratio of the heat capacity of the calorimeter, C, to the thermal conductivity of the link, G: $\tau = C/G$. The energy of the X-ray is determined by the magnitude of the temperature pulse recorded by the thermometer. In most work so far, a temperature-sensitive resistor, or thermistor, has been used as the thermometer. To keep the heat capacity of the system low, the device is cooled to cryogenic temperatures, typically ~ 0.1 K.

A fundamental consideration in estimating the energy resolution of such a device is associated with thermal fluctuations due to random exchange of heat between the calorimeter and the heat bath (Moseley, Mather, and McCammon 1984). A straightforward statistical mechanics calculation shows:

$$(\Delta E)_{therm} = (k T^2 C)^{1/2} .$$

Since the heat capacity of an insulator or semiconductor scales like T^3 at low temperature, the energy resolution scales like $T^{5/2}$ and cryogenic temperatures are required. However, the termodynamic fluctuations alone do not necessarily limit the energy resolution. Since the power spectrum of the X-ray signal is identical to that of the thermal noise and decreases at high frequencies, the signal-to-noise ratio is independent of frequency and arbitrary precision can be achieved by going to very high bandwidth in the pulse-processing electronics chain (McCammon et al. 1986). Unfortunately, the Johnson noise of the thermistor introduces an upper limit to the bandwidth. A sophisticated analysis (Moseley, Mather, and McCammon 1984) shows that the eventual resolution is given by a factor ξ times the thermodynamic limit, where ξ is a strong function of the responsivity, $\alpha = - d \ln R / d \ln T$, of the thermistor. For the semiconductor thermistors which have been used so far, $\xi \sim 2$.

Motivated by the desire to eliminate the Johnson noise component, some other thermometer schemes are currently being investigated. The GSFC group is looking at a kinetic inductance technique (Rawley et al. 1989), which utilizes the temperature dependence of the London penetration depth in a superconductor close to the transition temperature. This can be monitored inductively using a superconducting meander strip. Silver et al. (1989) of LLNL and LBL are investigating the use of temperature-sensitive dielectrics, which may provide capacitive readout, free of Johnson noise.

There are a number of practical limitations on the microcalorimeter imposed by the requirement to keep the heat capacity low. First, the device must be small: the XRS prototype devices are ~ 0.5 x 0.5 mm^2 x 10 µm. These can be arrayed to cover larger areas. Second, epoxy bonds must be avoided. This has been solved by implanting dopants directly into a semiconductor substrate to create the thermistor. Finally a suitable absorber must be found. The semiconductor itself is unsuitable because electron-hole pairs created by the initial ionization event can remain "trapped" at impurities, thereby removing energy from the thermal "sea". Metals do not support trapped states, but have high heat capacity. The best results have been achieved with a zero band-gap material, HgTe, which has no trapped states and moderate heat capacity (Kelley 1990).

Associated with the development of the XRS experiment for AXAF, the GSFC/Wisconsin has made great strides in this area over the past few years. Their latest result is shown in Figure 6 which is a measured pulse height spectrum of the MnKα line near 5.89 keV. The resolution is ~ 11 eV, implying a resolving power of 535. The measurement was made with one element of an array of silicon calorimeters overcoated with HgTe and using thermistor readout at 0.1 K. If recent progress is a guide, further improvements seem likely.

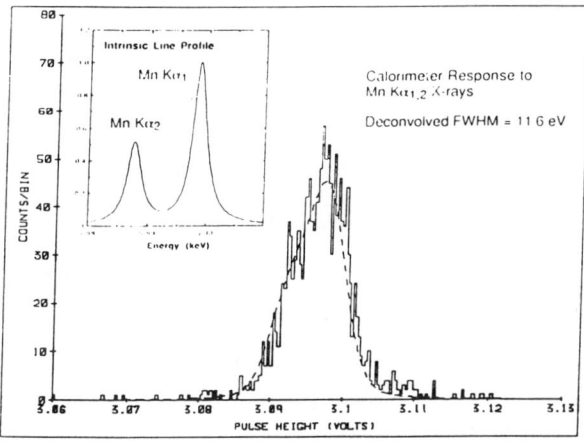

Figure 6: A pulse height spectrum of the Mn Kα$_{1,2}$ line obtained with the latest version of the GSFC/Wisconsin monolithic calorimeter. (Taken from a recent progress report on the development of the XRS).

6. Superconducting Tunnel Junctions.

Another cryogenic detector which has been actively studied recently is the superconducting tunnel junction (STJ; Kraus et al. 1986, Twerenbold 1986). An STJ consists of a pair of superconducting layers separated by an insulating layer (usually a natural oxide of the superconducting material), which is thin enough to permit quantum mechanical tunneling of electrons. If a magnetic field (~ 100 G) is oriented parallel to the junction, then the normal Josephson current associated with the tunneling of Cooper pairs can be suppressed. Thus, only thermally broken pairs or "quasi-particles" are free to cross the barrier. If a small bias potential, $V < 2\Delta/e$, where 2Δ is the energy required to break a Cooper pair, is applied to the junction, then a current of unpaired quasi-particles can be read out. Absorption of an X-ray photon yields an excess of broken pairs which shows up as a current pulse.

The STJ operates essentially as a photoconductive device, similar to conventional semiconductor detectors. However, the number of electrons produced per photon event, N, is much higher in an STJ since the energy required per electron, Δ, is only ~ few meV. The ultimate energy resolving power of the spectrometer scales as $N^{1/2}$. For an ideal STJ, the resolution should be ~ few eV, comparable to that which might be achieved with the microcalorimeter. For the STJ, however, ΔE is proportional to $E^{1/2}$, whereas in the calorimeter, ΔE is fixed across the band, so the junction may even be better at low energies.

There have been a few initial X-ray experiments with STJs using both tin and niobium junctions (Twerenbold 1986, Kraus et al. 1986, Gare et al. 1989). The best resolution obtained to date was ~ 41 eV. In many of these experiments, two pulse height peaks are obtained for any given line, corresponding to photons absorbed in the top and bottom layers of the junction. The charge collection efficiency is not the same, in general, for the two layers, so the pulse height peaks can be individually resolved.

There are some fundamental limitations on the size of an STJ detector imposed by two separate effects. First, because the tunneling probability is inversely proportional to the distance of a quasi-particle from the junction interface, the superconducting films must be made thin. In general, they are not thick enough to stop X-rays with high quantum efficiency. As indicated above, in the tests carried out so far, X-rays are absorbed on both sides of the junction. Second, because of capacitive amplification of electronic noise in the circuit, the junction area must be kept small $< 1 \text{ mm}^2$ (Twerenbold 1987). This means they would need to be arrayed in a manner similar to that used for the microcalorimeters.

However, both of these problems may now have been solved by the invention of a technique called "quasi-particle trapping" (Booth 1987, Kraus et al. 1990). This involves the use of multiple superconducting materials with different transition temperatures. Quasi-particles created in the higher T_C material can migrate into the lower T_C materials if they are placed in close electrical contact. Scattering off phonons then causes them to lose energy and get trapped in this lowest T_C region. They simply do not have enough energy to be quasi-particles in the other material. If the high T_C material is used as the absorber and the low T_C material is used to make the junction, quasi-particles get forced to stay in the tunneling zone indefinitely, which vastly increases the tunneling probability. The active volume of the detector can then be much larger than the junction volume. If a number of small junctions are placed around the absorbing area, the quasi-particle charge division observed in the tunneling pulses can also provide spatial information as to the position of the X-ray event (Kraus et al. 1990).

7. Conclusion.

As a conclusion, I think it is clear that a diverse array of technology exists even in the current era to provide high spectral resolving power coupled with moderate-to-high efficiency in the X-ray band. Reflection gratings seem to hold real promise at low energies, while calorimeters and STJs seem best suited at higher energies. However, as emphasized earlier, in the final comparison, the detailed trade-offs between one approach and another will depend on the particular implementations, perhaps even moreso than on the fundamental properties of the devices themselves. Without detailed information regarding the launch capabilities, constraints, and scientific drivers of the 21st Century, it is foolish to guess which technologies will be most useful. In all probability, all will find application eventually.

I am indebted to Claude Canizares, Dan McCammon, and Chuck Hailey for advice and suggestions connected with this review. This work was supported, in part, by grants from the NASA Innovative Research Program and the University of California Campus-Laboratory Collaborative Research Program.

REFERENCES

Alexandropoulos, N.G., and Cohen, G.G., 1974, *Appl. Spectroscopy*, **28**, 155.
Barbee, T., Jr., and Underwood, J.H., 1983, *Opt. Commun.*, **48**, 161.
Beuermann, K.P., Brauninger, H., and Trumper, J., 1978, *Appl. Optics*, **17**, 2304.
Booth, N.E., 1987, *Appl. Phys. Letters*, **50**, 293.
Brinkman, A.C., Aarts, H.J., Branduardi-Raymont, G., Hailey, C.J., Jansen, F.A., Kahn, S.M., de Korte, P.A.J., and Zehnder, A., 1989, *Proc. of the S.P.I.E.*, **1159**, 495.
Brinkman, A.C., van Rooijen, J.J., Bleeker, J.A.M., Dijkstra, J., Heise, J., de Korte, P.A.J., Mewe, R., and Paerels, F., 1985, *Proc. of the S.P.I.E.*, **597**, 232.
Canizares, C.R., Clark, G.W., Bardas, D., and Markert, T., 1977, *Proc. of the S.P.I.E.*, **106**, 154.
Canizares, C.R., Schattenburg, M.L., and Smith, H.I., 1985, *Proc. of the S.P.I.E.*, **597**, 253.
Cash, W., 1988, *Proc. of the S.P.I.E.*, **830**, 204.
den Boggende, A.J.F., de Korte, P.A.J., Videler, P.H., Brinkman, A.C., Kahn, S.M., Craig, W.W., Hailey, C.J., and Neviere, M., 1988, *Proc. of the S.P.I.E.*, **982**, 283.
Gare, P., Engelhardt, R., van Dordrecht, A., Peacock, A.J., Lumley, J., Pereira, C., Busfield, M., and Twerenbold, D., 1989, *Proc. of the S.P.I.E.*, **1159**, 433.
Giacconi, R., et al., 1979, *Ap. J.*, **230**, 540.
Holt, S.S., Kelley, R.L., Moseley, S.H., Mushotzky, R.F., and Szymkowiak, A.E., 1988, *Proc. of the S.P.I.E.*, **597**, 267.
Kahn, S.M., 1990, in *IAU Colloq. No. 115, High Resolution X-Ray Spectroscopy of Cosmic Plasmas*, ed. P. Gorenstein, in press.
Kelley, R.L., 1990, these Proceedings.
Kraus, H., v. Feilitzsch, F., Jochum, J., Mossbauer, R.L., Peterreins, Th., and Probst, F., 1990, *Phys. Letters B*, submitted.
Kraus, H., Peterreins, Th., Probst, F., v. Feilitzsch, F., Mossbauer, R.L., and Zacek, V., 1986, *Europhys.Letters*, **1**, 161.
Markert, T.H., Powers, T.R., Levine, A.M., McCullum, C.B., Mohr, J.J., and Canizares, C.R., 1988, *Proc. of the S.P.I.E.*, **982**, 245.
McCammon, D., Juda, M., Zhang, J., Kelley, R.L., Moseley, S.H., and Szymkowiak, A.E., 1986, *IEEE Trans. on Nucl. Sci.*, **33**, 236.
Moseley, S.H., Mather, J.C., and McCammon, D., 1984, *J. Appl. Physics*, **56**, 1263.
Rawley, G.L., Kelley, R.L., Moseley, S.H., and Szymkowiak, A.E., 1989, *Proc. of the S.P.I.E.*, **1159**, 414.

Schattenburg, M.L., Canizares, C.R., Dewey, D., Levine, A.M., Markert, T.H., and Smith, H.I., 1988, *Proc. of the S.P.I.E.*, **982**, 210.
Schnopper, H.W., and Byrnak, B.P., 1985, *Proc. of the S.P.I.E.*, **597**, 301.
Seward, F.D., Chlebowski, T., Delvaille, J.P., Henry, J.P., Kahn, S.M., Van Speybroek, L., Dijkstra, J., Brinkman, A.C., Heise, J., Mewe, R., and Schreiver, H., 1982, *Appl. Optics*, **21**, 2012.
Silver, E.H., Labov, S.E., Goulding, F., Madden, N., Landis, D., and Beeman, J., 1989, *Nucl. Inst. and Meth.*, **A277**, 657.
Stark, J.P.W., and Culhane, J.L., 1978, *M.N.R.A.S.*, **184**, 509.
Taylor, B.G., Andresen, R.D., Peacock, A., and Zobl, R., 1981, *Sp. Sci. Rev.*, **30**, 479.
Twerenbold, D., 1986, *Europhys. Letters*, **1**, 209.
Twerenbold, D., 1987, *Nucl. Inst. and Meth.*, **A260**, 430.
Vaiana, G.S., Reidy, W.P., Zehnpfennig, T., Van Speybroek, L., and Giacconi, R., 1968, *Science*, **161**, 564.

DISCUSSION

Richard Kelly

The energy resolution of ~ 11 eV that you showed was limited not by heat capacity but by the non-optimal value of the thermistor resistance at 0.1 K and FET amplifier noise. The heat capacity of the device is dominated by the thermistor, and not by the epoxy adhesive used to attach the HgTe.

Claude Canizares

That was an excellent review. I just want to add that, when it comes to observing the line-rich thermal sources like galaxies, clusters, and SNRs, some of the methods have extra problems. The gratings (both T and R) degrade rapidly for sources like these. The microcalorimeters at present are limited in focal plane area. Here the crystals do better, and their effective area × solid angle are comparable to or even exceed those of the microcalorimeters.

Kahn

Yes, I meant to make this point, although I was limited by lack of time. It's interesting to look at the tokamak community. They want to measure the same lines we do, and they mostly still use crystal spectrometers even though they are well aware of the developments in gratings and non-dispersive devices of recent years. The reason is that they need to accumulate photons over a large solid angle. When you look at very diffuse sources, the multiplex advantage of the other devices is no longer as relevant.

Martin Elvis

What are the prospects for imaging microcalorimeters and tunneling junction devices?

Kahn

For the microcalorimeters, one is limited in the basic size of the detector by heat capacity considerations. Therefore, the only way to make an imaging detector is to make a large array of individual detectors, along the lines that GSFC has done for AXAF. Several years ago there were suggestions that imaging microcalorimeters could be usable using "ballistic" phonons created in the initial event. I don't think this has been demonstrated so far. For the tunnel junctions, I think quasi-particle trapping will eventually lead to an imaging device. If an array of small low-T_c junctions are placed around the edges of a higher-T_c absorber, the division of the quasi-particle charge among the different junctions should provide spatial information. However, this has to be worked out!

Webster Cash

Your comparison between in-plane and off-plane reflection grating spectrographs is not fundamental in nature in that it picks a fixed "maximum" ruling-density. As we approach the 21st century this density will doubtless increase. Also, the density comment is inappropriate in that we can observe quite effectively in high order. Peak order efficiency and order separation, which drives one to first order at longer wavelengths, are not important considerations when a CCD can be used to reconstruct the spectrum.

When freed from ruling density considerations, one finds off-plane designs far superior. They yield eight times higher resolution at a given graze angle. The have higher efficiency and lower scatter. Azimuthal optical effects can be used to gain at least another factor of four in resolution in many applications.

Kahn

I certainly didn't mean to imply that there are "fundamental" reasons to choose an in-plane orientation over off-plane; I only meant to point out their different properties. I think the final comparison depends on the implementation, as I emphasized at the beginning of my talk. There may well be major increases in ruling density over the next 20 years or so, and those developments could favor off-plane designs.

As far as the use of higher orders goes, our measurements of prototype gratings for XMM have shown that the true reflection efficiency compared to the theoretical efficiency drops off significantly as you go to higher orders. That is undoubtedly related to "groove-shaped" imperfections; the higher orders are more sensitive. As you have suggested, you may get some of the efficiency back by summing over orders, but it isn't obvious that that will always be the case. The use of reflection gratings in this way is still rather new. As the technology improves, I don't disagree that off-plane designs may turn out to be superior.

LOW-TEMPERATURE DETECTORS FOR X-RAYS

Gilbert G. Fritz

E.O. Hulburt Center for Space Research
Naval Research Laboratory
Washington, D.C.

ABSTRACT

Temperatures below 10 K permit the design of x-ray detectors with energy resolution a factor of 10-100 better than existing semiconductor devices. The low heat capacity of materials at these temperatures, as well as the special properties of superconductors, can be exploited to produce non-dispersive x-ray spectrometers. Several approaches are examined, and the prospects of constructing arrays and using high Tc superconductors are briefly assessed.

INTRODUCTION

Interest in low-temperature radiation detectors has increased dramatically in the last few years. For x-rays, the promise is of spectral resolution improved dramatically over the 100 eV of conventional semiconductor detectors operating at 100 Kelvin, such as the Solid State Spectrometer flown on the Einstein Observatory. Energy resolution in the range 1-10 eV, which is theoretically possible and now appears within reach, offers the fascinating possibility of replacing x-ray spectrometers in some applications with a single, high-efficiency, non-dispersive detecting element having a resolving power as high as several thousand. The principal advantages of low-temperature operation are lower noise, lower heat capacity, and the availability of the special properties of superconductors. Several concepts for x-ray detection have emerged, which basically exploit either 1) the equilibrium or bolometric response of a small grain of material, or 2) the very small energy gaps (milli-eV) in superconductors. Specific types of devices include calorimeters, bolometers, tunneling junctions, superconducting transition detectors, and vortex-flow detectors; in some cases there are significant design variants within one type of device. Each approach has some special advantage and is being studied by at least one research group, but each also has at least one significant limitation that may or may not become an insurmountable obstacle. There is also considerable interest in building these detectors in arrays, and the approaches differ significantly in the ease with which this might be accomplished. Finally, although most work to date has concentrated on conventional superconductors below 10 K, there is some hope that the new high-Tc superconducting materials will be useful for x-ray detectors.

CALORIMETERS AND BOLOMETERS

Calorimeter is a general term used to refer to a class of devices in which input energy (in this case, x-radiation) produces heat, which is measured. Bolometers were originally devices in which a temperature change was measured by means of a change in electrical resistance, but the term has become broadened to include other sensitive thermometers. For these devices, the temperature rise for a given energy input is inversely proportional to the heat capacity of the detector, so that maximum sensitivity is achieved by using 1) small (mass) devices, 2) low specific heat materials, and 3) the lowest possible temperatures, since specific heats can vary as steeply as T^3.

A major breakthrough in calorimeter detectors for x-rays has been made by the Goddard Space

Flight Center/University of Wisconsin group (summarized by Kelley 1990). Energy resolution of about 10 eV has been achieved in detectors that use an implanted thermistor in silicon as the thermometer element. In addition, attempts to bond a small silicon thermistor to a block of a different x-ray absorbing material have now been successful. This allows the optimization of the absorber for high x-ray efficiency (high Z) and low heat capacity. Difficulties with these microcalorimeters include the necessity of cooling to temperatures <100 milliKelvin, fragility arising from the necessity to suspend the detecting element on micron-size leads for thermal isolation, and technical problems of assembling the detectors in arrays.

Other possibilities exist for the calorimetric measurement of x-ray energy. The large change in the dielectric constant of certain materials with temperature can be utilized (Silver 1990), as can the change in inductance ("kinetic inductance") of a meander pattern of superconducting lines (Kelley 1990). X-ray detection via magnetic effects in superconducting grains will be discussed after a brief digression on the basic properties of superconductors.

SUPERCONDUCTING POTPOURRI

Zero resistance, the Meissner effect, a very low energy gap, and quantum tunneling of single particles and Cooper pairs across a barrier, are all characteristics of superconductors that can be exploited for x-ray detection, and are illustrated below:

1. ZERO RESISTANCE

CRITICAL TEMPERATURE Tc

2. MEISSNER EFFECT: EXCLUSION OF MAGNETIC FIELD

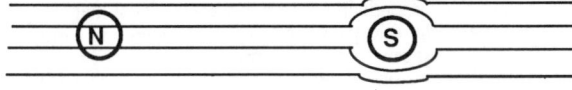

TYPE I: FLIP-FLOP
TYPE II: SLOW PENETRATION OF MAGNETIC FIELD

3. ENERGY GAP - ANALOGY TO SEMICONDUCTORS

SEMICONDUCTOR	SUPERCONDUCTOR
Create electron-hole pairs	Cooper pairs ⟶ Quasiparticles
Few eV in Si, Ge	Few meV or less

4. QUANTUM TUNNELING OF QUASIPARTICLES ACROSS INSULATING BARRIER (JUNCTION)

SINGLE PARTICLES
CAN BE PHOTON OR PHONON ASSISTED

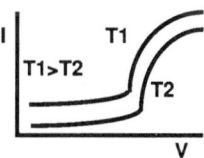

5. JOSEPHSON JUNCTION

PAIRS OF PARTICLES

SQUID (SUPERCONDUCTING QUANTUM INTERFERENCE DEVICE)
MOST SENSITIVE MAGNETOMETER

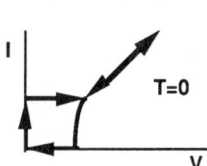

PENETRATION DEPTH BOLOMETER

Type II superconductors exhibit partial penetration of an external magnetic field into the material at temperatures near, but below, the superconducting transition temperature. This effect can be used to create an x-ray bolometer, as shown schematically below:

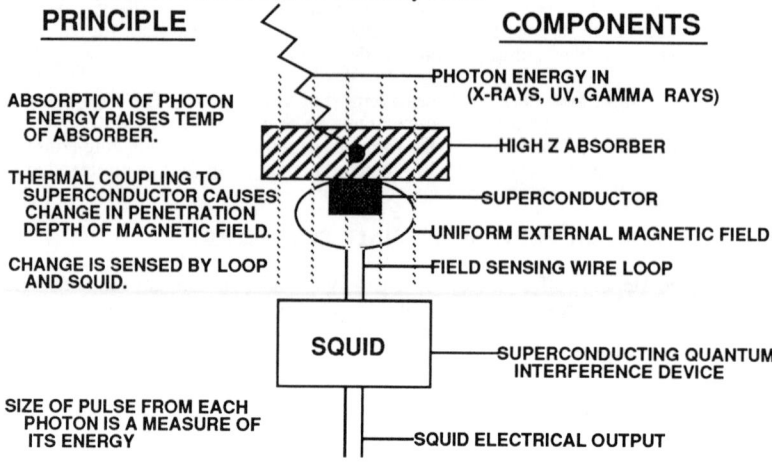

Important features of the design are 1) the use of an x-ray absorber with high Z, low heat capacity, and suitable size, 2) the use of a separate superconducting grain to vary the local magnetic field with temperature, and 3) the use of a magnetic pickup coil and SQUID (Superconducting Quantum Interference Device). The SQUID is the most sensitive magnetometer known, and is the key to obtaining high energy resolution from the detector. In fact, a commercially available SQUID coupled to a simple pickup loop has the sensitivity to detect temperature changes corresponding to <10 eV energy input to an absorber of 30x30x3 microns size. Custom SQUIDs and optimized pickup coil geometry can improve this by 1-2 orders of magnitude. The energy sensitivity translates directly into energy resolution for x-rays provided the photon energy is completely converted into heat within the response time of the system (milliseconds). The high potential sensitivity of this approach could allow the use of a single SQUID to read out an array of detecting elements, or multiple SQUIDs to read out energy and position, and could also allow operation at more moderate temperatures (1 K) than the microcalorimeter. Disadvantages include a limited dynamic range of x-ray energies, extreme non-linearity of response which requires precise temperature stabilization, and a high sensitivity to external electric and magnetic noise.

TUNNELING JUNCTIONS

Tunneling junctions consist of two layers of superconducting material separated by a thin barrier. Photons with energies greater than the binding energy, typically a few milli-eV, break Cooper pairs in the superconductor and produce quasiparticles that can tunnel through the barrier if a small voltage bias is applied. This results in a measurable current pulse whose amplitude should be proportional to the photon energy, with a theoretical limit on the energy resolution determined by the statistics of the number of quasiparticles produced, or about 5 eV for 6 keV x-rays in a typical superconductor. Advantages of the tunneling junctions are 1) production as monolithic devices by thin film deposition and etching; 2) operation at moderate temperatures (1 K); 3) simple, two-terminal devices (with four connections required); and 4) use of conventional low-noise electronics. However, the best energy resolution to date has been only about 40 eV (Twerenbold and Zehnder 1987), and the physical model of the junctions appears to be quite complicated, with the detailed interaction between phonons and quasiparticles assuming a major role. In addition, a number of practical problems persist, such as controlling the granularity of the superconducting films and the uniformity of the tunneling barrier, and maintaining low leakage currents in the junctions. A number of groups are now attacking the various problems, and the potential for the devices remains quite high. Since they are made by techniques similar to integrated circuits, the prospects are excellent for constructing tunneling junctions in arrays.

GRANULAR SUPERCONDUCTORS

A number of ideas have emerged that make use of the properties of grains of superconducting material. One possibility is to measure the time required for the transition of a grain out of the superconducting state (due to absorption of a photon) and back again (due to heat loss to the surroundings), as discussed by Kurfess *et al* 1990. This technique is likely to be useful for hard x-rays, but is not likely to have the energy resolution needed in the soft x-ray region. A quite different approach is to deposit superconducting grains in a thin film together with an insulating matrix. The resulting single layer, dubbed CERMET because it is usually a ceramic/metallic composite, can be patterned into a detector and operated much like a tunneling junction. The response to pulsed infrared light shows both a slow, bolometric component and a fast, quasiparticle component. While there is some controversy over the response mechanism, it is clear that a coupling between the individual grains is occurring, with magnetic vortex flow being possibly the dominant factor. Although the sensitivity of the CERMET detectors can be quite high, the random and irregular nature of the superconducting grains may mean poor energy resolution for x-rays.

PROSPECTS FOR HIGH Tc DETECTORS

The discovery of high temperature superconductors has raised the hopes of using these new materials in x-ray detectors. Tunneling junctions, for instance, which are normally operated at about 0.1Tc, would require only liquid He-4 for cooling (4 K). Granular thin films of the high Tc material have already shown some evidence of non-bolometric response, which is necessary for x-ray detection. However, the higher operating temperatures also result in higher noise, higher heat capacity, and higher band-gap energies, all of which work against achieving high energy resolution. Therefore, the prognosis is that the best x-ray detectors will be those operating at or below 1 K.

REFERENCES

Kelley, R. 1990, these proceedings.
Kurfess, J.D. *et al*. 1990, these proceedings.
Silver, E. 1990, these proceedings.
Twerenbold, D. and Zehnder, A.J. 1987, *J.Appl.Phys.* **61**,1.

DISCUSSION

William Priedhorsky

Why are bolometer detectors difficult to array?

Fritz

The basic problem is to use the center of the focal plane efficiently, i.e., with as high a fractional coverage as possible. For bolometers, the difficulty is in bringing out the multiple connections for each detecting element (pixel) and maintaining close thermal coupling for each element. With small arrays (a dozen pixels) various packing tricks can be used, but when the arrays grow to CCD size (hypothetically), the problem becomes extremely difficult.

Jonathan Grindlay

Are you pursuing the actual development of SQUID readout techniques in a laboratory program at NRL?

Fritz

This is being done on a low-priority basis. Jim Kurfess is giving a talk on grain-flipping (out of and then back into the superconducting state) for hard X-ray detection, and we expect to carry out some experiments, using SQUIDS, on magnetic penetration into superconductors, in connection with this work.

Steven Kahn

For a while, there were reports of junctions produced in high T_c materials by a "break" technique wherein the ceramic is deposited on a piezo and then deformed to make a small crack. They claim to have observed Josephson effects. Are those junctions useful for X-ray spectroscopy?

Fritz

There are some high T_c devices that have shown Josephson tunneling junction-like behavior, but my understanding is that they are very poor junctions. For instance, the leakage currents are generally so high as to preclude their use in detectors. It is, of course, one of the goals of the high T_c superconductivity program to produce such junctions, so the developments of the next few years may drastically change the present situation.

A NEW THERMAL SENSOR FOR X-RAY MICROCALORIMETRY

E. Silver, S. Labov and T. Pfafman

Laboratory For Experimental Astrophysics
Lawrence Livermore National Laboratory

F. Goulding, D. Landis, N. Madden and J. Beeman

Lawrence Berkeley Laboratory

ABSTRACT

Photon counting x-ray spectroscopy using microcalorimeters combines the high resolution power of a Bragg crystal with the broad bandwidth capability and high efficiency of charge sensitive detectors. To date, the developers of microcalorimeters take advantage of the strong temperature dependence of resistance in doped semiconductor crystals for converting the calorimeter temperature changes into useful voltage signals. Resistive sensors limit the ultimate sensitivity of microcalorimeters because they dissipate heat and add Johnson noise to the electronic conversion circuit. Potentially, these detrimental properties can be eliminated by exploiting the temperature dependence of the dielectric constant in ferroelectric materials. A mixed crystal quantum ferroelectric shows great promise for a dielectric thermal sensor. We present results from our most recent measurements which include the first detection of individual 6 MeV alpha particles with a dielectric calorimeter operating at 1.2 K.

I. INTRODUCTION

Microcalorimeters, in general, promise to offer the broad bandwidth capability of photoelectric detectors, nearly 100% efficiency above 1 keV, and the high resolving power typical of a Bragg crystal. The implications for astronomical spectroscopy are tantalizing. Sophisticated high temperature plasma diagnostic measurements of a broad variety of galactic and extragalactic x- ray sources could be performed for the first time in a model- independent way. To date, this has only been possible for the sun and a few of the brightest supernova remnants. As calorimeters begin operation in x-ray observatories, it will be possible to perform high resolution spectroscopy on entire classes of x-ray sources which were too faint for the previous inefficient dispersive spectrometers.

Our new detector concept takes advantage of the large temperature dependence of the dielectric constant in ferroelectric materials. Calculations have shown that practical instruments operating with an energy resolution less than 10 eV may be possible at 300 mK. In section II we summarize general calorimeter operation and discuss the advantages of the dielectric calorimeter. In Section III we review how a dielectric calorimeter operates and predictions of its performance are presented. Section IV includes measurements of the dielectric constant and its temperature dependence above 0.3 K for a specific mixed crystal quantum ferroelectric. The first detection of 6 MeV alpha particles with this dielectric material operating at 1.2 K is also demonstrated.

II. CALORIMETERS

In principle, a calorimeter may be any device that exhibits a perceptible temperature increase after it absorbs a single photon or particle. For the temperature rise to be measurably large the device must possess a very small heat capacity. In an ideal microcalorimeter the photon/particle-induced temperature rise must be measured above background temperature noise caused by the random exchange of energy through the thermal link between the detector and its temperature sink. This phonon noise manifests itself in the calorimeter as thermal fluctuations with a spectral density per unit frequency of

$$\overline{\delta T^2}(f) = \frac{4kT^2}{G} \frac{1}{1 + 2\pi f (C_v/G)^2} . \quad (1)$$

Here, T and C_v are the temperature and heat capacity of the detector, respectively, G is the conductance of the thermal link, and k is the Boltzmann constant. Integrating over all frequencies yields the mean square value, $\overline{\delta T^2} = kT^2/C_v$ (Van der Ziel 1976).

The most common way to convert the calorimeter temperature changes into useful voltage signals takes advantage of the strong temperature dependence of resistance in doped semiconductor crystals. When biased with a constant current, a resistive thermal sensor produces a voltage that is a measure of the size of the temperature excursion. Unfortunately, the electronic conversion of the temperature introduces Johnson noise, JFET noise associated with the preamplifier, and mechanically-induced microphonic noise.

Generally, the thermistor Johnson noise dominates the contributions from the JFET and microphonics, and above some frequency it will also be larger than the phonon noise. In the dielectric calorimeter Johnson noise is virtually eliminated so that the energy resolution for photon counting spectroscopy is limited only by the phonon noise. The frequency response of the dielectric extends over a broader frequency range than resistive devices making way for the development of detectors that accomodate higher count rates. In a dielectric calorimeter used with a constant DC bias, the output signal can be controlled by an applied voltage without elevating the sample's temperature or introducing self-heating. Our calculations show that potentially, a dielectric calorimeter can provide an energy resolution below 10 eV at 300 mK, an operating temperature that is easy to achieve with conventional ^{3}He closed-cycle refrigerators.

III. USING A DIELECTRIC CALORIMETER

When a photon of energy E is absorbed by a dielectric calorimeter exhibiting an electrical capacitance C_c, its temperature will change. The temporal signature of this temperature perturbation is $\Delta T(t) = T(t) - T_b = T_o \exp^{-t/\tau}$ where T_b is the temperature of the cold stage, $T_o = E/C_v$ and $\tau = C_v/G$. The thermal pulse has a finite risetime but for simplicity, we begin by assuming the risetime to be zero. (The implications of finite risetimes has been discussed elsewhere (see Silver et al.1989). This temperature change alters the dielectric constant ϵ of the capacitor. If the voltage across the capacitor is held constant, then the charge on the capacitor will change by an amount, ΔQ,

$$\Delta Q = C_c V_{bias} \frac{1}{\epsilon} \frac{d\epsilon}{dT} \Delta T(t) , \quad (2)$$

where V_{bias} is the bias voltage applied across the capacitor and ϵ is the dielectric constant. When coupled to a charge sensitive preamplifier as shown in Figure 1 the current generated in the calorimeter capacitor, $I_c = dQ/dt$, is integrated to produce a voltage across the feedback capacitor, C_f. The output signal that results is

$$V_{out}(t) = \int_0^t \frac{I_c dt}{C_f} = \frac{C_c V_{bias}}{C_f} \frac{1}{\epsilon} \frac{d\epsilon}{dT} \frac{E}{C_v} \exp^{-t/\tau} . \quad (3)$$

Referring to the schematic, the feedback resistor R_f, is needed to maintain stable DC operation of the preamplifier, the bypass capacitor C_{bypass}, maintains the constant voltage on the detector, and C_s is the sum of the JFET input capacitance plus all additional stray capacitances associated with the gate of the JFET. R_c represents the resistive losses in C_c. The larger the value of R_c, the better the element acts as a pure capacitor.

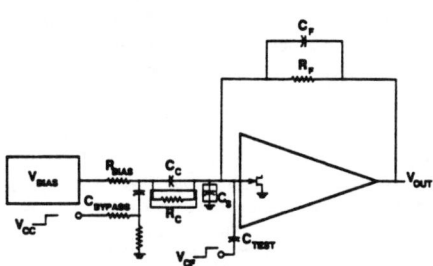

Figure 1. The schematic of the charge sensitive preamplifier used with the dielectric calorimeter

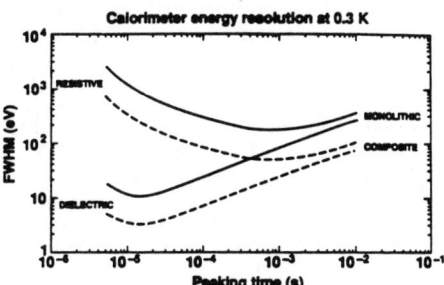

Figure 2. The predicted energy resolution (FWHM) for a dielectric and resistive calorimeter as a function of the peaking time of the post amplifier filter. The devices have equal heat capacity and thermal conductance.

The detector noise contributions have been treated in detail and may be found elsewehere (Silver et al.1989). The predicted energy resolution at 0.3 K for realistic monolithic and composite dielectric and resistive calorimeters of equal heat capacity and thermal conductance is compared in Figure 2. The FWHM resolution is plotted along the y axis and the bandpass of the amplifier system (Peaking Time) is plotted along the x-axis. A shorter peaking time implies a larger bandwidth. The curves show that the dielectric calorimeter has the potential for operating at 0.3 K with an energy resolution less than 10 eV for a monolithic design and approximately 4 eV in a composite configuration. The best that the equivalent resistive calorimeter could achieve is 180 eV and 50 eV, respectively.

IV. EXPERIMENTAL RESULTS

Material suitable for dielectric microcalorimetry must possess a dielectric constant that is strongly dependent on temperature at cryogenic temperatures. It must also exhibit low dielectric losses, possess small heat capacities, and it should have reasonable mechanical integrity. Ferroelectric crystals are potential candidates for capacitive sensors. Ferroelectrics exhibit electric dipole moments in the absence of an externally applied field. This intrinsic polarization or ferroelectric state exists below some transition temperature, T_c. We have focused our attention on the quantum ferroelectric $KTa_{1-x}Nb_xO_3$ (KTN). KTN is a well-known mixed crystal series whose ferroelectric transition temperature varies with the Nb concentration, x as $T_c = 276\sqrt{x - 0.008}$ K, applicable for x≤ 0.05 (Rytz et al.1983). For KTN with $0.012 \leq x \leq 0.03$, $C_v \propto T^3$ as for a Debye solid (Lawless et al.1985).

For a KTN sample with x =0.012, ϵ vs T data are plotted in Figure 3. For x =.012, ϵ peaks sharply at 18 K. We were not set up to accurately measure ϵ between 4.2 K and 77 K. However, we did observe that the value of C_c passed through a maximum value. In Figure 3a, ϵ at 4.2 K is 15000 and drops sharply to 11500 at 2 K and to 9900 at 0.3 K. These values are in excellent agreement with the data obtained by Rytz (Rytz et al.1983) who measured ϵ above 4.2 K. The linear dependence below 5 K is shown in Figure 3b. We find $\frac{1}{\epsilon}\frac{d\epsilon}{dT}$ at 1.7 K and 0.3 K are 11%/K and 13%/K, respectively and the parallel resistance, R_c is $\geq 5 \times 10^{13}\Omega$.

We fabricated a prototype KTN calorimeter for testing at 1.2 K and irradiated it with pulses of infrared light. Sample dimensions were 0.6mm x 0.6mm x 1.4 mm. It was suspended on 2 copper wires, each with a diameter of 25 μm and length of 5mm. The thermal conductance was estimated to be $1 \times 10^{-6} W/K$. The capacitance was 26 pf at 1.2 K and the heat capacity of the KTN alone was estimated to be $\approx 5 \times 10^{-9} J/K$ (Lawless et al.1985). The temporal profiles of a thermal pulse produced by an infrared LED are shown in Figure 4 for different values of applied voltage. The response for $V_{bias} = 100$ volts is shown in Figure 4a and for $V_{bias} = 0$ in 4b. The precursor transient that lasts for 200 μs arises from capacitive pickup between the pulsing LED and the input of the preamplifier. In this experiment 6 MeV alpha particles were also detected for the first time with a dielectric calorimeter. A typical signal from the postamplifier which filters the signal and noise is shown in Figure 5. We are presently evaluating these results to determine the exact signal to noise ratio. The noise has been measured in detail, and we are currently analyzing the contributions from the Johnson noise, phonon noise, and JFET noise. Independent measurements of the heat capacity are underway and a KTN sample with a smaller percentage of niobium is also under investigation.

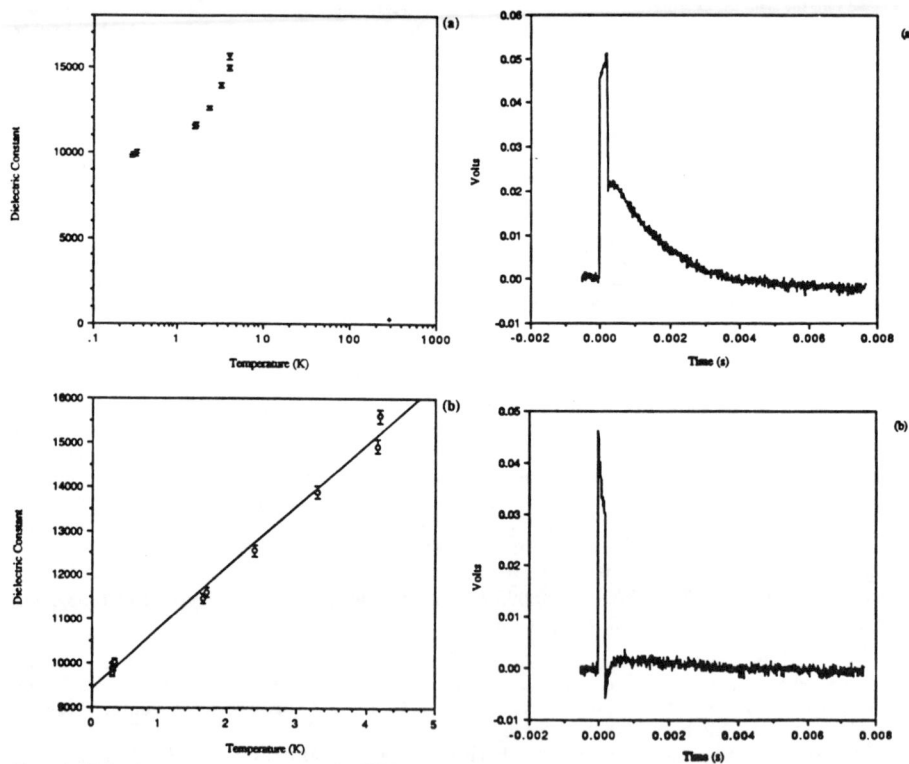

Figure 3. Dielectric constant vs. temperature for KTN sample with x= 0.012.

Figure 4. Dielectric calorimeter response to a 200 μs infrared LED pulse. In (a) the bias voltage is 100 V and in (b) the bias is 0 V.

The authors wish to thank R. Rytz for furnishing the KTN dielectric and W. Hansen for his help in processing it. This work was performed under the auspices of NASA Grant NAGW-1686.

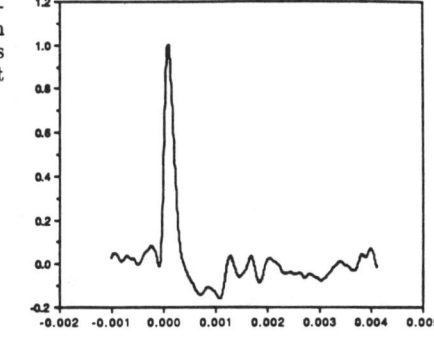

Figure 5. Detection of a 6 MeV alpha particle with a KTN (x=0.012) dielectric calorimeter. The signal has been filtered with a 200 µs peaking time.

VI. REFERENCES

Van der Ziel, A. , 1976, *Noise in Measurements* (Wiley, New York), 45.
Silver, E.H., Labov, S.E., Goulding, F., Madden, N., Landis, D., and Beeman, J., 1989, *Nucl. Instr. and Methods,* A **277**, 657.
Rytz, D., Chatelain, A., and Hochli, U.T., 1983, *Phys. Rev. B*, **27**, 11, 6830.
Lawless, W.N., Arenz, R.W., Rytz, D., and Buffat, P.A., 1985, *Proceedings of the Sixth International Meeting on Ferroelectricity,*; 1985, *Jap. J. Appl. Phys.,***24**, Supplement 24-2, 263,.

DISCUSSION

(Unknown)

Any hope of using these as a gamma-ray detector?

Silver

The use of alpha particles with energies of 6 MeV to diagnose performance of these detectors at 1.2 K should not mislead you into thinking that these calorimeters are good prospects for gamma-ray detectors. 6 MeV alphas have enough energy to thermally excite these detectors at 1.2 K; the heat capacity of this calorimeter is significantly high at 1.2 K. Remember that if heat capacity scales with temperature to the third power, a 6 keV X-ray will in principle produce the same size signal at 0.1 K as a 6 MeV particle at 1.2 K. Alphas are just convenient sources for this purpose, with a very short range for absorption.

For these detectors to stop a 6 MeV gamma-ray, they would have to have sufficient stopping power, which implies a longer detector. Since the heat capacity is proportional to volume, the detector would necessarily have to be long and thin to maintain the same heat capacity and, hence, energy resolution. How long and how thin would have to calculated.

MULTISTEP FLUORESCENCE GATED PROPORTIONAL COUNTERS

B. D. Ramsey and M. C. Weisskopf

NASA Marshall Space Flight Center

A new type of proportional counter is being developed at Marshall Space Flight Center (MSFC). The device, a multistep fluorescence gated detector combines superior energy and spatial resolution with a high degree of background rejection suited to x-ray astronomy for applications up to 100 keV. Its utility is demonstrated by the fact that detectors of this basic design are integral parts of the CASES, POF, and EXOSS missions discussed elsewhere in these proceedings.

The Basic Device

Figure 1 illustrates the basic concept. Photons incident on the detector are photoelectrically absorbed in a drift region, and the resulting electrons move, under the action of a small electric field towards a set of plane parallel "preamplification" grids. The grids are operated at a higher potential so that charge multiplication at low gas gain may take place. This signal is used both to trigger the system and measure the energy of the incident photon. A small portion of the charge produced between the grids is also transferred through to an imaging stage to record the position of the event.

There are several advantages to this "multistep" approach. First, the energy resolution resulting from parallel geometry is excellent, even at high gas gains. Second, the energy resolution of the imaging stage is dramatically improved by the coupling to the preamplification stage. This latter feature is of significance for the fluorescence gating technique used to reduce background (Ramsey and Weisskopf, 1986, 1987).

The detector utilizes xenon as the primary gas to obtain high efficiency at high energies. Further improvements in energy resolution are provided through the use of Penning ad-mixtures, such as isobutylene or trimethylanine, where the excitation energy of the xenon gas atoms is used to ionize the molecules of these quench gases. The result is an improvement in electron yield over more conventional quench gases which both lowers the operating voltages and improves the energy resolution (Ramsey and Agrawal, 1989).

The imaging section is a conventional multiwire proportional counter and has two orthogonal cathode planes of wires sandwiching an anode plane. The cathodes are segmented in that small, adjacent groups of wires each have a separate readout. Charge transferred from the preamplification grid through appropriate bias voltages is further amplified by the avalanche at the anode plane. This, in turn, induces a charge distribution on the cathodes. These cathode signals may then be used to determine the centroid of each distribution and thus determine the position of an event.

This type of imaging is essential for efficient fluorescence gating, employed for additional background suppression above and beyond the five-sided anticoincidence shown in Figure 1. Fluorescence gating exploits the high probability that photons, above the 35 keV xenon K-edge, fluoresce the xenon gas. When the fluorescent photon is absorbed within the detector's drift region a pair of separate, but highly correlated, events are produced for each incoming x-ray. This double event then becomes a signature for a true x-ray interaction and can be used to discriminate against the charge particle induced background which has a very low probability of inducing fluorescence. Because of exponential absorbtion a large protion of fluorescent pairs materialize close together and so fine spatial resolution is necessary to resolve them as two separate events.

The use of fluorescence gating also improves both the spatial and energy resolution above the xenon K-edge. The former is improved by avoiding the smearing due to measuring an energy weighted mean position of events where fluorescence and reabsorption take place. The energy resolution is improved as one can know, rather precisely, the energy of one of the two events--the fluorescent photon. Both of these advantages are illustrated in Figures 2 and 3, with data obtained from a 10 cm x 10 cm prototype (Ramsey and Weisskopf, 1987).

Current Status

Recently, as part of NASA's spring 1989 supernova campaign and in collaboration with Harvard University, we have flown a 30 cm x 30 cm single stage detector to investigate the efficiency of the fluorescence gating under actual flight conditions. Full details of this device were discussed by Ramsey, Weisskopf, and Joy, 1989. Data from the flight were just recently sent to us by the NSBF and are currently under analysis, especially with regards to background rejection. We did discover that the xenon gas was contaminated with radioactive krypton, but since the principal decay products are electrons, this appears to have little impact on the pair background due to the low fluorescence yield for this type of interaction. Based on a crude and conservative preliminary analysis, the pair background was below 10^{-4} counts/cm^2 s keV at the xenon k-edge and rose to a peak of about 2×10^{-4} counts/cm^2 s keV at about 70 keV. The source of this peak, observed both on the ground and in flight, is somewhat puzzling and is currently under investigation.

Future Developments

The next steps in the development of these detectors are illustrated in Figure 4. In the near term we will incorporate the preamplification stage into our current flight instrument, in addition to any other changes indicated by the results of the data analysis. In parallel to this, we are investigating, in the laboratory, the practical consequences of attempting to image the visible light emitted during the avalanche of the electrons in the preamplification stage. We are interested in this for a number of reasons, the principal being that this approach offers a method of being able to spatially resolve closely spaced pairs. This advantage will become even more important as we investigate the consequence of operating at pressures much higher than the current two atmospheres in order to increase the overall absorption efficiency. Finally, in collaboration with Harvard University, we are investigating a hybrid detector which combines a gas-filled first stage utilizing only the preamplification grid with a phoswich second stage. The light from both detectors will be imaged by photomultiplier tubes located behind the phoswich. In this concept, low energy photons would interact in the gas detector with the phoswich acting in anti-coincidence while at high energies the roles would be reversed.

References

Ramsey, B. D., and Weisskopf, M. C. 1986, Nucl. Instr. and Meth. in Phys. Res., **A248**, 550.

Ramsey, B. D., and Weisskopf, M. C. 1986, IEEE Trans. on Nucl. Sci., **NS34**, 3.

Ramsey, B. D., and Agrawal, P. C. 1989, Nucl. Instr. and Meth. in Phys. Res., **A278**, 576.

Ramsey, B. D., Weisskopf, M. D., and Joy, M. K. 1989, Proceedings of the SPIE, **1159**, 246.

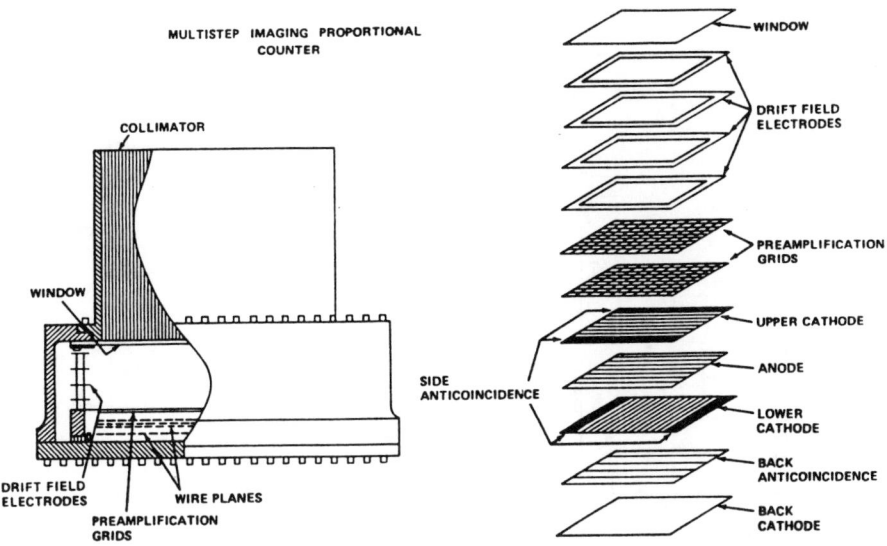

Figure 1. Sectional Schematic of the Multistep Detector.

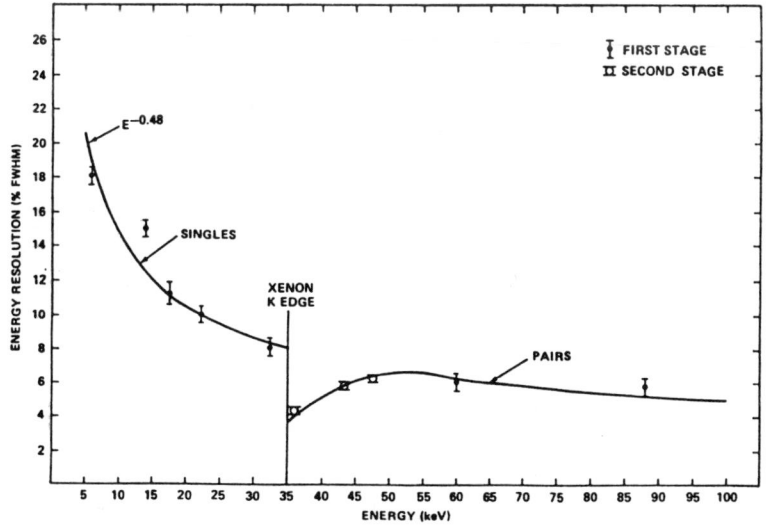

Figure 2. Energy Resolution as a Function of Energy for the Prototype Detector.

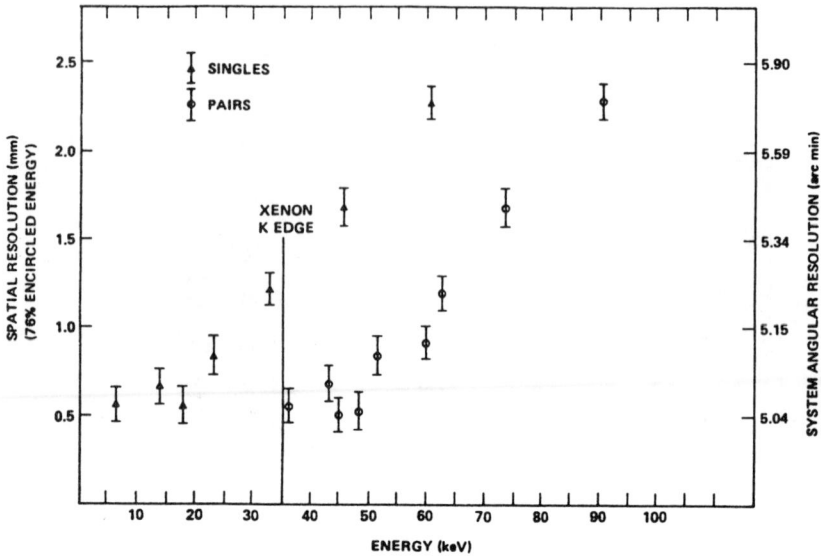

Figure 3. Spatial Resolution as a Function of Energy for the Prototype Detector.

Future Plans

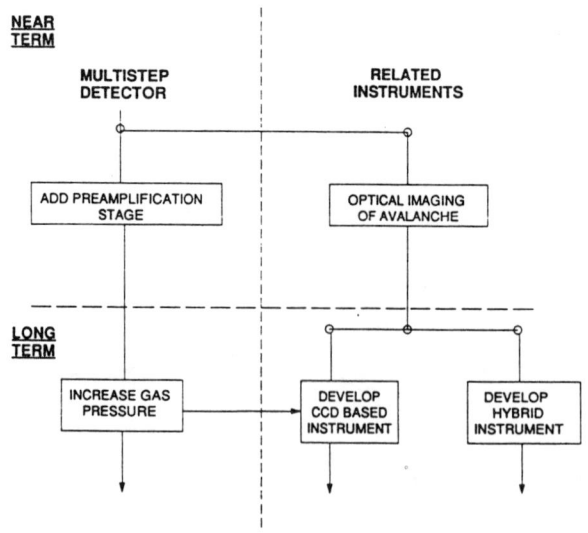

Figure 4. Near and Long Term Developments.

HIGH-RESOLUTION GAMMA-RAY IMAGING FROM THE MOON

William A. Mahoney

Jet Propulsion Laboratory 169-327
4800 Oak Grove Drive
Pasadena, CA 91109

ABSTRACT

An observatory is suggested for exploiting unique lunar features to perform sensitive, sub-arcsecond cosmic x-ray and gamma-ray imaging. The observatory would be built in an evolutionary manner and would eventually include several different position-sensitive detector systems which together would cover a broad energy range and address a wide variety of astrophysical problems. High angular resolution would be obtained by using a mobile crane on the flat lunar mare regions to move a coded aperture mask for source tracking with detector/mask separations of up to 5 kilometers.

I. Introduction

Astrophysical issues to be addressed by hard x-ray and gamma-ray observatories beyond the turn of the century will require experiments with larger collecting areas and orders of magnitude better angular resolutions than those presently planned. Sensitivity requirements dictate collecting areas of about 10^4 cm^2 or larger, depending on the instrument and the energy range of interest. Angular resolutions of an arcsecond or better will be required to conclusively identify sources with known objects and to map the geometry of compact gamma-ray sources.

The Moon would provide an ideal location for an observatory meeting these requirements. Since it is not practical to focus hard x rays or gamma rays, imaging is achieved with position-sensitive detectors and a means of chopping or modulating the source flux. Methods for modulating the flux include coded aperture masks, modulation collimators, and Fourier transform systems. With these techniques, the angular resolution is inversely proportional to the separation between the detector and the mask. The vast stretches of flat lunar terrain, such as the mare regions, would provide a stable platform where detector/mask separations of several kilometers could be conveniently maintained and monitored. A mask carried on a mobile crane with a vertical range of 50 m and a horizontal range of 5 km in any direction from the detector would provide for sub-arcsecond angular resolution while allowing access to nearly the entire sky. Continuous movement of the mask would allow tracking of a particular source for up to 14 days, although the angular resolution would decrease as the mask was moved closer to the detector. Such long, uninterrupted observations are not possible in low earth orbit. The mask could also function essentially as a "zoom lens". A source whose position was initially poorly known could be located with increasingly higher precision by first placing the mask near enough to the detector to ensure that the source was within the field-of-view. The mask would then be moved away in steps to locate the source to the desired accuracy.

In addition to providing for an extremely flexible observing program with high angular resolution, a number of complementary instruments could be built to cover a very broad energy range. The instruments should be of a modular design so they could be expanded and upgraded. With the assistance of personnel stationed at the nearby lunar outpost, masks and detector configurations could be changed to optimize observations with different instruments. This would allow the lunar observatory to be utilized in much the same manner as earth-based observatories are now operated.

II. Scientific Requirements

While a lunar gamma-ray observatory would address a wide variety of astrophysical problems, the following discussion will emphasize those issues involving high-resolution imaging as this is the area where the proposed concept offers the greatest improvement in capability. Although several dozen gamma-ray sources are known, few have been conclusively identified with objects observed at other wavelengths. The identifications that do exist have typically been established by temporal variations rather than by direct imaging. Of the 14 gamma-ray sources observed above 80 keV by the HEAO A-4 experiment (Levine et al. 1984), only 7 have been tentatively identified with known objects. Of these, 3 are extended extragalactic objects (Cen A, 3C273, and NGC 4151), and it is not known where, within the the extended region, the gamma rays originate. While COS B detected 25 sources above 100 MeV (Swanenburg et al. 1981), only four have suggested identifications, including 3C 273 and a complex molecular cloud region known as ρ Oph. Meaningful searches for the counterpart to the gamma-ray sources at other wavelengths using large modern telescopes will likely require angular locations of the gamma-ray sources to about 15 arcseconds even in uncrowded regions of the sky (Sciama 1971). Searches in densely populated regions will require even better localization.

The galactic center region provides a striking example of the need for arcsecond gamma-ray imaging. A map of the central 24" x 24" region at 3.5 μm, observed with an angular resolution of approximately 0.2", is shown in Figure 1 (Tollestrup, Capps, and Becklin 1989). There are over 30 infrared sources evident within this region. An important question is the location of the true dynamic center of the Galaxy. It has been suggested that the center is the very compact nonthermal radio source known as Sgr A*, which may be a massive black hole (Brown 1982; Lacy, Townes, and Hollenbach 1982). Sgr A* is indicated on Figure 1 by a cross, but it is not coincident with any of the infrared sources. High-resolution gamma-ray spectroscopy has revealed a source of narrow line emission at 0.511 MeV from the annihilation of positrons in the general direction of the galactic center. This 0.511 MeV line emission may also be associated with center of the Galaxy, however, this association remains highly controversial and will likely remain so until the source location is accurately known. It is clear from Figure 1 that a localization to an arcsecond or better would be required to conclusively tie the source to Sgr A*, for example, without confusion with other nearby objects.

Except for the diffuse galactic emission, no cosmic gamma-ray source has been spatially resolved. However, based on intensity variations, a number are known to be highly compact. For example, the 0.511 MeV emission from the vicinity of the galactic center varies over a 6-month time scale, and may vary even faster. Light travel time arguments thus constrain the angular extent of the source to 4 arcseconds or less, provided it is located at the distance of the galactic center. Clearly the only hope for resolving the source and measuring its spatial distribution is through imaging with arcsecond resolution.

The relativistic jets of active galactic nuclei (AGN) and their apparent galactic analog SS 433 provide another example of the need for high-resolution gamma-ray imaging. In the case of SS 433, by combining optical and sub-arcsecond radio observations, a detailed model of the kinematics of two opposite and precessing jets has been established (Margon 1982). However, far less is known about the physical conditions leading to the ejection of the jets. Owing to the high energy nature of the phenomenon (jet velocities of 0.26 c), and the fact that SS 433 is a strong x-ray source (Seward et al. 1980), it is highly likely that a map with angular resolution of 1 arcsecond would reveal detailed gamma-ray structure, possibly time variable as is the case with radio measurements (Hjellming and Johnston 1981). This would surely lead to a better

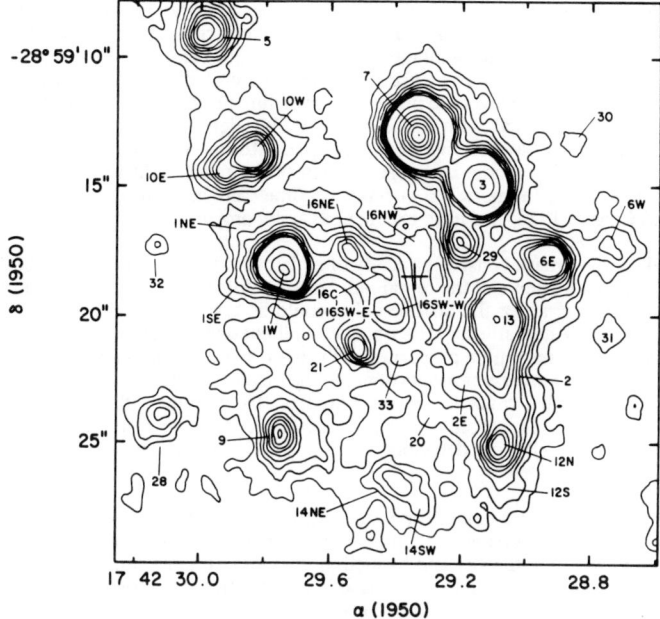

Figure 1. Contour plot of 3.5 μm emission in the central 24" x 24" region of the Galaxy (from Tollestrup, Capps, and Becklin 1989). The compact radio source, Sgr A*, is indicated by a cross. The proposed lunar observatory would provide gamma-ray maps with comparable angular resolution.

understanding of the physical processes operating in the source region, including the acceleration of the jets, interactions within the jets, and interactions between the jets and the interstellar medium. Similar arguments apply to the jets in AGN.

Finally, high-resolution gamma-ray imaging of the Sun would provide a breakthrough in the understanding of processes occurring in solar flares, many of which have direct analogs in astrophysics (Colgate 1988). Basic questions involve the location, size, and number of sites of particle acceleration and the propagation of solar particles, especially protons and ions. Gamma-ray emission from solar flares is dominated by nuclear interactions. Therefore gamma-ray imaging in combination with high energy resolution will provide measurements of the isotopic composition of both the accelerated particles and the ambient medium.

III. Facility Description

The proposed lunar gamma-ray observatory would combine large, position-sensitive detector arrays with a distant mask to chop or modulate the source flux yielding gamma-ray maps with extremely high angular resolution. Although there are a number of techniques for modulating the source flux, the following discussion will use coded aperture masks as an example. The coded aperture concept goes back to a suggestion by Dicke (1968) for overcoming the limitations of the simple pinhole camera through the use of multiple pinholes for x-ray imaging. The pinhole camera offers the possibility of high resolution imaging without the use of refracting or reflecting elements, but has the great disadvantage of achieving high resolution at the expense of a tiny sensitive area. Dicke's idea was to use a random array of many pinholes to obtain a large collecting area, and then to recover the image by correlating the pattern of the random pinhole mask with the x-ray data, recorded with a position-sensitive detector. More recently Gunson and Polychronopulos (1976) have shown how the noise introduced into the image by the random pinhole pattern can be overcome by proper design of the mask. Masks based on the theory of cyclic difference sets (Baumert 1971) have become known as uniformly redundant arrays (Fenimore and Cannon 1978) and have excellent imaging properties.

A coded aperture mask consists basically of a plate divided into a number of cells. Typically, about half the cells are open and the remaining cells are filled with a material opaque to gamma-rays. A practical material for the opaque cells would be tungsten, 2 cm of which is completely opaque below 400 keV and at least 80% opaque at all energies. The cell size and shape must be matched to the position-resolution properties of the main sensors. For a given mask configuration, the angular resolution is inversely proportional to the distance between the detector and the mask. For example, a mask with 1 cm cells, located 5 km from the detector with a positional resolution of 1 cm would give an angular resolution of 0.4 arcsecond. Angular resolutions approaching 0.04 arcseconds would be possible for detectors with a positional resolution of 1 mm and masks having 1 mm cells.

An artist's rendition of a lunar observatory for high-resolution gamma-ray imaging is shown in Figure 2. The principal elements include a position-sensitive gamma-ray detector capable of being pointed anywhere in the sky and located in a large, flat region of lunar terrain. The coded aperture mask is shown suspended from a mobile crane which can move it in any direction from the detector array to distances approaching 5 km.

A. Site Location. In order to allow viewing access to as much of the sky as practical, the observatory should be located near the lunar equator. If the observatory was offset from the lunar equator by 5°, only the region within 5° of one lunar pole would be inaccessible. It would also be desirable to avoid site locations near the lunar limb. Cosmic sources near the lunar horizon could be located with the highest angular precision and it would be preferable not to have the Earth nearby. Since the mobile crane should be capable of moving the mask almost anywhere within 5 km of the detector array, a flat region free of large craters and boulders is needed. These requirements could probably all be met with the observatory on the Mare Tranquillitatis, site of the Apollo 11 landing and a strong candidate for the site of a lunar outpost. The region consists of flood basalts and, aside from some small craters, is nearly a gravitational equipotential. Thus the crane would maintain a nearly constant height with respect to the detector array as it moves, aside from effects resulting from the curvature of the lunar surface, which reaches about 7 m at a distance of 5 km. It might be necessary to initially smooth the area over which the crane moves to eliminate rocks and to fill in small holes. It would also be beneficial to locate the observatory in a region of low natural radioactivity to minimize the instrumental background count rate. The natural radiation will almost certainly have been mapped by an orbital survey mission prior to the establishment of any lunar base. Detailed mapping by a surface rover could supplement this if necessary.

B. Detector Array. A large number of x-ray and gamma-ray detector systems could be used including proportional counters, large scintillators, germanium detector arrays, and spark chambers. Together they would cover a broad energy range and would address a wide variety of astrophysical problems. All of the detectors would require areas approaching 10^4 cm^2, and in some cases areas in excess of 10^5 cm^2 would be needed. An example of an experiment consisting of an array of position-sensitive germanium detectors is shown in Figure 3. Such a system would

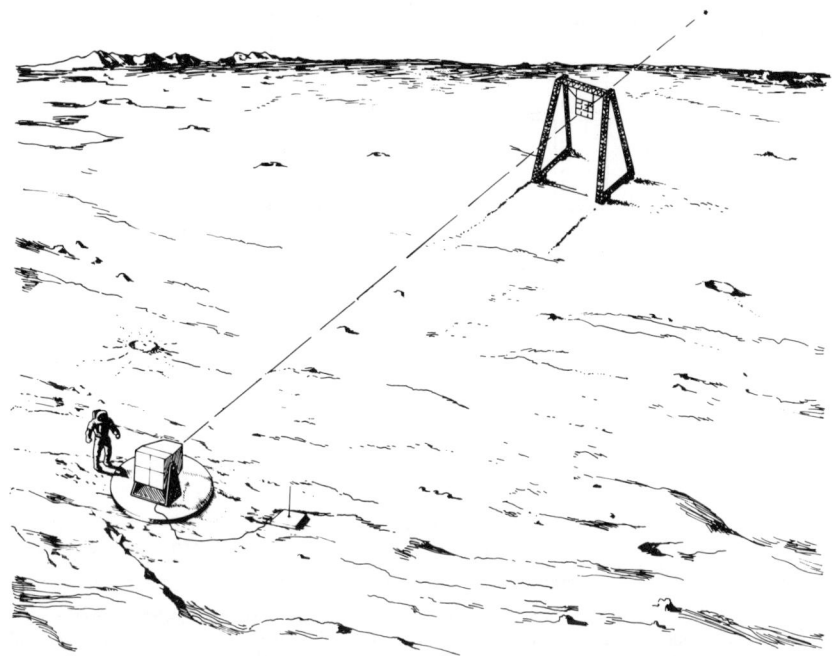

Figure 2. Artist's sketch of a lunar gamma-ray observatory consisting of a large, position-sensitive detector array and a mobile mask. The mask could be moved in any direction out to a distance of about 5 km, providing angular resolutions of 0.4 arcseconds or better.

combine the high energy resolution capabilities of germanium with extremely good angular resolution to study the nuclear component of the gamma-ray emission. Each germanium detector would have the largest volume practical at the time, probably about 400 cm^3. All would be segmented as indicated and mounted horizontally to provide positional information on the location of the gamma-ray interaction to about 1 cm. By the time of a lunar observatory, finer segmentation in two dimentions will likely be possible. Background suppression and crude collimation would be provided by an active shield of bismuth germanate at least 10 cm thick. The detector array shown would be of modular design to allow for future expansion. It would be mounted on a stable platform with an orientation system capable of pointing it toward any region of the sky. Cooling for the germanium detectors could be provided either by mechanical refrigerators or by radiative coolers. The detector system shown in Figure 3 has a mass of about 2000 kg and a size of about 1.5 m by 1.5 m by 1 m, exclusive of any radiative coolers. It is expected that the full instrument would eventually consist of at least 4 of the modules shown in Figure 3. Since the detector is fully steerable, it could be operated without the mask, although with limited imaging capabilities.

C. Mask. A typical coded aperture mask would be a square approximately 2.5 m on a side with a mass of about 1000 kg. The opaque cells might be made of tungsten with a 2-cm thickness

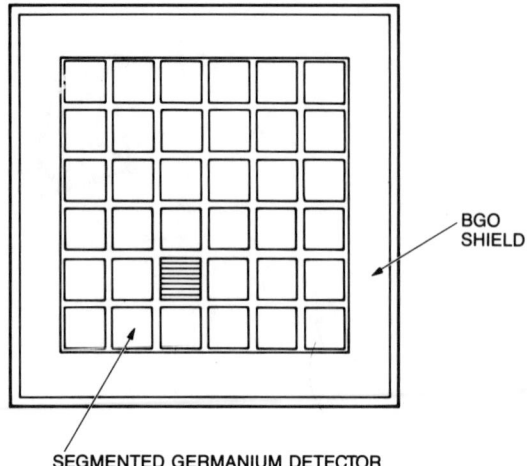

Figure 3. An example of a gamma-ray detector using an array of position-sensitive germanium detectors. When used in conjunction with a coded aperture mask, this system would combine high angular resolution with superb energy resolution.

and lateral dimensions of about 1 cm. The mask would be maintained perpendicular to the viewing direction to within about 1° and its location would be controlled by the mobile crane to within about 10 cm. Knowledge of the location in the directions perpendicular to the viewing direction would have to be known at all times to within about 0.1 of the cell spacing, or about 1 mm for the mask described above. This could be achieved by a network of several microwave interferometer ranging stations distributed around the observatory site. Approximately 100 updates per second should be sufficient to remove any effects from oscillations or other motions of the mask.

The great advantage of the mobile mask is the flexibility it gives the observing program. A wide field-of-view with coarse imaging could be provided with the mask near the detector while a distant mask would give a narrow field-of-view with high angular resolution. When placed 5 km from the detector array, it would provide the highest angular resolution, about 0.4 arcsecond. Since a cosmic source moves at the rate of 0.5° per hour, a stationary crane at 5 km with a vertical range of 50 m would allow the source to be tracked for about one hour. While some gamma-ray observations, such as the imaging of a solar flare, could be completed in an hour, others require observing times of days. Such long observations could be achieved by continuously moving the crane toward the detector array while maintaining the source in the field-of-view of the mask. Uninterrupted observations as long as 14 days could be performed by starting to track a source as it rose above the lunar horizon and following it toward the detector, across the meridian, and ontward until it set. Although the angular resolution would decrease as the mask approached the detector, at the poorest it would be 40 arcseconds when the mask was directly over the detector at a height of 50 m.

D. Uniqueness of the Moon. When compared with a gamma-ray observatory in low earth orbit, the Moon opens up new capabilities and provides a number of advantages. The Moon has no atmosphere to stop or attenuate gamma rays, allowing the surface to be used for the observatory. This provides a stable platform for multi-element systems, such as the detector array and mask discussed here, easing the problems of alignment and coordinated pointings. Long baselines are readily available for detector/mask separations of 5 km. Such an observatory in low earth orbit would not be practical. A detector/mask system connected by a 5 km boom would be subjected to large tidal torques and would present serious construction and orientation problems. Conversely, if the two elements were on separate spacecraft, coordinated pointings and alignment would be difficult and would require continual thrusting. Finally, nearly half the orbital time would be spent pointing at the Earth. An observatory in high earth orbit would have large transportation costs and would lose the flexibility of manned access. The problem of the construction and alignment of very long structures, or of maintaining the alignment of two spacecraft would remain. On the Moon, locational information on the mask could be provided by a microwave ranging network fixed on the surface. Following an initial alignment and calibration, there would be no further worry about changing baselines.

The assembly of the large, heavy detectors needed for sensitive gamma-ray observations would probably be easier on the (1/6 g) lunar surface than in the completely weightless environment of earth orbit. The inhabitants of the nearby lunar outpost would provide a flexible resource for assembly, servicing, changeout of masks and detector elements, and for upgrading and expanding the prime sensors. There is also a good possibility that lunar materials could be used for shielding or for the construction of parts of the crane.

Uninterrupted observations of up to 14 days could be conveniently carried out in a constant background environment. Such observations could not be accomplished in low earth orbit.

The biggest negative aspect of a lunar observatory is the cost of transportation, and even this would be alleviated if lunar materials could be used for parts of the system. The instrumental background from primary cosmic ray interactions would also be several times higher than for the same instrument in low earth orbit, owing to the the lack of shielding from the earth's magnetic field. However, the background in low earth orbit varies with geomagnetic latitude and during encounters with the South Atlantic Anomaly whereas the lunar background is constant. An observatory in high earth orbit would be subjected to a cosmic-ray bombardment similar to that on the lunar surface.

Before construction of a lunar gamma-ray observatory could be initiated, a number of technological issues must be addressed. Large position-sensitive germanium detectors must be developed and tested with the appropriate coded aperture masks. Alternate cranes must be investigated, with emphasis on the structural design, materials, and power sources. An autonomous system for controlling the crane and locating the mask must be studied. The operation of both the mask and the detector array should be fully programmable to eliminate the need for real-time operations.

IV. Summary

A gamma-ray observatory on the Moon would open up a wide range of new capabilities, the most important of which is the ability to readily obtain high-resolution gamma-ray images. A position-sensitive detector and a coded aperture mask at a distance of 5 km would allow imaging with 0.4 arcsecond resolution or better. The Moon has no overlying atmosphere to attenuate gamma-rays, thus the surface provides a convenient, stable platform for maintaining and monitoring the alignment of the detector/mask system. A mobile crane operating on the vast smooth mare regions could track nearly any cosmic source for periods of up to 14 days, providing for an extremely flexible observing program.

The establishment of a lunar gamma-ray observatory would be evolutionary beginning with precursor orbiter and surface rover experiments to identify a flat site with minimum background radiation. Construction of the observatory would begin with the first prime detector module on a rigid platform. It would be fully steerable and would have a sensitivity surpassing any experiment now in the planning stages. Its operation would provide a valuable experience base to guide the design and operation of future experiments. Next the mobile crane and mask would be constructed, providing an observatory with full imaging capabilities. Finally, the initial instrument would be upgraded for increased sensitivity and new instruments and masks would be constructed and utilized

The site of the prime detectors would accommodate a number of instruments which, together, would cover a wide range of astrophysical objectives. Masks on the crane could be changed out as desired to optimize their effectiveness with a particular detector or for a particular scientific observation. With this flexibility in being able to service, upgrade, and modify the equipment for specific observations, the lunar observatory would function in a similar manner to present ground-based observatories.

The author appreciates the valuable comments and suggestions made by R. Cohen, J. Higdon, R. Radocinski, L. Varnell, A. Vaughan, and W. Wheaton. The research described in this paper was carried out by the Jet Propulsion Laboratory, California Institute of Technology, under contract with the National Aeronautics and Space Administration.

REFERENCES

Baumert, L. D. 1971, Cyclic Difference Sets, (New York: Springer).
Brown, R. L. 1982, Ap. J., **262**, 110.
Colgate, S. A. 1988, Solar Physics, **118**, 1.
Dicke, R. H. 1968, Ap. J. (Letters), **153**, L101.
Fenimore, E. E. and Cannon, T. M. 1978, Applied Optics, **17**, 337.
Gunson, J., and Polychronopulos, B. 1976, MNRAS, **177**, 485.
Hjellming, R. M., and Johnston, K. J. 1981, Ap. J. (Letters), **246**, L141.
Lacy, J. H., Townes, C. H., and Hollenbach, D. J. 1982, Ap. J., **262**, 120.
Levine, A. M., et al. 1984, Ap. J. (Suppl.), **54**, 581.
Margon, B. 1982, Science, **215**, 247.
Sciama, D. W. 1971, Modern Cosmology, (Cambridge: Cambridge University Press), p. 56.
Seward, F., Grindlay, J., Sequist, E., and Gilmore, W. 1980, Nature, **287**, 237.
Swanenburg, B. N., et al. 1981, Ap. J. (Letters), **243**, L69.
Tollestrup, E. V., Capps, R. W., and Becklin, E. E. 1989, Astron. J., **98**, 204.

DISCUSSION

Charles Hailey

If one varies the focal length to change the field of view, what accomodations are made to handle the altered collimation requirements?

Mahoney

The active collimator cannot be made long enough to limit the field of view to the area subtended by the mask. Thus, the collimation requirements remain the same as the distance separating the mask and collimator is varied.

Elihu Boldt

It's not clear that your configuration lends itself to tracking a source for two weeks. There appears to be a severe zenith-angle limitation.

Mahoney

The mask is suspended from a mobile crane which would actively control its vertical position to follow a source. The distance between the crane and the detector array would also be actively maintained. Together, these control systems should allow a source to be tracked continuously from the time it rose in the East, across the meridian, and westward until it set. Of cource, the angular resolution would vary continuously from about 0.5 arcseconds at the extreme ranges to just under an arcminute near the meridian.

(Unknown)

Your zeroth-order crane design for a lunar gamma-ray observatory doesn't seem very optimal. How about a vertical tower (say 1 km high) supporting the mask at the end of a boom, much like a fore-and-aft sail rig? This would give full altazimuth coverage.

Mahoney

Alternate crane designs, and other systems for moving the mask, need to be studied. It is quite likely the crane shown in the figure is not optimal.

Neil Gehrels

The lunar maria are significantly (factor 5 − 10 from Apollo data) more radioactive than the highlands. Also, a detector system sitting on the surface will be exposed to the full cosmic ray beam and the activated surface material.

Mahoney

It's correct that the cosmic-ray intensity is higher on the lunar surface than for a spacecraft in equatorial low earth orbit. As to the lunar radioactivity, it is anticipated that precursor orbital and possibly rover missions would be used to select a site with minimum background radiation. However, a flat lunar surface would be required to allow the crane to be moved as described.

Gerald Share

Wouldn't a Fourier Transform Telescope be better suited for high angular studies with the 1 cm spatial resolution you described?

Mahoney

I have no doubt that a coded aperture mask would work, and it appears preferable because of higher efficiency and less complexity. However, other techniques for imaging must be investigated.

A GAMMA-RAY IMAGING TELESCOPE BASED ON LIQUID XENON

Elena Aprile, Reshmi Mukherjee and Masayo Suzuki

Columbia Astrophysics Laboratory
Columbia University

ABSTRACT

Liquid xenon is an ideal ionization and scintillation medium for a position sensitive detector in the imaging of astrophysical gamma-ray sources. We discuss here the relevant features of a liquid xenon Time Projection Chamber that we have recently proposed as an imaging telescope for high energy astrophysics.

1. Introduction

In the challenging phase of gamma-ray astronomy beyond the Gamma-Ray Observatory, imaging capability will be the essential requirement for a telescope to accurately determine the location of gamma-ray sources, without problems of source confusion. Instruments based on rare gas liquids promise major advances not only in imaging capability but also in energy and position resolution as well as detection efficiency. Table 1 summarizes the most relevant properties of xenon and argon. Xenon is especially suited for the detection of gamma-rays because of its high density and high atomic number, which make it similar to NaI(Tl) in detection efficiency. It can be used either in an ionization calorimeter, with excellent energy and position sensitivity, or in a scintillating calorimeter. Xenon is in fact not only a good ionization medium but also an excellent scintillator, with abundant light emission in the ultraviolet region (175 nm) and a fast decay time constant (Kubota et al. 1982). As seen from Table 2, which shows a comparison of liquid xenon with other scintillators, the photon yield of liquid xenon for 1 MeV electrons is comparable to that of NaI(Tl) crystals, which has the highest photon yield for the gate time of longer than $10\mu s$. On the other hand the photon yield of liquid xenon for the gate time of nsec is superior even to that of the faster BaF_2.

The main challenge for a practical liquid xenon scintillating calorimeter is efficient light collection with photo-sensors working in the UV region. In this regard, recent developments of large area UV sensitive Si photo-diodes are very promising (Doke 1989).

To realize an imaging gamma-ray telescope based on liquid xenon, we have adopted the approach of an ionization calorimeter, implemented as a Time Projection Chamber (TPC) for 3-dimensional tracking. The imaging capability of a TPC will allow one to visualize complex gamma-ray events with multiple interactions, initiated by either Compton scattering or pair production. Since the detector is continuously sensitive to any ionizing event, no topological constraints limit the acceptance, thus enormously increasing the

Table 1

Properties of argon and xenon

Property	Argon	Xenon	Unit
Atomic number	18	54	
Atomic weight	39.95	131.3	g/mol
Boiling point	−185.9	−109.1	°C
Melting point	−189.4	−111.8	°C
Density (1 bar, 15°C)	1.6689	5.517	kg/m^3
$T_{critical}$	−122.43	16.59	°C
$P_{critical}$	48.65	58.40	bar
Properties of liquids			
Volume ratio	784.0	518.9	(gas/liquid)
Vap. enthalpy	163.4	99.3	J/g
Density	1.4	3.06	g/cm^3
v_d at 1 kV/cm	2.2	2.4	mm/μs
Mobility	525	2000	cm^{2}V^{-1}s^{-1}
Radiation length	13.5	2.6	cm
W-value	23.6	15.6	eV/pair
$(dE/dx)_{min}$	2.2	3.9	MeV/cm
Fano-factor	0.11	0.041	
WF	2.54	0.64	eV

Table 2

Comparison of Scintillators

	Liq. Xe	NaI(Tl)	BaF$_2$	BGO
Density (g/cm^3)	3.06	3.67	4.89	7.1
Rad. length (cm)	2.8	2.59	2.05	1.12
$(dE/dx)_{min}$ (MeV/cm)	3.79	4.84	6.60	9.02
Refractive index	1.58	1.775	1.56	2.15
Emission peak (nm)	175	410	220 / 310	480
Fast decay time (ns)	2.2	...	0.6	...
Slow decay time (ns)	27	250	620	300
Recombination time (ns)	45	...	...	...
W_{ph} (eV)	24.1	23.1	120	360
I (∞) (photons/GeV)	4.2×10^7	4.3×10^7	8.5×10^6	2.8×10^6
I (10 ns) (photons/GeV)	9.8×10^6	1.7×10^6	2.1×10^6	9.2×10^4
I (3ns) (photons/GeV)	3.4×10^6	5.1×10^5	2.0×10^6	2.8×10^4

detection efficiency. To obtain similar efficiency with the more common techniques used in gamma-ray astronomy, much larger detectors or longer observation times would be required. With an energy resolution of about 1% RMS at 1 MeV and a spatial resolution of about $250\mu m$ RMS on each coordinate, the liquid xenon instrument will perform not only as an excellent spectrometer but also as a Compton and Pair Telescope. The accuracy in the determination of the incident gamma-ray direction depends on the gamma-ray energy. For precise imaging of low energy gamma-ray sources, emitting below a few MeV, one would need much better spatial resolution than $250\mu m$, which is not practical. If one wants to focus on the low energy sources, a system made of liquid xenon TPC with a rotating coded aperture mask, could be considered as an attractive possibility. On the other hand, for imaging gamma-rays of several MeV, which will interact with liquid xenon through an initial Compton scattering, one can use the liquid xenon TPC as a conventional Compton Telescope, with the difference that the interaction points of the primary and scattered photons are seen in a single detector rather than in two separate ones, at a fixed distance. The angular resolution of the liquid xenon TPC used in the Compton mode is determined by the spatial and energy resolution. Simple estimates indicate a value of about 0.5 degrees at 1 MeV and better at higher energies. The ambiguity in source location that characterize the Compton measurement, can be completely removed if one can also determine the direction of the Compton electron. For gamma-ray events at high enough energy to allow tracking of the scattered Compton electron, as well as for events which start with pair production, the source of gamma-ray emission can be unambiguously identified. This capability is a unique feature of the liquid xenon instrument.

The imaging capability can be used not only to analyze the topology of true events, but also to reject charged and neutral background, reducing the requirement for a massive anticoincidence shield of the type that is needed for germanium and sodium iodide detectors. Imaging will also be useful to reject background events represented by upward moving gamma-rays which are rejected by time-of-flight measurement in a Compton Telescope of the double scatter type. In the liquid xenon telescope one will use the directionality associated with the reconstructed Compton electron or electron-positron pair. To stop lower energy gamma-rays a passive detector shield can be used. For a more detailed discussion of the expected performance of the liquid xenon TPC as a Compton-Pair gamma-ray Telescope we refer to Aprile et al. (1989).

2. Instrument Description

The proposed gamma-ray telescope consists of a liquid xenon TPC operated in the ionization mode and triggered by the scintillation light. A schematic diagram of the TPC system is shown in Figure 1. The device is continuously sensitive to any ionizing event. Gamma-ray interactions with the xenon will produce charged particles which will dissipate their energy by ionization and excitation. The total energy of the gamma-ray event as well as its spatial distribution will be measured from the ionization signals induced on the collection electrodes. To contain high energy gamma-rays and to be sensitive to small fluxes, we have considered a chamber with a depth of 26 cm, which corresponds to about

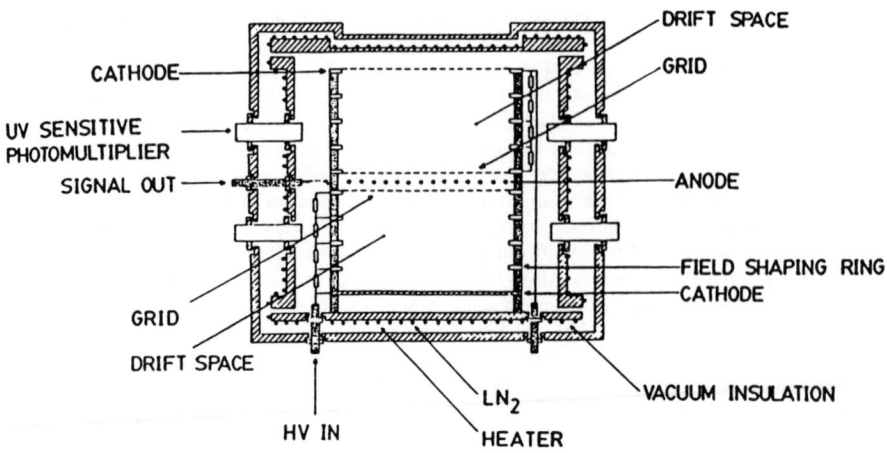

Figure 1. Schematic of the liquid xenon imaging chamber

ten radiaiton length, and a sensitive area of 1200 cm^2.

The total active volume of about 3×10^4 cm^3 is subdivided into 2 drift regions, each 13 cm long. Ionization electrons produced by an ionizing event can drift from both sides towards a readout electrode structure of about 40 cm diameter. The uniformity of the electric drift field is insured by a sequence of field shaping rings. High voltage is applied to the cathode planes made of thin mesh to minimize the interaction of γ-rays with passive materials. A liquid nitrogen heat exchanger can be used to maintain the liquid xenon temperature.

UV sensitive photomultipliers, a minimum of two for each drift region, are used to detect the primary scintillation light which is produced, along with ionization, by the passage of a charged particle through the sensitive volume, and which will be used for triggering the TPC. We have extensively studied the time dependence and the intensity of the scintillation light in liquid xenon, irradiated with alpha particles and fast electrons. For recent results we refer to Aprile *et al.* (1989). The dependence of the scintillaiton light on the specific ionization of the event will be used to discriminate against background, with pulse shape discrimination.

For two-dimensional imaging of the spatial development of the events in a TPC, different read out schemes can be adopted. The two most commonly used are multi-wire structures, in which the drifting electrons are passed from one sensing plane to the next determining one coordinate of the electron cloud at a time, or segmented anodes, where an interwoven pattern of strips in both directions is printed on one plate. We will initially test a segmented anode of the type used by Mahler *et al.* (1983) for a liquid argon TPC, but with much finer segmentation. The spacing between the sensing elements of the TPC will ultimately limit its spatial resolution. For the anode under consideration, a reduction of the pattern dimensions to the level of 0.5 mm, needed to achieve few hundred microseconds spatial resolution, appears feasible.

The charge signals induced on the sensing elements of the anode will be fed to low noise charge-sensitive preamplifiers and then multiplexed into wave form digitizers. These will be continuously active to provide a uniform sampling of the collected charge in the X- and Y- directions. The third coordinate, parallel to the electric field, is deduced with the known drift velocity from the measured drift time, as referenced to the event starting time provided by the light signal. Thus a complete 3-dimensional event image is realized.

To minimize charge losses in the detectable signal, which would directly affect the detector's resolution, the liquid filling the chamber has to be free from electronegative impurities, the applied electric field has to be sufficiently strong to reduce the electron-ion recombination process, and the transparency of the shielding grid has to be maximized. All these problems have been extensively studied at Columbia in the past few years and should facilitate the development of the proposed instrument.

The spectral response of liquid argon and xenon ionization chambers to gamma-rays, electrons and alpha particles has also been investigated by our group (Aprile *et al.* 1987; Aprile *et al.* 1988; Aprile *et al.* 1989). Figure 2 shows the measured energy spectrum of ^{207}Bi in liquid xenon, from which a resolution of about 14 keV RMS is obtained for the dominant 569 keV gamma-ray line. Experiments with liquid xenon are continuing to better understand the separate effects of recombination and electron attachment and to improve the energy resolution.

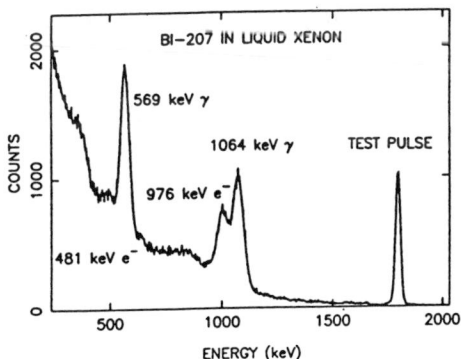

Figure 2. Energy spectrum of ^{207}Bi in liquid xenon

3. Summary

A new type of imaging instrument for gamma-ray astrophysics in the energy region of 1 – 30 MeV is being developed at Columbia. It is based on a liquid xenon detector, operated as a Time Projection Chamber, with 1% RMS energy resolution at 1 MeV and 250μm position resolution. At high energies, the unique capability of the TPC to image the direction of the scattered electron in a Compton interaction and to image pair-production events will translate into unambiguous source localization.

Acknowledgements

We are grateful to NASA for its support of this project under grant NAGW-1370. This is contribution number 413 of the Columbia Astrophysics Laboratory.

References

Aprile, E., Mukherjee, R., and Suzuki, M. 1989, *Proceedings of the SPIE 33rd Annual International Symposium on Optical and Optoelectronic Applied Science and Engineering*, **Vol. 1159**, 295.

Aprile, E., Mukherjee, R., and Suzuki, M. to appear in *Proceedings of the IEEE 1989 Nuclear Science Sumposium*, San Francisco, 15 – 19 January 1990.

Aprile, E., Ku, W., Park, J. and Schwartz, H. 1987, *Nucl. Instr. and Meth.*, **A261**, 519.

Aprile, E., Ku, W., and Park, J. 1988, *IEEE Trans. Nucl. Sci.*, **Vol. 35**, No. 1, 37.

Aprile, E. and Suzuki, M., 1989, *IEEE Trans. Nucl. Sci.*, **Vol. 36**, No. 1, 311.

Doke, T. 1989, (private communication).

Kubota, S., Hishida, M., Suzuki, M. and J. Ruan (Gen), 1979, *Phys. Rev.*, **B20**, 3480 and 1982 *Nucl. Instr. and Meth.*, **196**, 101.

Mahler, H.J., Doe, P.J., and Chen, H.H. 1983, *IEEE Trans Nucl. Sci.*, **NS-30**, 86.

HIGH RESOLUTION COMPTON TELESCOPE FOR THE 21ST CENTURY

W.N. Johnson, R.A. Kroeger, J.D. Kurfess
Naval Research Laboratory
Washington DC

ABSTRACT

The properties of Compton scatter telescopes are examined in the light of possible developments in high energy and spatial resolution detectors. Improvements in spectral resolution produce detector systems with sensitivities to gamma ray emission lines which improve proportionately with improvement in spectral resolution.

INTRODUCTION

Gamma ray instruments in the 21st century should provide much improved angular and spectral resolution over the instruments in use today. Current gamma ray source positions and uncertainties make multi-wavelength source identifications difficult while the instruments' spectral resolution hide or distort spectral features which provide insight into the underlying energy production and transport mechanisms in the source. We are examining one of the current gamma ray detection techniques, the Compton scatter telescope, for its applicability to new instruments which achieve the goals of improved angular and spectral resolution. Compton scatter telescopes have been used in gamma ray astrophysics for over ten years in balloon-borne experiments (see, for example, Herzo et al. 1975). The COMPTEL collaboration's experiment (Diehl 1988), V. Schoenfelder, Principal Investigator, will provide the first in depth demonstration of the Compton scatter telescope aboard NASA's Gamma Ray Observatory satellite which is to be launched in 1990. In general, these instruments provide imaging in the 1 - 10 MeV energy range with angular resolution of a few degrees and spectral resolution of ~10%. In association with the development of new X-ray and gamma-ray detection techniques, we are examining the properties of a Compton scatter telescope which incorporates detectors with both high spectral resolution and high spatial resolution. These properties and their effect on instrument sensitivity are discussed below.

COMPTON TELESCOPE BASICS

Figure 1 shows the general principles of a Compton scatter telescope. The upper detector plane (D1) is made of low Z material and a thickness to optimize the probability of a photon interaction via a single Compton interaction. The lower detector (D2) is designed to capture the scattered photon. In this case, the incident photon energy is the sum of the measured energy losses in the two detectors ($E_1 + E_2$) and the scattering angle θ is calculate from the Compton kinematics,

$$\cos\theta = 1 - mc^2(\ 1/E_1 - 1/(E_1+E_2)\ ).$$

The vector between the interaction positions in the upper and lower detectors, when projected onto the celestial sphere, defines the direction of the scatter photon. The incident photon, is offset from this direction by the angle θ defined above. Without a measurement of the direction of the Compton electron in the upper detector, the Compton process has an azimuthal uncertainty (about the scattered photon vector) in the arrival direction of the incident photon. This uncertainty, when projected onto the celestial sphere, produces a small circle of half-angle θ centered on the direction of the scattered photon such that an incident photon of the observed energy anywhere on the small circle could produce the observed interaction in the detector. The position of a celestial point source of gamma rays is therefore detected by the intersection of these circles from many gamma-ray events.

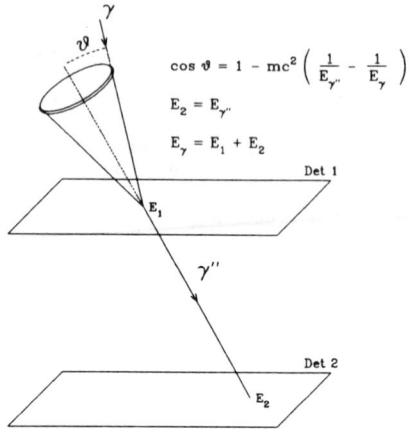

Figure 1. Compton Telescope Basics

PERFORMANCE

Spatial resolution and spectral resolution in Compton telescopes produce uncertainties in the scattered photon direction and uncertainties in the scattering angle, respectively, and consequently limit the angular resolution of the telescopes. This can be visualized as changing the small circle projections into annuli of half-angle θ and width $d\theta$. Figure 2 shows a one dimensional projection of the point spread function of a Monte Carlo simulated response to a γ-ray point source for a telescope with good position and spectral resolution. Figure 3 demonstrates the degradation in the angular resolution which occurs as the spectral resolution degrades. In the simulations displayed in Figure 3, the position resolution in the detectors was assumed to

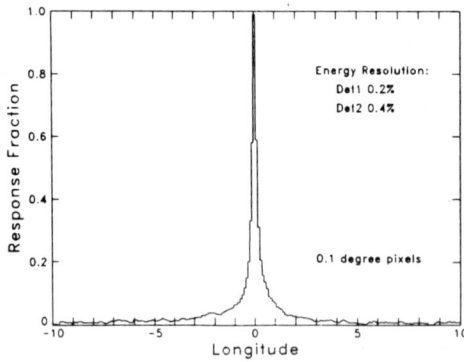

Figure 2. Point Spread Fuction for 1.3 MeV incident γ ray.

be much better than the spectral resolution so that the resultant point spread function was determined by the spectral resolution.

The coincidence requirement for energy losses in the telescope's two detectors produces telescopes with relatively low efficiency, typically 1 - 3%, but also reduces the detector background. An accurate measurement of the relative interaction times in the detectors ("time of flight" measurement) can provide additional reduction in background by discriminating against upward traveling photons and non-relativistic particles. Activation of material in and around the telescope with simultaneous emission of 2 or more photons has the coincidence characteristics which can produce a background problem for Compton telescopes. An example of this is neutron interactions with aluminum, ^{27}Al (n,α) ^{24}Na, with γs at 2.8 and 1.4 MeV.

Consequently, a Compton telescope made of detectors with good position and spectral resolution can provide a low background instrument with spectroscopy and imaging capabilities. One possibility for such an instrument would incorporate superconducting transition detectors which are currently being investigated at NRL (Kurfess et al. 1990).

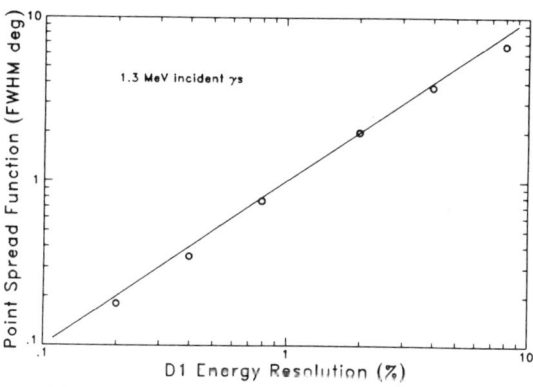

Figure 3. Effect of Detector Energy Resolution on Compton telescope angular resolution for 1.3 MeV incident γ rays. Monte Carlo simulation of the point spread function vs. energy resolution in the upper detector. The lower detector resolution was x2 the upper.

If it is assumed that the telescope background is dominated by the cosmic diffuse background, we find that a 10^6 sec observation with this Compton telescope, as summarized in Table 1, would have a sensitivity to gamma ray emission lines in the MeV range of

$$\text{Sensitivity } (5\sigma) \sim 2 \times 10^{-7} \ \gamma/\text{cm}^2\text{-s}$$

or, ~100-fold improvement in sensitivity of the COMPTEL instrument on GRO. This large increase is attributable to the fact that in Compton telescopes, the angular resolution improves as well with energy resolution, so that in background limited measurements of point sources, the emission line sensitivity improves at least in proportion to energy resolution. In conventional calorimeter detectors, the sensitivity improves with the square root of the energy resolution.

Table 1.
High Resolution Compton Telescope

Detectors:	Superconducting Transition Detectors
Energy Range:	0.2 - 10 MeV
Frontal Area:	~ 1 m^2
Spatial Resolution:	~ 1 mm
Energy Resolution:	~ 0.2% D1 ~ 0.5% D2
Resultant Angular Resolution:	~ 0.1 degree

A Compton telescope of detectors with millimeter position resolution and < 1% spectral resolution would provide an excellent spectroscopy survey instrument. Its broad field of view, fraction of a degree angular resolution, and good energy resolution would permit the investigation of the full range of astrophysical problems in the gamma ray range.

REFERENCES

Diehl, R. 1988, *Space Science Reviews*, **49**, 85.

Herzo, D., R. Koga, W.A. Millard, S. Moon, J. Ryan, R. Wilson, A.D. Zych, R.S. White 1975, *Nucl. Instru. and Meth.*, **123**, 583.

Kurfess, J.D. *et al.* 1990, (this publication).

DISCUSSION

Martin Elvis

How much of an improvement would this instrument give for broad-band continuum measurements?

Johnson

This instrument benefits from a smaller pixel size for continuum measurements. We expect improvements of a factor of 10 – 50.

Alan Bunner

What would be the useful energy range of a Compton telescope using superconducting transition detectors?

Johnson

200 keV – 10 MeV.

New Drift Chamber Technology
for
High Energy Gamma-Ray Telescopes

Stanley D. Hunter and Rajani Cuddapah
NASA Goddard Space Flight Center, Greenbelt, MD 20771

Abstract

In the era of high energy gamma ray astronomy (10 MeV – >30 GeV) following the results from the EGRET experiment on the Gamma Ray Observatory, additional improvements in angular resolution, sensitivity and possibly also energy resolution will be required. To achieve the angular resolution, energy determination and sensitivity goals to make continued advances in gamma ray astronomy, the next generation of gamma ray telescope is envisioned as a "2 meter class" instrument, a picture type detector with an aperture of approximately 2 m × 2 m and a depth of about 4 m. Drift chambers are an attractive type of detector for construction of this type of instrument. This paper discusses our work to develop a low power amplifier and discriminator, for use on large space qualifiable drift chambers.

Introduction

Information on high energy gamma rays, (10 MeV – >30 GeV), are derived by direct measurement of the electron and positron resulting from pair production. Previous gamma ray telescopes, SAS-2, COS-B, GAMMA-1 and EGRET (Thompson, 1986 and Hartman, 1988 and references therein) used an assembly of high voltage spark chambers, interleaved with high-Z metal foils which provide the pair production medium, to record a picture of the "Λ" formed by the ionization tracks left by the electron and positron produced when a high energy gamma ray undergoes pair production. Large area drift chambers are an attractive alternative to spark chambers for the development of a large area picture type detector, they have long been used at accelerators, but have not been viable for use in space because of the power required by their high speed electronics and the need to find appropriate materials and mechanical approaches for their fabrication.

In section one of this paper we consider the goals of a next generation high energy gamma ray telescope design and how we can achieve these goals using xenon gas drift chambers. Section two describes our design and construction of a low power drift chamber amplifier and discriminator. Section three covers the design of a quad time to amplitude converter.

I. Design Goals of a New Gamma Ray Telescope

To achieve the goals of improved sensitivity, angular resolution and possibly also energy determination needed to make continued advances in gamma ray astronomy, the next generation of gamma ray telescope is envisioned as a "2 meter class" instrument, a picture type detector with an aperture of approximately 2 m × 2 m and a depth of about 4 m. A new approach, using large area drift chambers as an alternative to spark chambers, is envisioned to construct such an instrument. To get high resolution in drift chambers, one must have a high speed amplifier and discriminator for each anode and the anodes must be closely spaced (not more than about 6 cm). A "2 meter class" instrument, containing of about 200 drift chambers, each with about 32 anodes, would have on the order of 6400 anodes. The development of low a power amplifier and discriminator are essential to the construction of this scale of instrument for space flight with a limited power budget.

II. Low Power Amplifier and Discriminator Design

We felt that a discrete design was preferable at this time to an integrated design because the circuit could easily be tailored to our specific needs in terms of power limitations, bandwidth and physical size. We began our amplifier development with a Japanese design (Suekane, 1986). We modified this design by changing the transistors to domestic near-equivalents and by removing the 50 Ω output driver stage. We also did a tradeoff between bandwidth and gain to match the response of our drift chambers. Our final four transistor surface mount design, shown in figure 1a, has a gain of 20, 100 MHz bandwidth and dissipates only 20 mW.

Our voltage level discriminator design is an in-house design based on earlier work. We modified it slightly to respond to the fast rise time of the amplifier and to use modern components. The final ten transistor surface mount design, shown in figure 1b, dissipates only 135 mW. Our amplifier and discriminator designs realize a power reduction of about 65% over commercially available designs (Lecroy Catalog, 1988, TRA402 and MVL407).

To test the amplifier and discriminator, we designed a printed circuit board containing three complete amplifier and discriminator circuits for use on specially constructed six inch drift chambers. These small chambers were constructed so that the electronics for the three anodes of each frame are outside the drift gas volume. The electronics could be tested without disturbing the vacuum system.

III. Quad TAC Design

Location of a particle's position in a drift chamber is deduced from the drift time of the ionization cloud from the track location to the anode. This time can be measured in several ways. We chose an analog TAC (Time to Amplitude Converter) approach using a field effect transistor to switch a charging voltage onto a capacitor. The start signal is generated by a event trigger system. The stop signal is the output of the discriminator. Our choice of a constant voltage TAC, rather than the more common constant current TAC, was based on simplicity of design, stability of the constant voltage source and "closed loop operation", the constant voltage is derived from the ADC (Analog to Digital Converter). The nonlinear voltage rise on the capacitor will be calibrated out along with several other nonlinearities associated with drift chambers.

For application on a "2 meter class" instrument we feel that it will be necessary to time up to four signals from each anode, the electron, the positron, a random cosmic ray and one extra channel for redundancy. Therefore, each anode requires four TACs. In our final design, figure 1c, the discriminator output is switched through a series of flip-flops to stop each of the four TACs sequentially. We designed a printed circuit board containing twelve channels of this quad TAC for use on $1/2$ m × $1/2$ m drift chambers. This twelve channel board is designed for computer controlled readout. Each individual TAC of the twelve quad TACs is addressable. Addressing a TAC, multiplexers the analog timing voltage onto the ADC and returns a valid-data-bit, derived from the stop signal, indicating that digital conversion is needed and timing data can then be read out.

IV. Summary

These low power amplifier and discriminator designs will be used on $1/2$ m × 2 m drift chambers which are currently being designed. The 2 m length of these chambers will simulate the electrical and mechanical characteristics of 2 m × 2 m chambers and will be used for testing of xenon as a drift gas.

References

Hartman, R.C., Proc. High Resolution Gamma Ray Cosmology Workshop (1988).
Suekane, et al., IEEE Trans. on Nucl. Science, **33,** No. 1, (1986).
Thompson, D.J. Nucl. Instr. and Meth. **251** 390 (1986).

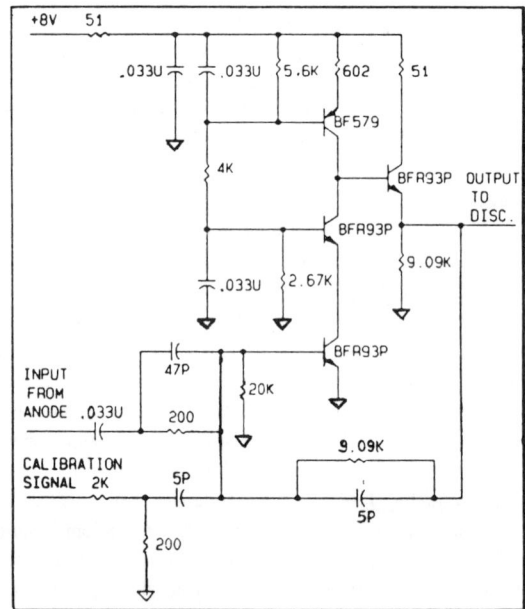

Figure 1. Low power drift chamber amplifier.

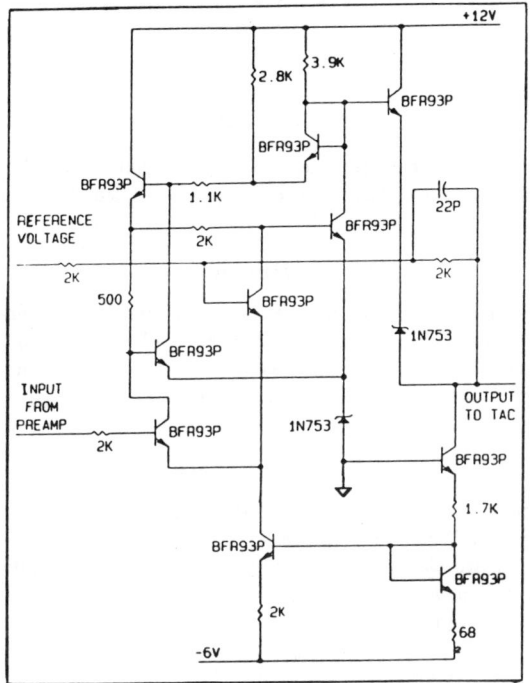

Figure 2. Low power drift chamber discriminator.

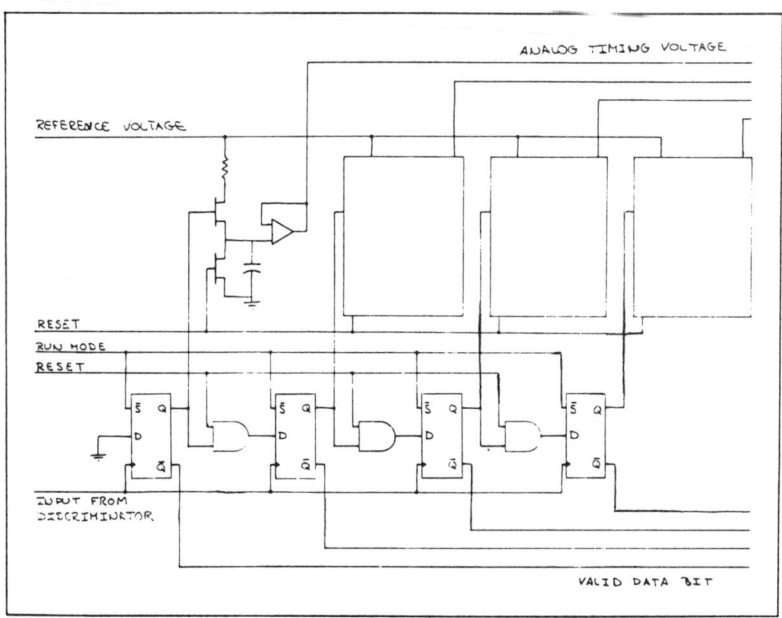

Figure 3. One channel of drift chamber quad TAC.

DISCUSSION

James Adams

As you know, most γ-rays in this energy range come from diffuse sources. What capability will this instrument have to survey diffuse emission? Will it only look at point sources?

Hunter

In the post-EGRET/GRO era it is felt that the whole-sky coverage, during the first year of the GRO mission, will have produced good observations of the diffuse galactic emission. If higher angular resolution observations of the diffuse emission are warranted, they could be done by mapping the regions of interest.

Gerald Share

About 15 years ago we (NRL) suggested using a xenon streamer chamber to make similar measurements. Would this still be a viable option relative to the drift chambers? One of our design features was to reduce γ production in the supporting materials and enclosure. In our design we had the enclosure made with plastic scintillator. How will you handle this background problem?

Hunter

It is desirable to reduce the amount of material surrounding the γ-ray detector. A "next generation" instrument which has a limited field of view can be made less sensitive to γ-ray background produced in the walls of the drift chambers. Another method, also used on EGRET, is to calibrate this source of background and thus be able to correct for it.

HIGH RESOLUTION GAMMA-RAY IMAGING AND SPECTROSCOPY

G. H. Nakano and J. R. Kilner

Lockheed Palo Alto Research Laboratory
Palo Alto, CA 94304

ABSTRACT

Rotational aperture synthesis techniques combined with high resolution Ge detectors have provided clearly resolved images produced with monoenergetic gamma rays. Both point and distributed sources are effectively imaged by this method using an energy window of about twice the FWHM detector resolution (2-4 KeV below 1 MeV). With the combined features of Compton suppression and background rejection provided by active shielding, significant improvement in the signal to noise ratio may be obtained for image reconstructions generated by a single narrow line, a group of lines, or a well defined region of interest in the gamma-ray spectrum. For example, a spaceborne or balloon-borne instrument of this type can yield images of sources in terms of specific line emission such as the 511 KeV annihilation radiation (i.e., from the galactic center region), 2.223 MeV neutron capture line, 1.807 MeV line from ^{26}Al, and other discete nuclear deexitation lines. Monte-Carlo simulations and laboratory measurements were performed to study various imaging and reconstruction techniques applied to rotation modulation collimator systems. Some of these findings are useful in guiding the design of future instruments.

INTRODUCTION

Since the gestation period of typical space experiments generally exceeds 10 years, it is not too soon to consider next generation instruments that will satisfy observational requirements in astrophysics in the early 21st century. It is clear that the need for observations with both high energy resolution and fine spatial resolution in the hard x-ray and γ-ray region is essential for the advancement of high energy astrophysics. In fact, such measurements were recently declared the top priority imperatives by the Gamma-Ray Program Working Group (1988). The technology to perform these observations in space has been developed over the years and is presently quite mature. Yet, it is more than sobering to note that, to the best of our knowledge, no Ge spectrometer has been flown in space since HEAO-3 which was over 10 years ago and the next opportunity to do so will apparently take another 10 years.

High energy astrophysical processes and their resulting nuclear interactions generate copius amounts of gamma radiation that act as carriers of detailed information on the nature and mechanisms that produce these quanta. Precise identification of the γ-ray lines by their energy shift and line shapes in terms of kinematic, thermal, and/or gravitational effects, by their polarization properties, and by their associated non-thermal continuum spectra are some of the basic data that may be acquired by high resolution spectroscopy. The recent high resolution detections of the 847 KeV and 1238 KeV line emissions produced by the decay of the ^{56}Co from SN1987A in the LMC (Sandie et al. 1988; Mahoney et al. 1988; Rester et al. 1988; Teegarden et al. 1989) attest to these points. These data have not only confirmed the standard theory of nucleosynthesis but also provided important information constraining existing models (Shigeyama et al. 1987; Pinto and Woosley 1988; Chan and Lingenfelter 1988). Such measurements and detailed characterization of the γ-ray flux provide the most direct evidence relating to the processes at work in compact objects such as neutron stars, black holes, active galactic nuclei, as well as nova and supernova explosions. Comprehensive investigations into such regimes will not only advance the state of our knowledge in astrophysics but may have far reaching consequences regarding our current theories of nuclear matter, quantum gravity, and general relativity.

GAMMA-RAY IMAGING

In the past several years, we have developed a rotation modulation collimator (RMC) imaging system based on techniques of Mertz et al. (1986) and have performed Monte-Carlo simulations (Kilner and Nakano 1989) to study effects due to statistical noise and aberrations resulting from realistic operating conditions. These findings have proven to be useful in the design of new and larger imaging spectrometers. Improved imaging techniques derived from these studies are tested in the laboratory, when feasible, with the use of the WINKLER RMC imaging spectrometer (Fisher et al. 1989).

A photograph of the WINKLER imaging system is shown in Figure 1. This instrument incorporates a rotation modulation collimator system with 9 consecutive space frequencies (n=9) combined with a high resolution γ-ray spectrometer. The collimators, placed in front of each of the 9 Ge detectors, consist of a pair of grid masks mounted as a cylinder and are made to rotate synchronously about their individual axes. The separation distance of the grids define the FOV. Generally two detectors are required for each frequency order to obtain both the even and odd Fourier components. However, WINKLER acquires each component in alternate rotation periods or a series of rotation periods. At the end of each rotation, the upper grids are made to change their phase so that both Fourier components are acquired in two rotations.

Figure 1. The WINKLER (n=9) RMC Imaging High Resolution γ-Ray Spectrometer

Figure 2 shows reconstructions of two point sources, one at 59.6 KeV from ^{241}Am and the second at 356 KeV from ^{133}Ba, observed simultaneously. This image was taken before proper parallax corrections were applied which significantly sharpens the peak and deminishes the side lobes; however, it does serve to illustrate the point of resolving different sources by their spectral features.

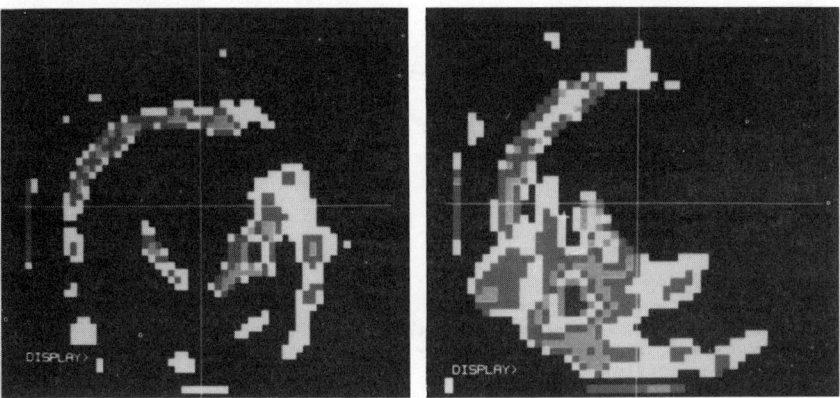

Figure 2. Image displays of two point sources taken simultaneously, at 59.6 KeV on the left and at 356 KeV on the right.

The angular resolution of the RMC system depends of the full FOV angle and the highest order (space frequency) n; the nominal point spread function is given by:

$$\delta\theta = 0.705\ \theta/n \quad \text{(FWHM)}, \qquad \text{where } \theta = \text{FOV angle.}$$

An important result that impacts on the desired spatial resolution and the number of detectors required for a given RMC imager is illustrated in Figure 3. Point source reconstructions are presented for two different size RMC arrays, each having the same number of signal counts (1000 cts.) but the image on the left is derived with an array of order 18 and the image on the right is obtained from an array having four times as many detectors, or with order 72. The image on the right is magnified (zoomed) by a factor of four so that the width of the point source should be the same in each plot. It is concluded that for a constant number of counts, a reconstructed image of any order n, when magnified proportional to n, remains unchanged within statistics, including the rms noise and side-lobes. Hence, the signal to noise ratio is independent of the number of detectors in an instrument with the same total area. Since there is no optimum array size from statistical considerations, we are free to select whatever array size is required to achieve the desired resolution and sensitivity. Instrument complexity will still be a primary constraint in the design of a practical system.

A number of artifacts are produced in the reconstructions of images taken under real or non-idealized conditions. Some of these effects are summarized in Table 1. Different algorithms have been devised that effectively remove artifacts which develop at or about the image center due to rotation axis misalignment and time varying background effects. Aberrations due to differences in detector size or efficiency are easily handled by normalization. Although parallax corrections are generally not an issue in astronomical observations, such corrections have been applied to laboratory test data and have proven to be an important procedure for certain applications involving close range imaging. Various enhancement techniques for image reconstructions such as the maximum likelyhood method and the inclusion of positivity constrains are presently being studied.

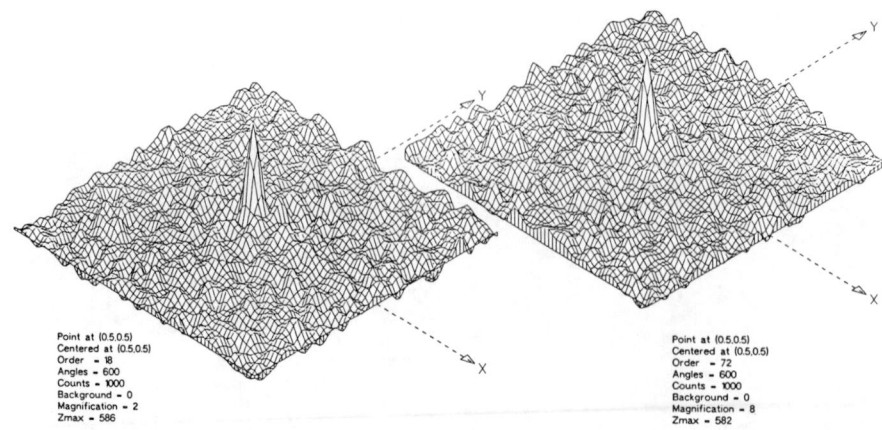

FIGURE 3. Reconstructed images of a single point made with RMC arrays of different sizes but with the same number of detected photons. The plot on the right is for an array with four times as many detectors as for the array on the left and is also magnified by four.

TABLE 1. SUMMARY OF ARTIFACTS PRODUCED IN NON-IDEALIZED IMAGES

Mismatched detectors -- Can be handled by normalization.

Rotation axis misalignment -- Produces central artifact that can be subtracted exactly.

Time varying background -- A "leveling" algorithm has been developed to correct central artifact.

Grid Transparency -- Signal to noise ratio decreases rapidly with incresing transparency.

Parallax -- Correction applied for laboratory test data.

Background spectrum -- Inherent on-source/off-source subtraction technique with RMC.

One detector/order -- $\pi/4$ phase shifted collimators provide both odd and even components of the signal; central artifact can be removed.

Signal to noise vs. number of detectors -- No optimum array size from statistical considerations.

Skipping orders -- A technique allowing high spatial resolution (few arcsec); produces sidelobes around a point source that are effectively removed with the Clean algorithm.

The RMC system provides an inherent on-source/off-source technique to obtain a background subtracted spectrum for a point source. Once the point source is located and is off-center, geometry can be applied to determine the rotation angle at which each collimator best views and best obscures the point source. Data about these angles serve to define the on-source and off-source data. This technique may be exploited in a simple way for the galactic center observations of the 511 KeV annihilation radiation line. Since this line is also a principal background line, the spatial separation of the point source signal and background may be especially useful. A precise knowledge of the background line in this data set should also be an advantage in mapping the diffused 511 KeV component.

At higher photon energies, the partial transparency in the tungsten collimator grids rapidly reduces the modulated signal. The present collimator grids attached to WINKLER are 5 mm thick and are effective below ~500 KeV, but significantly thicker grids would be required to obtain images at higher gamma-ray energies. However, thicker grids produce increasing vignetting effects that places restrictions on the FOV angle, thereby, forcing traded-off considerations. Such a trade can be advantageous for solar observations where the total FOV may be as small as 0.5 deg and still contain the total disk.

A 45 deg phase shift is incorporated in each set of modulation collimators providing both even and odd components of the source signal, making it is possible to use only one set of collimators per frequency order instead of the nominal two collimator-detector sets per order. Simulations of this imaging configuration have produced successful reconstructions for an RMC system having fixed rotation axes as in WINKLER. However, this system has not been tested under various conditions that may possibly complicate matters. Although imaging system using two collimators per order is certainly more conservative in this regard, the simplification of using one set of collimators per order offers significant advantages. Further studies will be required before such a system can be implemented.

Order skipping (e.g., using every 4th or 8th order) is an important technique that offers a method of attaining high spatial resolution in practical systems with a limited number of detectors. Of course, the alternative, is to use a large number of collimator-detector sets that would provide both increased resolution and sensitivity. Future imaging spectrometers will benefit by incorporating both options to achieve an optimum design. Figure 4A and its corresponding contour plot in 4B show a point source reconstruction with n=9 and 10,000 counts. Figure 4C and 4D show the image of the same scene with n=36, using every fourth order and a total of 9 detectors. Note the presence of four sidelobe rings which are a consequence of order skipping with the number of rings equal to the multiple of skipped orders. These sidelobes are readily removed to noise level by the application of the CLEAN algorithm or other standard enhancement techniques. This lower image also corresponds to that for n=72 magnified by two; similarly, the upper reconstruction corresponds to an image with n=72 magnified by eight. With the collimator grid diameter of 5.5 cm and a separation distance of 5 meters, the FOV is sufficient to include the entire solar disk. Spatial resolution of 10-20 arcsec can be achieved.

HIGH RESOLUTION IMAGING SPECTROMETERS FOR THE 21ST CENTURY

Table 2 lists some instrument parameters of future RMC imaging high resolution spectrometers, assuming the availability of 8 cm diameter x 8 cm multi-segmented Ge sensors in the next few years. This assumption is not unreasonable since some of the large n-type Ge detectors presently available range up to 7 cm. These parameters are based on simple extentions in the array size of a proven instrument and thus, represents the current state of the art. Array sizes of 18, 36, and 72 detectors with frontal areas of 905, 1810, and 3620 cm^2, respectively, are considered. The angular resolutions are given for a nominal FOV of 15° and 0.5° in terms of RMC orders equal to either the total number or one-half of the number of detectors in each system. Order skipping by multiples of 4 and 8 is employed only in the case of the 0.5° FOV as indicated. In addition, the estimated efficiency in the non-imaging mode is compared to that of the HEAO-3 instrument and a rough estimate of the 3σ narrow line sensitivity for an observing time of 10^6 second is given for each system. The latter estimates include advantages obtained by the use of segmented detectors (Varnell et al. 1990) and an effective antishield. Obviously, any major advances in imaging techniques or breakthroughs in detector technology in the intervening years would be implemented where possible to enhance these observational parameters.

It is seen that even the rather modest 18 detector system will provide sensitivities of about 1x10^{-6} photons/cm^2 s for the non-imaging case and angular resolutions of 1.2° or 0.6°. This instrument already satisfies the specifications for future γ-ray spectrometer recommended recently by the Gamma-Ray Program Working

210 High Resolution Gamma-Ray Imaging and Spectroscopy

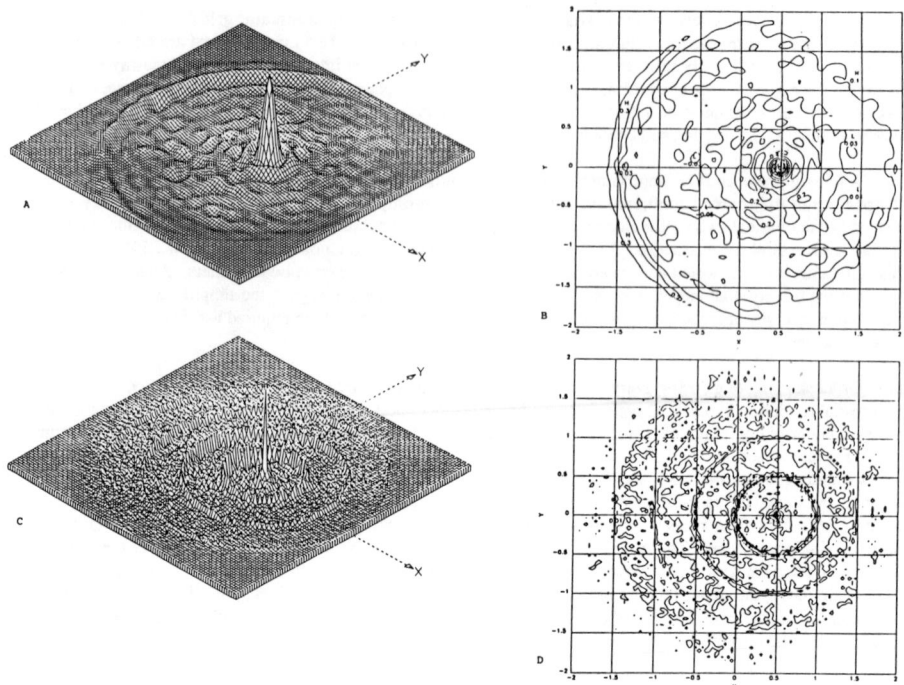

FIGURE 4. Single point response of a nine detector RMC with order skipping using every 8th order (n_{max}=72). Magnification: x8 (top) and x2 (bottom).

TABLE 2. PARAMETERS OF FUTURE RMC HIGH RESOLUTION SPECTROMETERS

No. Det.	Area (cm^2)	Eff. factor* (N.I.)	3J Sens. (ph/cm^{2}s)	RMC Order (n)	ANG. RES. 15° FOV	0.5° FOV	Order Skipping	n_{max}
18	905	50	1 x 10^{-6}	9	1.2°	35"	4	36
				18	35'	18"	8	72
36	1810	100	7 x 10^{-7}	18	35'	18"	4	72
				36	17'	9"	8	144
72	3620	200	5 x 10^{-7}	36	17'			
				72	9'			

* Compared to HEAD-3, sensitivity 2×10^{-4} ph/cm^{2}s ($\Delta T = 10^6$s)

Group. The larger systems just scale by factors of two. Since the maximum throughput in the imaging mode is 25% due to the two layer grid system, our design of spaceborne instruments have generally included mechanisms that deploy the RMC collimators in and out of the FOV, reflecting a philosophy of maximizing the sensitivity. It should be noted, however, that the sensitivity is not reduced by a factor of two but something closer to $\sqrt{2}$ since the RMC system can reduce the observing time to about 1/2 that required for on-source/off-source measurements. Thus, the trade-off between higher sensitivity and added mechanical risk is not clear. There is also a limit to the maximum effective frequency order that can be employed, particularly in the case of order skipping where edge effects will ultimately constrict the FOV. The fabrication of finer and finer tungsten grids may also a limiting factor.

CONCLUSIONS

The RMC imaging system described here does not require position sensitive detectors; thus, it is well suited to be coupled with high resolution Ge detectors which together constitute a very effective instrument for astrophysical observations. In its present form, the array of Ge detectors is mounted in a common cryostat in a close-packed configuration providing a natural sensitivity as a Compton scattering γ-ray polarimeter which is most sensitive in the 100 KeV to 1 MeV range. Polarization measurements can reveal important information relating to emission mechanisms and structures in the accreting disk of collapsed objects and of the magnetoshere of young radio pulsars. The combined observational features of the high resolution RMC spectrometer should indeed satisfy many of the important measurements that have been identified as future imperatives by a number of science advisory committees. A similar but smaller version of such an 18 detector RMC system (Nakano et al. 1989) was proposed but not selected for the Space Station Attached Payloads Program. However, if the observational requirements in astrophysics are to be served, a larger fully optimized version of this type of instrument should be considered in the near future as a national facility for a high resolution γ-ray observatory on the Space Station.

This paper was supported in part by the Lockheed Independent Research Program. We also wish to thank all the scientists and engineers that contributed to the successful development of the WINKLER RMC imaging system.

REFERENCES

Chan, R. W., Lingenfelter, R. E. 1989, Nuclear Spectroscopy of Astrophysical Sources (AIP Conf Proc 170), ed. N. Gehrels and G. H. Share (NY:AIP).
Fisher, T. R., Hamilton, Hawley, J. D., Kilner, J. R., Murphy, M. R., Nakano, G.H., 1989 SPIE Vol. 1159 EUV, X-Ray, and Gamma-Ray Instrumentation for Astronomy and Atomic Physics.
Gamma Ray Program Working Group Report, Gamma Ray Astrophysics to the Year 2000, 1988.
Mahoney, W. A., Varnell,L. S., Jacobson, A. S., Ling, J. C., Radocinski, R. G., and Wheaton, W. A., 1988, Ap. J. 334, L81.
Mertz, L. N., Nakano, G. H., and Kilner, J. R., 1986, J. Opt. Soc. Am. A3, 2167.
Nakano, G. H., Chase, L. C., Kilner, J. R., Sandie, W. G., Fishman. G. J., Paciesas, W. S., Lingenfelter, R. E., Woosley, S. E., G.H., 1989 SPIE Vol. 1159 EUV X-Ray, and Gamma-Ray Instrumentation for Astronomy and Atomic Physics.
Pinto, P. and Woosley, S. E. 1988, Ap. J., 329, 820.
Rester, A. C., Coldwell, R. L., Dunnam, F. E., Eichorn, G., Tromka, J. I., Starr, R., and Lasche, G. P. 1988, Ap. J. 342, L71.
Sandie, W. G., Nakano, G. H., Chase, L. F., Fishman, G. J., Meegan, C. A., Wilson, R. B., Paciesas, W. S., Lasche, G. P. 1988, Ap. J. 334, L91.
Shigeyama, T., Nomoto, K., Hashimoto, M., Sugimoto, D. 1987, Nature 328, 320.
Teegarden, B. J., Barthelmy, S. D., Gehrels, N., Tueller, J., Leventhal, M., and MacCallum, C. J. 1989, Nature 339, 122.
Varnell, L. S. et al. 1990, This Workshop.

DISCUSSION

Gerald Share

(1) Have you performed a simulation using a realistic detector background and, say, a typical strong source such as the The Crab or perhaps the reported galactic center point source? (2) Have you considered using an active mask to reduce background?

Nakano

(1) No, not specifically, but it should be done and perhaps we will do it soon. We don't expect a serious problem for very strong sources, i.e., the Crab and the galactic center point source when it's on. (2) We have thought about systems amenable for implementing active masks, but our present RMC system is not one of them.

DIRECTIONS IN GAMMA-RAY SPECTROSCOPY

Neil Gehrels and Robert M. Candey[1]

NASA - Goddard Space Flight Center
Greenbelt, MD 20771

I. INTRODUCTION

In this paper, we will describe current and future instrumentation for gamma-ray spectroscopy in the spectral range from 10 keV to 10 MeV. We will emphasize new technologies for Germanium (Ge) spectrometers and emerging detector technologies. Sodium Iodide (NaI) detectors with their moderate energy resolution, typically 50 keV at 1 MeV, have been the work horse of astrophysical gamma-ray spectroscopy since the 1960's. The NaI instruments on the Gamma-Ray Observatory (GRO) will be the culmination of this technology. Ge detectors, with their high resolving power, typically 2 keV at 1 MeV, will undoubtedly be the technology of choice for future gamma-ray spectroscopy missions. Some of the emerging detector technologies have tremendous potential, but face significant technical hurdles before they replace Ge for spectroscopy detectors.

The paper begins with a brief discussion in Section II of science objectives, with an emphasis on capabilities beyond those of the Nuclear Astrophysics Explorer (NAE). A list of instrument requirements is given. Section III presents technologies under development for an NAE-era spectrometer. Section IV discusses Ge spectrometers beyond the NAE, and Section V presents other types of future detector technologies. Table 1 gives a partial list of current and future spectrometers, to orient the reader.

Table 1
Gamma-Ray Spectrometers

Quasi-Steady-State Sources				Gamma-Ray Burst Sources		
Instrument	Launch	Detector/Shield	Sensitivity* (ph/cm^2-s)	Instrument	Launch	Detector
HEAO-3	1979	Ge / Cs I	3×10^{-4}	GRO/BATSE	1990	NaI
Balloon Instruments	1988	Ge / NaI Ge / BGO	3×10^{-4}	WIND/TGRS	1992	Ge
GRO/OSSE and COMPTEL	1990	Na I / Cs I	3×10^{-5}	TTTS	?	Ge
NAE	1999?	Ge / BGO	3×10^{-6}			
Lunar Instrument	?	Ge / BGO / dirt	2×10^{-7}			

*Narrow line sensitivity (3σ) at 1 MeV in 10^6 s

[1] Atlantic Research Corporation, 8201 Corporate Drive, Suite 350, Landover, MD 20785

II. SCIENTIFIC OBJECTIVES AND INSTRUMENT REQUIREMENTS

Gamma-ray lines are tracers of unique and fundamental physics which cannot be studied in any other wavelength band. The production mechanisms are radioactive decay, nuclear excitation, electron-positron annihilation with its 511 keV line, and cyclotron radiation in 10^{12} Gauss fields. Spectroscopy can be used in the same way it is in the optical band, to study elemental abundances, temperatures, and motion of the emitting material.

The excitement and challenge of gamma-ray spectroscopy is illustrated in Figure 1 by the Ge balloon observation of Supernova (SN) 1987A made by the balloon-borne Gamma-Ray Imaging Spectrometer (GRIS) (Tueller et al. 1988). The flux of the 847 keV line uniquely identifies the emitting nucleus as ^{56}Co; the width and centroid of the line give information about the structure of the ejecta. The disagreement between this measured line and the predicted (10 HMM model of Pinto and Woosley (1988)) line shape is one of the key pieces of evidence that have led people to suggest clumping in the structure of the ejecta at early times.

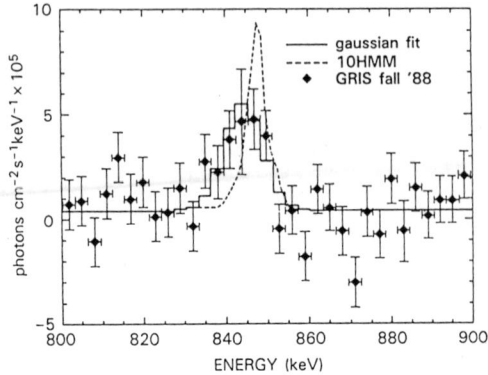

Figure 1. Supernova 1987A 847 keV line.

The challenge of the GRIS observation is that SN 1987A was the closest supernova observed in 400 years. What kind of instrument is required to study the more distant and more common events? For Type I supernovae, the Nuclear Astrophysics Explorer (NAE) (Matteson et al. 1988, Matteson 1990) will be able to observe events out to the Virgo Cluster. The gamma-ray lines from Type II events, like SN 1987A, are much more difficult to detect due to the massive ejecta that covers the radioactive material in the center of the star. Figure 2 shows the flux curve measured from SN 1987A, with the flux at 3 Mpc shown on the right axis. One would need a sensitivity of 2×10^{-7} ph/cm^2-s, an order of magnitude better than the NAE, to be able to detect this line out to 3 Mpc. The reason 3 Mpc was chosen is shown in Table 2, which lists numbers of galaxies in the three nearest groups. With five large galaxies and eight medium galaxies closer than 3 Mpc, one would expect a Type II supernova every ~5 years (Tammann 1982), or two events within a 10 year mission. So it would take an order of magnitude better sensitivity than that of the NAE to see another Type II supernova.

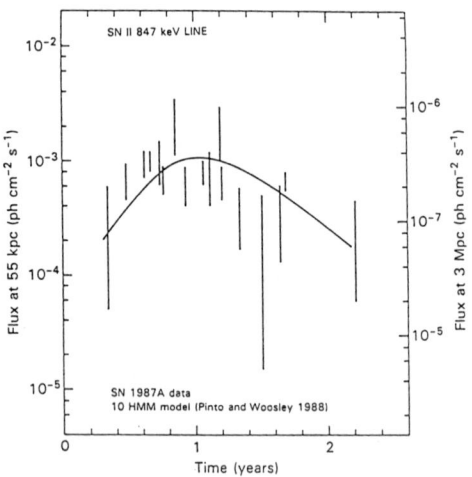

Figure 2. Supernova 1987A 847 keV line.

Figure 3 shows the predicted emission (Ramaty, Kozlovsky, and Lingenfelter 1979) from the interstellar medium being excited by cosmic ray protons. The line intensity predicts have recently been modified by Higdon (1987). The narrow lines are due to cosmic ray excitation of dust and the broader lines are due to excitation of gas. This is a very weak emission; the NAE, with its 10° field-of-view (FOV), will be able to detect the ^{16}O

Table 2

Group	Number of Large Galaxies	Number of Medium Galaxies	Distance (Mpc)
Local	2	2	< 1
Sculptor	1	4	~ 3
M81	2	2	~ 3
	5	8	

line from the interstellar medium, which will be an important measurement. To really study the spectrum, one again needs a factor of 10 or more sensitivity to see 5 or 10 of these lines and to start seeing the broad and narrow components.

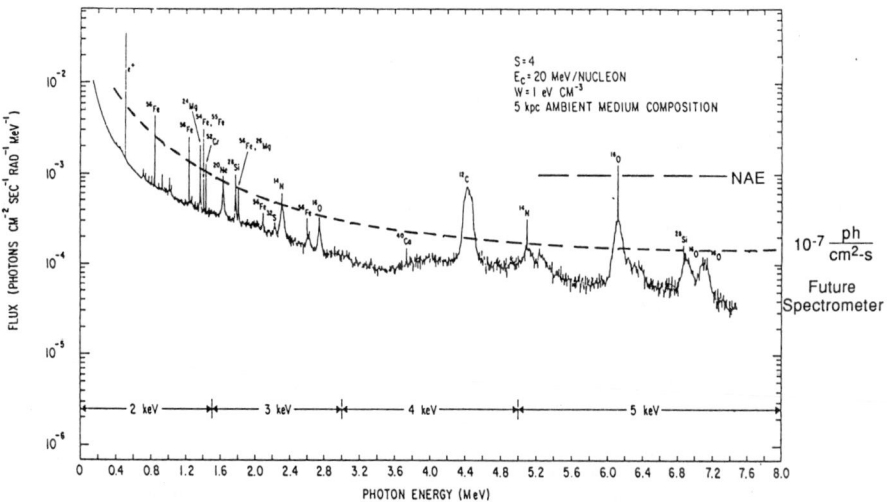

Figure 3. Cosmic rays interacting with interstellar gas and dust (Ramaty, Kozlovsky, and Lingenfelter 1979).

Table 3 lists the gamma-ray spectrometer requirements for quasi-steady-state and transient source instruments. A few sample science objectives are listed for different sensitivity levels for the steady-state spectrometers. For the transient spectrometers, we need high resolution to resolve the narrow cyclotron lines (see, e.g., Fenimore 1988), enough sensitivity to study the evolution of the spectra during the bursts, and long-term missions.

Table 3
Gamma-Ray Spectrometer Requirements

(Quasi) Steady-State Sources	
Energy Range	10 keV – 10 MeV
Energy Resolution	~ 1 keV ($\Delta E/E \sim 1000$) or better
Angular Resolution	~ 1° (post GRO) to resolve Galactic structures
Sensitivity (ph cm^{-2} s^{-1})	~ 10^{-5} detect Type I supernovæ study Galactic plane
	~ 10^{-6} study Type I supernovæ to Virgo ^{44}Ti – historic Galactic supernovæ detect interstellar medium interactions detect novæ
	~ 10^{-7} detect Type II supernovæ study novæ study interstellar medium interactions
Transient Sources	
Energy Resolution	~ 1 keV ($\Delta E/E \sim 1000$)
Number of Bursts with Line Detections	10 per year discover lines 100 per year compact-object studies
Wide Field-of-View	
Time-Resolved Spectroscopy	

III. GERMANIUM DETECTOR TECHNOLOGY

How do you make a 10^{-6} ph/cm^2-s sensitivity multi-year spaceflight gamma-ray spectrometer? Not by simply scaling-up the HEAO-3. The HEAO-3 had four Ge detectors (400 cm^3 total volume) and a sensitivity of 10^{-4} ph/cm^2-s (Mahoney *et al.* 1980). Since these are background-dominated instruments, the sensitivity scales like the square root of the background and the square root of the detector volume. Just scaling-up the HEAO-3 would require 10^4 times more Ge, which is not feasible. Achieving this kind of sensitivity level requires a number of technological advances that have been made over the past decade and are continuing to be made. These are listed in Table 4 and briefly discussed below.

Table 4
New Technologies for Ge Detectors

Large n-type Ge detectors	– High efficiency – Resistant to radiation damage
BGO shields	– Reduce shield leakage
Segmented detectors	– Reduce beta activation – Eliminate neutron scattering
Ge isotope enrichment	– Reduce beta activation – Reduce line background
Beryllium cryostats	– Reduce Compton scattering – Reduce Uranium background – Reduce 511 keV line background
Mechanical coolers	– Long mission life
Coded masks	– Clean signal modulation

Since the purpose of many of these technologies is to decrease instrumental background, we first show in Figure 4 the background components for the GRIS balloon instrument measured during a flight over Australia in 1988. The measured points are displayed as triangles and the calculations are shown by the lines. The calculations were carefully done as the instrument was being designed and were published prior to the flight (Gehrels 1985). The agreement between the sum of the calculations (solid line) and the measured data is quite good over the entire energy range, indicating a good understanding of the background components. The known background components are the flux coming in through the aperture of the instrument, elastic neutron scattering of the Ge nuclei, activation of the Ge nuclei themselves, and leakage of gamma rays from the atmosphere or the spacecraft coming through the shield (15 cm of NaI).

1. Large n-type Ge Detectors

There has been a tremendous improvement in quality and size of available Ge detectors since the 5.5 cm diameter crystals of the HEAO-3. We have already flown 7 cm diameter Ge detectors in our balloon instruments. They are n-type Ge detectors which are 50 times (Pehl *et al.* 1979) more resistant to radiation damage than the p-type detectors on HEAO-3.

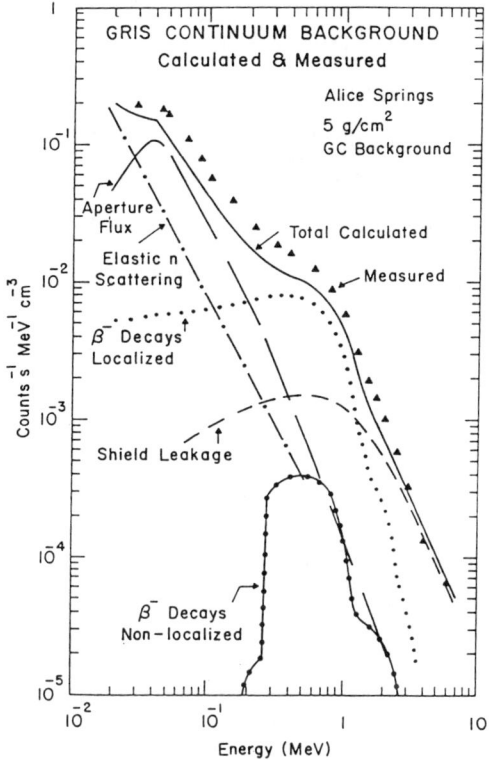

Figure 4. GRIS calculated and measured continuum background.

2. BGO Shields

Another technology advance over the HEAO-3 that is well in hand is the use of Bismuth-Germanate (BGO) for shielding material surrounding the Ge detectors. BGO is much denser (ρ = 7.1 g/cm^3 BGO vs. 3.7 g/cm^3 NaI) and has a higher Z than NaI, so one can make a thicker effective shield for less weight due to its more compact geometry. The shield for the UC-France balloon instrument is a BGO shield and has been successfully flown.

3. Segmented Detectors

Segmented detectors, pulse-shape discrimination, and Ge isotopic enrichment are used to reduce beta decay and elastic neutron scattering background. A segmented detector has its outer contact divided into strips and electronically monitored. The high resolution signal is still taken from the inside contact as with non-segmented detectors. The reason for doing this is to distinguish between good gamma-ray events and background. For a good event in the MeV range, a gamma-ray coming in scatters in the detector (Compton scattering) and deposits ionization in two or more sites. The beta decay background which is so important in the MeV energy range (Figure 4) is due predominantly to the single site localized beta decays. The beta decay produces a neutrino and an electron; the neutrino escapes and the electron stops within about a millimeter, within one segment. Keeping only multiple segment events rejects the background events. The beta decay background is reduced by a factor of more than 10. The first balloon test flight is planned for 1990.

4. Germanium Isotope Enrichment

Ge isotope enrichment is a idea that has only been around in the gamma-ray spectroscopy community since mid-1989 (Gehrels 1989). Natural Ge has four different isotopes, two fairly abundant ones and two of lesser abundance (^{72}Ge – 21%, ^{73}Ge – 8%, ^{74}Ge – 37%, ^{76}Ge – 8%). The background we cannot eliminate with segmentation is primarily beta decays of ^{74}Ge. It recently has become possible to isotopically separate Ge by gas centrifugation; the Soviets are doing this now and the U.S. plans to start at Oak Ridge National Laboratory this year. Ge is turned into a gas, the isotopes are separated by gas centrifugation, and selected isotopes are turned back into a metal and made into detectors. We plan a balloon test flight of an isotopically-enriched Ge detector in 1991.

5. Mechanical Coolers

HEAO-3 had a solid cryogen cooler for cooling the Ge detectors that lasted 9 months. A significant amount of science was done in those 9 months but future space missions require longer observing times. Recent mechanical coolers, in particular the British Aerospace cooler, have multi-year lifetimes. Over the next two years, the British Aerospace cooler will be tested with large Ge detectors. It is the cooling mechanism of choice for detectors like the NAE.

The NAE will use all of the above technologies to achieve a sensitivity of ~ 3×10^{-6} ph/cm^2-s, as shown in Figure 5. The mission is described by Matteson (1990) in this volume.

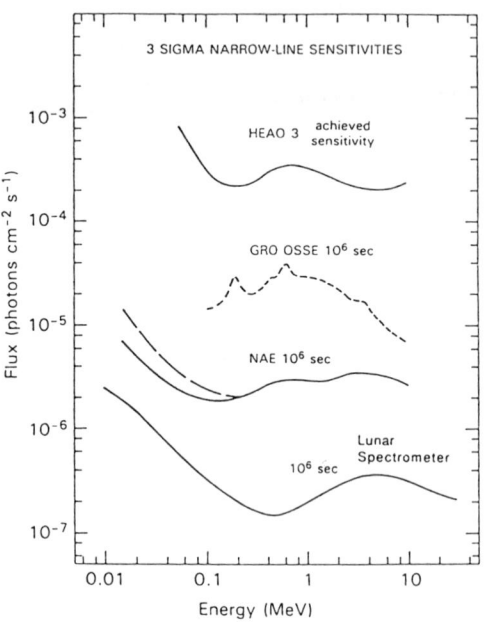

Figure 5. Narrow line sensitivities (3σ).

IV. BEYOND THE NAE

Moving beyond the NAE requires an instrument of a few 10^{-7} ph/cm^2-s. Again, it is difficult to simply scale up the previous mission. To get an order of magnitude better sensitivity than the NAE, we need 100 times more Ge (= 650 future 8 cm detectors) resulting in a very heavy spacecraft and a complex detector array. A better way would be to increase the Ge volume (say 5 times) and decrease the background (say 20 times) by doubling the shield and decreasing the field-of-view (2°). It is the obvious thing to consider for the next step, but this kind of background decrease, although it looks good on paper, will probably not work. There are other components of the background that are below the NAE background but would certainly dominate when one goes a factor of 20 lower. Probable sources for this background are background lines, Compton continuum from the lines, ß+ decays, and other things we have not even thought of at this point. We must decrease the primary radiation environment at the detectors, i.e., reduce the number of cosmic rays and neutrons that are going through the detectors. A reasonable-sized BGO shield does not stop those; ~ 500 g/cm^2 of BGO shielding, or something like the Earth's atmosphere, is required to prevent cosmic rays from getting to the detectors. That kind of shielding around an NAE-size Ge array would weigh ~ 28 000 kg.

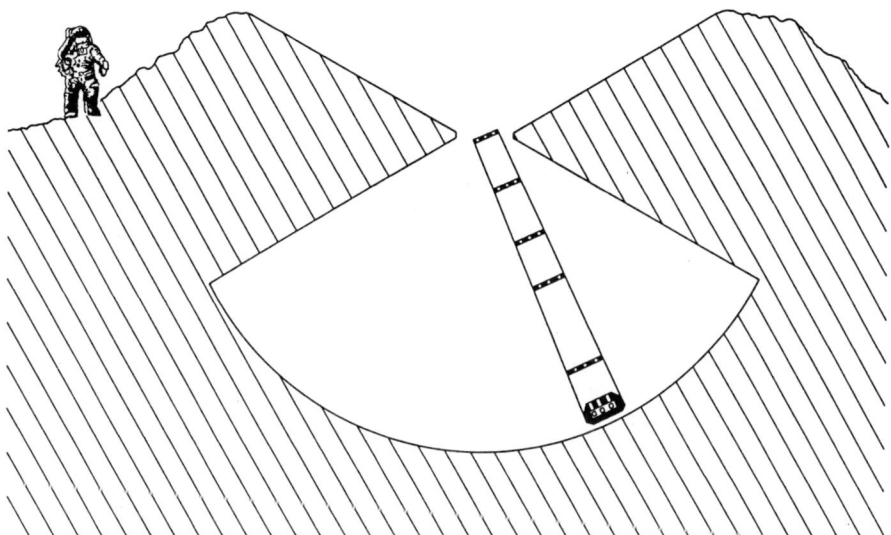

Figure 6. Lunar-Based Gamma-ray Spectrometer (LBGS).

One alternative is to use the lunar soil for shielding. We have performed a first-order engineering concept study at Goddard defining this instrument and calculating its background and sensitivity; the study was done in collaboration with J. Matteson. Figure 6 shows what this kind of instrument might look like. We assumed a 19-element detector array (vs. 9 for NAE), somewhat larger detectors (8 cm diameter vs. 6.8 cm), and a thicker shield (15 cm BGO vs. 10 cm). The primary shielding would be ~ 2 m of lunar regolith on top of the cavern. The detectors would move around in this azimuthally-symmetric instrument and point to most positions in the sky. The fraction of the sky observable at any one time is only 25%, but 87% is visible (all but the ecliptic poles) over a lunar day, assuming the instrument is located near the equator. The aperture holes define a 1° FOV. Including the fact that the cosmic ray fluxes are higher on the Moon (no magnetosphere) than in Earth orbit, and integrating the neutron and proton fluxes coming through the roof of the cavern, we find that the total particle environment at the detectors would be down by more than an order of magnitude compared to the NAE. With that background decrease, we calculate an order of magnitude sensitivity improvement over the NAE for a 10^6 second observation, as shown in Figure 5.

V. GAMMA-RAY TRANSIENT SPECTROMETERS

There are a very limited number of high resolution observations that have been made of gamma-ray bursts. The next high-resolution space mission will be the Transient Gamma-ray Spectrometer (TGRS) to be flown on the WIND spacecraft in 1992. It will use a radiative cooler which views dark space to keep the detector cold. Mars Observer will also have a radiatively-cooled Ge detector. Future instruments will require Ge arrays for high sensitivity. One example is the Total Throughput Transient Spectrometer (TTTS) proposed for the space station that is described by Hurley (1990) in this volume.

VI. EMERGING DETECTOR TECHNOLOGIES

1. Argon and Xenon

The most promising emerging detector technologies at this point are noble gas and liquid detectors, including high pressure Xenon (Xe) detectors, liquid Argon (Ar) and liquid Xe detectors. A large number of groups are working on these technologies. High pressure Xe detectors are being used for hard X-ray observations, but liquid Xe detectors are best for gamma-ray range spectroscopy. Liquid Xe detectors have many desirable properties: a high detection efficiency, a significantly higher Z than Ge, and a good density (3 g/cm^3). Large unit sizes (1000 to 2000 cm^2) and three-dimensional spatial resolution to better than a millimeter are possible. The ultimate energy resolution could eventually be within a factor of two of Ge. The Fano factor is low for liquid Xe, allowing high resolution capability in principle. The drawback to liquid Xe at this point is that the ultimate energy resolution is probably not achievable. The best resolution currently achieved is 6% and the goal for the next few years is 3% (30 keV at 1 MeV). This is likely due to fundamental problems in charge collection that may not have solutions. Aprile, Mukherjee, and Suzuki (1989) at Columbia have proposed a liquid Xe instrument using a time projection chamber (1600 cm^2) in which the ionization is collected at the grids, giving the positions of the interactions through the chamber. It is a very interesting technology, but from a spectroscopic point of view, liquid Xe will not replace Ge in the near future.

2. High-Z Solid-State Detectors

Two high-Z solid-state detectors are Mercuric Iodide (HgI$_2$) and Cadmium Telluride (CdTe). We recently did a detailed review of these technologies (Gehrels et al. 1988) and found them not to be very promising at this point. A list of characteristics is given in Table 5. The nice feature about these detectors is that they can operate at room temperature. HgI$_2$, with its high Z, is the most interesting of these two. Ultimately HgI$_2$ could have energy resolutions similar to Ge but again, as with liquid Xe, the resolution is probably not achievable. The actual measured resolutions are in the 10's of keV at 1 MeV (Table 5). The detectors are only available currently in small sizes.

3. Phonon Detectors

Phonon detectors offer the potential of extremely high energy resolutions, since the energy required to create phonons in the detector is milli-eV instead of eV for solid-state detectors. However, this technology is in its infancy and no practical gamma-ray detectors are conceived of yet. Phonon detectors are planned for X-ray spectrometers (Kelley et al. 1988), but are limited to very small sizes (millimeters). Possible use of detectors that collect ballistic phonons before they thermalize (Sadoulet et al. 1989, Cabrera 1989) may eventually allow larger cm-size detectors to be made.

Table 5
Detector Comparisons

Parameter	CdTe	HgI$_2$	Ge	LXe
Operation T (K)	300	300	77	195 (5 atm)
Density (g/cm^3)	6.1	6.4	5.4	3.0
Highest Z	52 (Te)	80 (Hg)	32 (Ge)	54 (Xe)
Energy resolution at 1 MeV				
Ultimate (keV)	1.5	1.4	1.0	2.0
Achieved	~40	~30	2.0	60*

*Actual measurement is 34 keV at 569 keV = 6%

REFERENCES

Aprile, E., Mukherjee, R., and Suzuki, M. 1989, in *Proc. of the SPIE 33rd Ann. Int. Symp. on Optical and Optoelectronic Applied Science and Engineering*, submitted.

Cabrera, B. 1989, private communication.

Fenimore, E. E. *et al.* 1988, *Ap. J.*, **335**, L71.

Gehrels, N. 1985, *Nucl. Instr. and Meth.*, **A239**, 324.

Gehrels, N., Crannell, C. J., Forrest, D. J., Lin, R. P., Orwig, L. E., and Starr, R. 1988, *Solar Physics*, **118**, 233.

Gehrels, N. 1989, *Nucl. Instr. Meth.*, submitted.

Higdon, J. C. 1987, *20th Int. Cosmic Ray Conf.*, **1**, 160.

Hurley, K. 1990, this volume.

Kelley, R. L., *et al.* 1988, in *X-Ray Instrumentation in Astronomy II* (SPIE Conf. Proc. 982), ed. L. Golub (Bellington, WA: SPIE), 982-25, p. 210.

Mahoney, W. A. *et al.* 1980, *Nucl. Instr. Meth.*, **178**, 363.

Matteson, J. L., Teegarden, B. J., and Mahoney, W. A. 1988, in *Nuclear Spectroscopy of Astrophysical Sources* (AIP Conf. Proc. 170), ed. N. Gehrels and G. H. Share (New York: AIP), p. 417.

Matteson, T. L. 1990, this volume.

Pehl, R. H., Madden, N. W., Elliott, J. H., Raudorf, T. W., Trammell, R. C., and Darken, L. S., Jr. 1979, *IEEE Trans. Nucl. Sci.*, **NS-26**, No. 1, 321.

Pinto, P., and Woosley, S. E. 1988, *Ap. J.*, **329**, 820.

Ramaty, R., Kozlovsky, B., and Lingenfelter, R. E. 1979, *Ap. J. Supp.*, **40**, 487.

Sadoulet, B., Cabrera, B., Maris, H. J., and Wolfe, J. P. 1989, in *Proc. of the 3rd Int. Conf. on Phonon Physics and 6th Int. Conf. on Phonon Scattering in Condensed Matter*, submitted.

Tammann, G. A. 1982, in *Supernovæ: A Survey of Current Research*, ed. M. J. Rees and R. J. Stoneham (Boston: Reidel), p. 371.

Tueller, J. *et al.* 1988, in *Nuclear Spectroscopy of Astrophysical Sources* (AIP Conf. Proc. 170), ed. N. Gehrels and G. H. Share (New York: AIP), p. 439.

DISCUSSION

William Mahoney

With respect to isotopically enriched germanium detectors, aside from the cost, I am concerned that while some background lines could be reduced, others would be increased in intensity. For example, a detector made from Ge^{70} would result in the production of more Ga^{69}, which has a number of background lines above 1 MeV.

Gehrels

Your comment is correct. Although the total number of Ge background lines will generally be reduced in an enriched detector, particular lines and families of lines will increase in intensity. This effect is calculated in a paper by Gehrels recently submitted to NIM. The principal benefit of enriched detectors is the reduction of the continuum background. The effect on background lines is secondary.

George Ricker

CdTe and HgI_2 are not being as extensively developed by semiconductor producers as Ge and Si. Therefore, there is less of a "coattails effect" ($\sim$ 10^8 – 10^9 have been spent by the industry on Ge and Si). It seems very unlikely that even the combined efforts of NASA, ESA, ISAS, and IKI will provide materials development money within even orders-of-magnitude of these numbers by the early 21st century. Hence, it is hard to be optimistic to expect that *theoretical resolutions* Gehrels cited will be achieved in practice, unlike what has happened with So and Ge.

James Kurfess

I question the optimistic background reduction techniques which have been presented. In the cosmic ray environment, 0.01 – 0.02 spallation products, many radioactive, will be produced in each gram of material. It may be difficult to reduce this component by a large factor. Can HEAO-3 data set limits on this component of the background?

Gehrels

All spallation products have been included in the calculations described in my talk. The total β^- decay background component shown in the figures is actually the sum of $\sim$ 60 beta decay spectra. The background reductions presented are based on these calculations. The HEAO-3 background was calculated by the same technique and came out very close ($\sim$ 20%) to the measured spectrum.

Richard Kelly

I believe that P. Luke at LBL has already demonstrated the principle of measuring the energy of the e-h pairs as heat.

IMAGING GERMANIUM TELESCOPE ARRAY FOR GAMMA-RAYS
(IGETAGRAY)

Charles J. Hailey and Klaus P. Ziock

Laboratory for Experimental Astrophysics,
Lawerence Livermore National Laboratory

Fiona A. Harrison

Department of Physics
and Space Sciences Laboratory,
University of California at Berkeley

Judith Fleischmann

Department of Astronomy,
University of Massachusetts

ABSTRACT

The Germanium Drift Chamber (GDC) is a gamma-ray detector with excellent energy and one-dimensional spatial resolution. Due to recent developments in coded aperture optics, it is feasible to couple one-dimensional coded apertures and GDCs in a special array geometry producing a telescope with true two-dimensional imaging. This Imaging Germanium Telescope Array for Gamma-rays (IGETAGRAY) has made a comparable field of view and sensitivity to true two-dimensional systems, but simplified engineering requirements. IGETAGRAY will make possible high sensitivity spectroscopy of the gamma-ray sky.

I. INTRODUCTION

A major challenge for soft gamma-ray astronomy in the 21st century will be to produce imaging spectrometers with high sensitivity and very good energy and angular resolutions. The current generation of non-imaging germanium spectrometers has difficulty even localizing point source emission, much less mapping the extent of spatial emission in regions like the galactic center or clusters of galaxies.

II. THE DETECTOR CONCEPT

Several years ago a new class of detector called the semiconductor drift chamber (SDC) was developed (Gatti and Rehak 1985). The SDC consists of a silicon $p^+n^-p^+$ sandwich which is fully depleted by an electric field. When an incident photon produces a hole-electron cloud, the holes are collected on the top and bottom p^+ electrodes. A second electric field transverse to the depleting field, produced by a metal gate structure, drifts the electrons to an anode at the edge of the SDC. Because the electron drift time to the anode is a function of the original interaction point-anode distance, the SDC is position sensitive in one dimension. Initial results on the SDC with gamma-rays were encouraging (Rehak *et al.*, 1985). Shortly after the introduction of the SDC, a germanium drift chamber (GDC) was constructed (Luke, Madden and Goulding 1985). Germanium has better stopping power for gamma-rays. The GDC has the added innovation of generating its electron drift field via an intrinsic doping gradient in the n-type material, which was phosphorous doped, rather than through more complex metal electrodes. A position resolution of better than 500 μm is obtained with a prototype detector of about 9 cm^2 active area. Energy resolution is about 1 keV.

Despite its promise, applications of the GDC to soft gamma-ray astronomy have not been pursued. In part, this is due to its intrinsically one-dimensional nature. While two-dimensional versions of the GDC can be envisioned, they are more complicated and expensive and no research has been done on them to date.

III. ROTATED ONE-DIMENSIONAL CODED APERTURE TELESCOPES

The GDC's prospects have been improved by a recent development in coded aperture imaging. This development allows an array of one-dimensional detectors to be used for two-dimensional imaging (Harrison et al., 1989). Each detector in the array is matched to its own one-dimensional coded aperture (OCA) to form a telescope module. The total instrument, which we call the imaging germanium telescope array for gamma-rays, IGETAGRAY, consists of the array of co-aligned modules. The imaging direction of each module is rotated with respect to that of all of its neighbors, allowing a two dimensional image to be reconstructed from the array of one-dimensional images collected by the instrument. IGETAGRAY, represents a significant improvement over conventional coded aperture/detector schemes.

With a spatial resolution better than 500 μm, the GDC can obtain an angular resolution of ≤ 30 arc-seconds with modest focal lengths. It is not necessary to develop more costly two-dimensional variants of the GDC given the imaging scheme described above.

As explained in Harrison et al., (1989) compared with an ideal two-dimensional system, the rotated OCA offers superior off-axis collimator response. Moreover, because the ratio of mask pixel size to detector spatial resolution does not change with energy, high sensitivity can be maintained over a broad energy bandwidth.

High angular resolution gamma-ray telescopes require masks and collimators with small pixels and very high aspect ratios; two-dimensional versions are difficult to fabricate. Rotated OCA's are significantly simpler to fabricate than a two-dimensional system. The use of a redundant detector array ensures reliability because the loss of a few GDC in a large array does not affect the ability to accurately reconstruct an image. Finally, the use of the simple one-dimensional GDC provides a reduction in cost per unit area over a two-dimensional position sensitive germanium detector, so telescopes can be scaled to large areas.

IV. RESULTS OF IGETAGRAY TESTS

We are in the early stages of evaluating IGETAGRAY. Using a GDC obtained from Lawrence Berkeley Laboratory the spatial dependence of detection efficiency, energy and spatial resolutions and position linearity are being measured.

In our first tests, the 59.5 keV line from a ^{241}Am point source was utilized. The "shadow" pattern of an OCA on the GDC is shown in Fig. 1a. The 0's and 1's mark the position of opaque and transparent elements in the Pb mask, respectively. The pixel size is 2 mm. The reconstructed image of the source is shown in Fig. 1b.

V. FUTURE DIRECTIONS AND CONCLUSION

Future tests will utilize arrays of rotated OCA and GDC to evaluate on- and off-axis performance for multiple point sources and extended sources. Extensive simulations demonstrate the validity of the rotated OCA/detector array concept and it is currently being

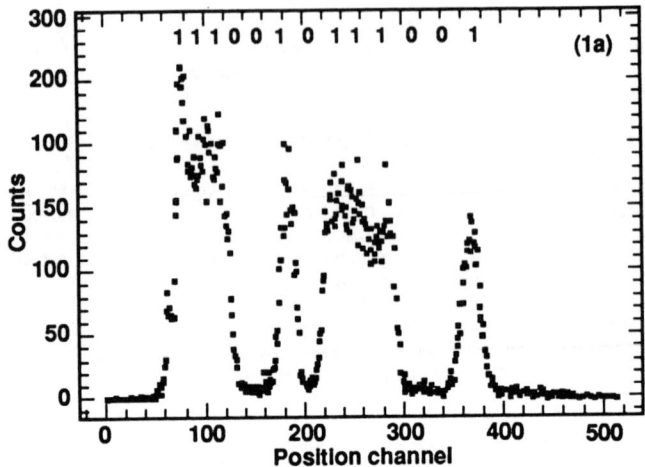

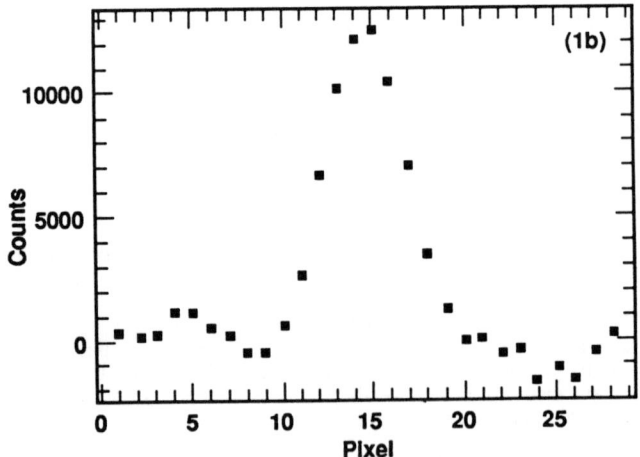

Fig. 1a: Shadow image of a one-dimensional coded aperture mask on the Germanium Drift Chamber. The source is the 59.5 keV line from ^{241}Am. The 0s and 1's indicate the position of opaque and transparent elements of the mask, respectively.

Fig. 1b: The reconstructed image of the shadow pattern of Fig. 1a.

incorporated into a balloon payload using a different type of position sensitive detector (Harrison et al. , 1989; Seiffert et al. , 1989).

A variety of technical issues must be addressed concerning the optimization of the GDC. The GDC physical dimensions, doping gradient, doping density and operating temperature vary, in an interdependent fashion, the electron drift time, spread in electron collection times for different gamma-ray absorption depths, the amount of electron trapping, the effective area and the required bias voltage. The charactistics related to the electron mobility, the drift time, the spread in collection times, and the amount of trapping; respectively determine the detector dead-time, spatial resolution and energy resolution. Finally, work must be done on simulating the performance of the IGETAGRAY with realistic GDC/OCA/shielding geometries to appraise the total telescope sensitivity.

IGETAGRAY represents a novel scheme for obtaining the high performance imaging spectroscopy which will be necessary to answer the important questions in soft gamma-ray astronomy in the 21st century.

We wish to thank Fred Goulding and Richard Pehl of Lawrence Berkeley Laboratory and Klaus O. H. Ziock of the University of Virginia for loan of the germanium drift chamber and dewar respectively. This work was performed under the auspices of the U.S. Department of Energy by Lawrence Livermore National Laboratory under contract No. W-7405-Eng-48.

REFERENCES

Gatti, E., and Rehak, P. 1984, *Nucl. Instr. Meth.*, **225**, 608.
Harrison, F. A., Hailey, C. J., Kahn, S. M., and Ziock, K. P. 1989, in *Proceeding 1159, EUV, X-ray and Gamma-ray Instrumentation for Astronomy and Atomic Physics*, ed. C. J. Hailey and O. H. W. Siegmund (Bellingham: SPIE), p.36.
Luke, P. N., Madden, N. W., Goulding, F. S. 1985, *IEEE Trans. Nucl. Sci.*, **NS-32**, 457.
Rehak, P., Gatti, E., Longoni, A., Kemmer, J., Holl, P., Klanner, R. Lutz, G, and Wylie, A. 1985, *Nucl. Instr. Meth.*, **235**, 224.
Seiffert, M. Lubin, P. M., Hailey, C. J., Ziock, K. P., Harrison, F. A., and Kahn, S. M. 1989, in *Proceeding 1159, EUV, X-ray and Gamma-ray Instrumentation for Astronomy and Atomic Physics*, ed. C. J. Hailey and O. H. W. Siegmund (Bellingham: SPIE), p. 344.

DISCUSSION

Jonathan Grindlay

Did you say the detector concept could be generalized to 2-D? I continue to worry about this loss of effective area in a multi-1-D telescope versus the same total detector area in a 2-D detector.

Hailey

The detector can be generalized to 2-D by the use of segmented electrodes in the non-drift direction or, perhaps, by use of more clever electric field geometries. But it is not really neccesary. According to our simulations, the multi-1-D telescopes have about the same effective area and field of view as a full 2-D system.

Steven Hunter

Instead of using a single thick detector, with its associated doping problems, have you considered using several layers of thinner detectors?

Hailey

No, we have not considered layering thinner detectors. I believe the doping can be optimized to give good performance in a single thick Ge chip.

A Large Area, Low Cost, Gamma-Ray, Imaging Spectrometer

K. P. Ziock and C. J. Hailey

Laboratory for Experimental Astrophysics
Lawrence Livermore National Laboratory

Abstract

We are developing a low-cost per unit-area, imaging, low-energy (.1 - 10 MeV) gamma-ray spectrometer. This device has excellent position resolution (~ 2 mm @ 511 keV), near unity quantum efficiency and an energy resolution typical of alkali halide-type detectors. The low unit cost is achieved through separation of the energy and position sensing functions in an alkali halide-based gamma-ray detector.

As we enter the next century, high energy astrophysics will progress from the youthful field of today to a more mature branch of astrophysics. This change will be achieved when the large area, high angular resolution instruments required to obtain meaningful images of the gamma-ray sky are in use. In particular, we look forward to instruments capable of imaging with sensitivities and angular resolutions allowing direct comparison of astronomical maps of the high energy sky with those taken at other wavelengths. While such instruments will surely be developed during the next century, as in all endeavors, cost will play a role in the time to commissioning, the performance and the number of instruments available. It is with this in mind that we are developing the Fiber-Fed Imaging Spectrometer (FIFIS).

Due to a lack of imaging optics for low energy gamma-rays, the typical telescope at these energies is based on a coded aperture design (Fenimore, 1989). Such instruments consist of a mask of transparent and opaque pixels located a distance in front of a position sensitive detector. Sources in the field of view project a shadow pattern of the mask on the detector -- the precise pattern depending on the source locations. One recreates an image of the sources from the known mask pattern and the shadow image obtained by the detector.

This imaging technique has several design implications for gamma-ray telescopes and the position sensitive detectors they require. First, the angular resolution of a coded aperture instrument is determined by the ratio of the pixel size to the mask-detector spacing. The limit on pixel size is set, to first order, by the position resolution of the detector. To achieve reasonable angular resolution in an instrument of manageable length requires high position resolution. For instance, to achieve ~ arcminute angular resolution in an instrument limited to several meters length, requires ~ 1 mm position resolution. Such a length limit is desirable due to the increase in costs associated with increased instrument length. A second implication of the coded aperture technique is that the collecting area (and sensitivity) scale with the detector area. Typically, detectors have a high cost per unit area rendering the cost of a 1 to 10 m^2 active area instrument prohibitive. Unfortunately, such large areas are required to obtain sensitivities comparable to those attained at other wavelengths.

Currently, the most versatile low energy gamma-ray detector consists of an alkali halide scintillator coupled to an imaging phototube (Hailey *et al.*, 1989). FIFIS is an extension of this detector with its unit area costs reduced through separation of the energy and position resolving functions. Optically isolated bars of scintillator are mounted on a standard (non-imaging) phototube (see Fig. 1). A separate fiber optic for each bar couples it to an imaging device. The majority of the light from an event is collected by the non-imaging phototube and is used to determine the energy of the event. The event location is obtained by comparing the unique map of fiber positions at the imager with the imager output. The entire device provides alkali halide-type energy resolution and a position resolution fixed by the width of the bars.

A Large Area Imaging Spectrometer

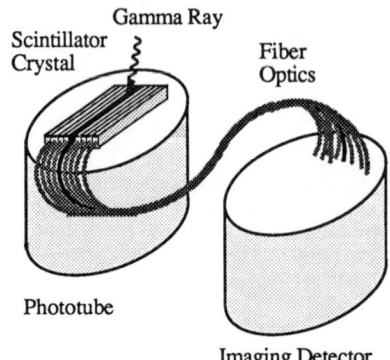

Figure 1. A schematic diagram of FIFIS. A gamma-ray's energy is determined from the phototube output, its position from the location of the active fiber end at the imager.

The cost advantage of FIFIS is realized by multiplexing a single imager to many non-imaging phototubes. As the former cost ~ five times the later, a significant decrease in cost per unit area results. This savings is further compounded by the reduction in support electronics required by a non-imaging phototube. For instance, each imaging phototube used in our current generation of imaging gamma-ray detectors requires four complete channels of electronics; a non-imaging phototube requires only one. This not only reduces the cost and complexity of an instrument, it also reduces its weight and power consumption -- important considerations in a device which must be lofted to the top of the atmosphere. As one imager can be multiplexed to handle ~ 30 non-imagers, the combined savings more than offset the costs associated with the fiber optics.

To achieve a detector with two dimensional position resolution, one could further subdivide the scintillator bars into segments. However, due to advances in imaging techniques (Harrison *et al.*, 1989), the additional complexity and costs are unwarranted. As a one-dimensional detector, this device is well suited to the Gamma-Ray Arcminute Telescope Imaging System (GRATIS) currently under construction (Seiffert *et al.*, 1989). GRATIS uses the images from multiple rotated one-dimensional telescopes together with the MEM image reconstruction algorithm to create two-dimensional images.

There are several issues which we are addressing to develop FIFIS. The primary concern is that sufficient light is transmitted to the imaging device to avoid position dependent quantum efficiency variations. Such a position bias can occur because the amount of light collected by the fiber varies with the event position in a bar. If too little light is transmitted to the imager to generate a trigger, then a valid event will be missed.

A second concern is the pixel size of the imaging device. We are currently exploring the use of Hamamatsu imaging phototubes which possess a position resolution proportional to the size of, and inversely proportional to the intensity of, the light cloud at the photocathode (Hailey *et al.*, 1989). Thus, the more light transmitted by the fiber optic, the more pixels in the imager. The pixel size is important to determine the allowed spacing of fibers from the same non-imager. By establishing an electronic coincidence, one can pack many fibers from different non-imagers into the same pixel. Thus, the ultimate number of bars which a single imager can handle is given by the size of the fiber optics. Consequently, one wishes to use the smallest fiber optic available which collects a sufficient amount of light to avoid bias in the quantum efficiency. (Note that the overall counting rate the imager can handle also sets an upper limit to the number of bars it can service.)

The position resolution of FIFIS is determined by the width of the scintillator bars. Since the light is collected from the bar edges, their width must be sufficient to transmit enough light from all locations within the bar for reliable energy and position determination. The width required varies with bar height and length. The bar height determines the highest energy gamma-rays which are efficiently stopped. However, light collection efficiency decreases as thicker bars are used. This means that low energy events, which preferentially occur near the top surface of the bar, have poorer energy resolution and may go undetected by the imager. Thus, bar thickness also determines the low energy threshold of the detector.

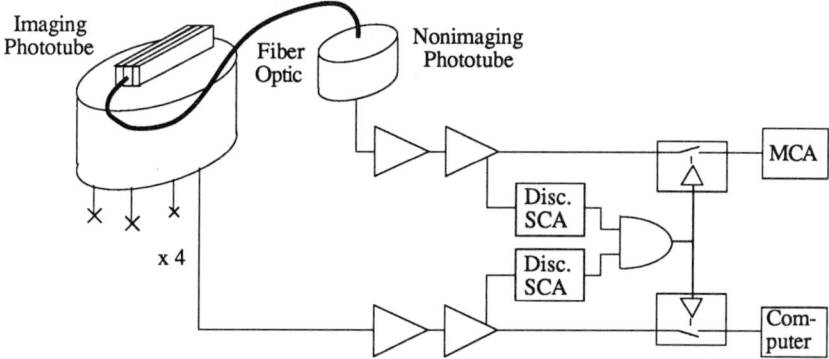

Figure 2. Experimental Apparatus. Note that the roles of the imaging and non-imaging phototubes are reversed.

We have conducted preliminary experiments of the FIFIS concept using the apparatus shown in Fig. 2. Three .3 x 3 x 4 cm CsI (Na) bars are mounted on an Hamamatsu R2486-02 imaging phototube. The bars are coupled to the tube with optical grease and optically isolated from each other with 7.6 µm Al foil. The other end of the fiber bundle is connected to an EMI 9954B phototube. *Note that this is the opposite configuration used for a working FIFIS system!* The roles of the imaging and non-imaging tubes have been reversed to allow imaging of events in the bar, and to use the single electron resolution of the EMI tube to measure the number of photoelectrons generated by light from the fiber optic. An energy dependent, electronic coincidence is established between the two phototubes to identify valid events.

Using this apparatus, we have verified the basic concept behind FIFIS and its extension to high energy gamma rays. In particular, a ^{22}Na source has been used to irradiate all of the bars with the 511 keV positron annihilation line. The plots in Fig. 3 show the number of counts as a function of position summed along the length of the bars. The data was collected with an energy window centered on the 511 keV peak in the imaging phototube. In Fig. 3A the coincidence requirement is turned off and one sees all three bars. In Fig. 3B the coincidence requirement is set and only the central bar appears.

There are several interesting features in the data. Although the bars are only separated by 7.5 µm there appears to be a large gap between them. This is an artifact of the centroiding nature of the Hamamatsu phototubes. For each event, the phototube determines the centroid of the light cloud obtained at the photocathode. The large aspect ratio of the crystal bar prevents the light cloud from spreading and the centroid of the light distribution is always observed near the center of the bar.

A second feature of interest is the small peaks between the bars observed in Fig. 3A. These are due to events where the gamma-ray energy is shared between neighboring bars. The primary mechanisms for such events are Compton scattering of the gamma-ray into a neighboring bar and photoelectron transit of the interbar optical barrier. Confirmation of this comes from the absence of the peaks at lower photon energies (350 keV), and from their reduction in Fig. 3B. Their absence at lower energies is due to the reduction of both the Compton cross section and the range of the primary photoelectrons, so that few multibar events occur. Their absence in Fig 3B is explained by the small number of photoelectrons (2.5) detected per event by the EMI tube. Whenever energy is shared by two bars, the probability of seeing the event at the EMI tube vanishes and the events are suppressed. This supports the idea that one could run the fiber optics from neighboring bars to different imagers and use a threefold coincidence to suppress such events. Unfortunately, it also means that too few photoelectrons are being collected. This is seen by the reduction of the peak

232 A Large Area Imaging Spectrometer

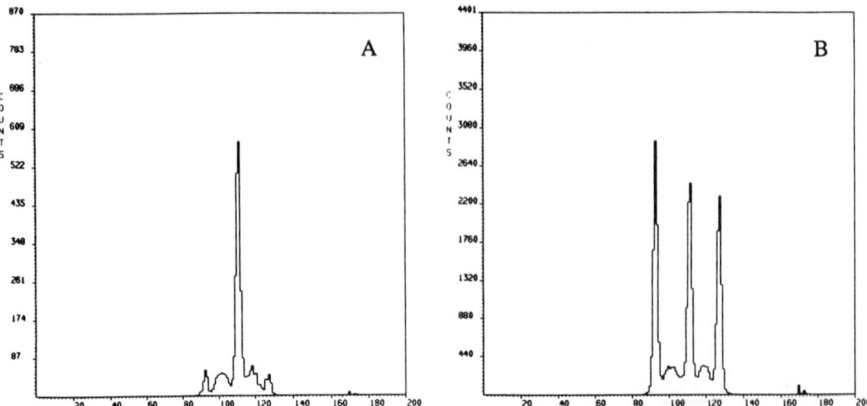

Figure 3. Counts as a function of position summed along the CsI (Na) crystals, taken as described in the text. In Fig 3B the FIFIS coincidence is required. The bar center-to-center spacing is 3 mm.

height in Fig. 3D when compared with Fig. 3B. Insufficient light is seen through the fiber optic used, to reliably obtain a trigger from the EMI tube. Solutions of this problem include changing the fiber optic size, type and attachment point; collecting light from multiple locations on each crystal and changing the crystal size. Other measurements with smaller crystals did not show this effect, even at lower energies (Ziock and Hailey, 1989).

In summary, we have demonstrated an imaging spectrometer with position resolution of 3 mm and energy resolution of 12.5% at an energy of ~ 511 keV. The technique upon which this detector is based gives a low cost per unit area. In addition, the detector design reduces the weight and power requirements in an imaging system through a significant reduction in the electronics required. This makes FIFIS an ideal candidate for use in the large area telescopes envisioned for the next century.

We wish to acknowledge a grant from the University of California Campus-Laboratory Collaborative Research Program. Portions of this work were performed under the auspices of the U. S. Dept. of Energy by Lawrence Livermore National Laboratory under Contract No. W-7405-ENG-48.

References:
Fenimore, E. 1989, See review in this proceeding and references therein.

Hailey, C.J., Harrison, F., Lupton, J.H., and Ziock, K.P. 1989, Nucl. Inst. Meth., **A276**, p. 340.

Harrison, F.A., Hailey, C.J., Kahn, S.M., and Ziock, K.P. 1989, in SPIE Proceeding **1159**, ed. C.J. Hailey and O.H.W. Siegmund (Bellingham: SPIE) p. 36.

Seiffert, M. Lubin, P.M., Hailey, C.J., Ziock, K.P., Harrison, F.A., and Kahn, S.M. 1989, in SPIE Proceeding **1159**, ed. C.J. Hailey and O.H.W. Siegmund (Bellingham: SPIE) p. 344.

Ziock, K.P., and Hailey, C.J. 1989, in **1159**, ed. C.J. Hailey and O.H.W. Siegmund (Bellingham: SPIE) p. 280.

DISCUSSION

Gerald Share

There appears to be a trade-off between getting enough light out through the fiber optics at low energy and making the crystals big enough for improved high-energy response.

Ziock

This is true and forms one of the major design considerations for FIFIS. However, it is more a question of careful design to achieve the energy range desired than one of limiting the viability of the detector. We have only begun to perform the careful studies required to optimize the light into the fiber, which can be increased by such factors as the crystal preparation, fiber placement, fiber type, or simply by increasing the fiber size.

Paul Gorenstein

What percentage of light goes to imaging and what percentage to energy?

Ziock

Only a small fraction of the light will go to imaging. We require ~ 10 detected photoelectrons by the imager at the lowest energy at which FIFIS operates. If we take this to be 100 keV, where we normally obtain ~ 800 photoelectrons for NaI, then $\sim 1.25\%$ of the light goes to imaging.

A TOTAL THROUGHPUT TRANSIENT SPECTROMETER FOR GAMMA RAY SOURCES

Kevin Hurley

University of California
Space Sciences Laboratory
Berkeley, California

INTRODUCTION

The vast majority of all the cosmic gamma ray sources appear suddenly, flicker for a few seconds, and then lapse into a long quiescent state. These are the cosmic gamma-ray bursters. It has become a cliché to state that they are perhaps the outstanding astrophysical mystery of the past two decades, and are still poorly understood despite intense theoretical and experimental efforts. I will not claim that the experiment which is described below will solve this mystery − not because it cannot (it could), but because it probably would not fly for at least another decade. Here I will make the assumption (optimistic) that missions in the intervening years such as the Gamma Ray Observatory and the High Energy Transient Experiment will have resolved the question of the origin of cosmic gamma-ray bursts, and I will make the prediction (hazardous) that the majority of observable bursts are associated with galactic neutron stars. In that case, in a decade or so from now, we will be asking a very different set of questions.

OBJECTIVES

In the unique astrophysical laboratory associated with those neutron stars which emit gamma-ray bursts (and this is probably the most populous class of observable neutron stars), numerous and complex physical processes are occuring: energy release, transport of energy to a plasma, energization of plasma particles, and the radiation of power, primarily in gamma-rays. By studying the energy spectra of bursters in detail (i.e., with excellent time and energy resolution) we can determine (or at least strongly constrain) not only the emission mechanisms, but also the physical conditions in, on, and around neutron stars. Specifically, some of the questions which we will probably be asking ten years or more from now (and which we will be able to answer) are the following. [For more background on these subjects, refer to the conference proceedings edited by Liang and Petrosian (1986), Woosley (1984), and Burns et al. (1983).]

What is the magnetic field configuration for gamma-bursters, and how is it related to the source geometry? There is now good evidence that fields $\approx 10^{12}$ G are present in some sources; there is also evidence that bursters are old neutron stars. Have the dipole fields decayed, leaving higher order multipoles, or do dipole fields align with rotation axes? Evidence for magnetic fields comes, of course, from the observation of cyclotron features, and these are known to depend on the magnetic field-photon source-observer geometry. The effects are subtle, but are beginning to be observed for X-ray pulsars such as Her X-1, where phase-resolved spectra show line energy shifts. Other predictions include intensity-line energy correlations and line width-observing angle correlations, as well as line intensity variations dependent upon whether the photons from the observed source are transmitted through a plasma, reflected from it, or embedded in it.

If intense magnetic fields are present in the emitting region, what is the polarization of the gamma-rays? For example, there is evidence from spark chamber measurements that Vela pulsar gamma rays are almost 100% polarized. However, polarization measurements of medium energy gamma rays have not yet been carried out for any astrophysical object. The detection of polarized gamma-rays from gamma bursters would clearly be a major step in establishing the emission mechanism in the presence of a strong field.

There is good evidence that the high energy photons from bursters are emitted in some cases from a relativistic pair plasma. Line features have been observed in numerous cases at energies around 400 keV, which are interpreted as gravitationally redshifted electron-positron annihilation. Yet it is also recognized that line broadening and blueshifting should occur in a medium whose implied temperature is this high; this makes it difficult to understand how one could observe a line at all, much less a redshifted one. Possible explanations include multi-temperature sources, spatially distinct sources with different temperatures, and single sources whose temperatures are changing faster than the time scale for spectral measurements. A closely related issue is whether the hard and soft components of burst spectra can really be generated in the same environment, since high energy photons (now observed up to 100 MeV) should not escape in great numbers from regions whose magnetic field strengths are as high as those implied from the cyclotron features.

If accurate redshifts can be derived from spectral measurements, what do they tell us about the neutron star equation of state? Some might answer that gravitational redshifts cannot be inferred at all from these data, for the reasons outlined above; however, other explanations of these spectral features seem so contrived that serious consideration must be given to the simplest idea, namely that a mass-to-radius ratio of about 1 $M_\odot$/10 km prevails in the emitting region, and that we may indeed understand something about the conditions inside by observing the radiation outside. It may well happen that this view will not hold up when good quality high time and energy resolution observations become available, and that the real explanation is more complex. That in itself would represent a significant step, but it is unlikely to come about unless measurements of the caliber described below become available.

What is the composition of a neutron star crust, and/or of the material which is accreting onto it? There is already tantalizing evidence for the observation of a true nuclear line in the energy spectrum of a gamma-ray burst. If interpreted as a redshifted Fe line, it is likely to originate in the crust, which is expected to be iron-rich. There has been speculation for over a decade that material accreted onto the surface of a neutron star may excite the crust or become excited itself and radiate gamma-ray lines. Measurements of them would be almost unique indicators of elemental composition, with only a possible uncertainty due to gravitational redshift.

What are the timescales for plasma energization mechanisms, and what are the emission mechanisms? It can be shown that practically all current instruments are incapable of detecting transient events shorter than a few milliseconds long, and that scintillators are poor instruments for defining energy spectra. It is quite possible that in the phase space defined by time and energy resolution, we are failing to detect an entire population of very short events, and are not detecting important spectral features (broad or narrow).

Measurements such as the ones described above must be carried out with simultaneous high time and energy resolution. The two cannot be separated, particularly in the case of gamma-ray bursters, whose time histories and energy spectra are already known to vary on timescales down to the limits of all current measurements.

INSTRUMENTATION AND SENSITIVITIES

The Total Throughput Transient Spectrometer is shown in Figure 1 for a Space Station configuration. It consists of 12 high purity segmented germanium detectors (6.5 cm dia. × 7 cm long) in a BGO anticoincidence shield. The Ge detector array has a field of view of 135° FWHM; each detector analyzes photons in the 10 keV–16.3 MeV range, while the BGO shield covers the 50 keV–100 MeV range. The Ge detectors have an energy resolution which varies from about 1–4 keV over their energy range. The total on-axis detection area is 398 cm^2; the total detection volume is 2800 cm^3.

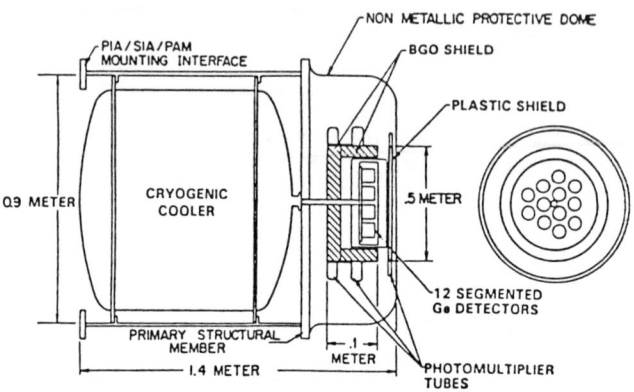

Figure 1. Schematic drawing of the Total Throughput Transient Spectrometer in a Space Station configuration.

The amount of onboard processing is kept to a minimum. Every interaction is time-tagged to 3 μs and energy-analyzed in 8192 channels before transmission to the ground, where burst searches are carried out. The throughput of the instrument is total in the sense that there is no degradation of temporal or energy resolution due to binning before transmission; >90% of the interacting photons are analyzed and transmitted for >90% of the bursts detected. Rates over 2×10^5 counts/s can be handled with no distortion, and even a "March 5-type" burst can be accurately analyzed, using two specially fabricated Ge detectors with small upper segments.

The resulting telemetry rate, 300 kb/s, can be accommodated on the ground by a PC-based system writing to VHS tapes, and later archiving the data to WORMs. The total experiment mass is 550 kg, and the power requirement is 300 W. A Lockheed cooler provides about one year of operation (Nast 1986).

Monte-Carlo simulations have been performed to analyze the expected background, and the resulting sensitivity to transient events. Unlike many hard X-ray and gamma-ray instruments, TTTS does not operate in a background-limited mode, due to the intensity of most bursts. The main background is from atmospheric and cosmic diffuse photons, with β-decays providing only a small contribution. We estimate the 3σ narrow line sensitivity to be between 0.01 and 0.1 photons/cm^2 s depending upon the energy, and the 3σ continuum detection flux limit to be around 10^{-7} erg/cm^2 s > 30 keV. Figures 2a and 2b show simulated burst spectra. We expect approximately 100 large cosmic gamma-ray bursts per year, and about 300 small events. The polarization sensitivity reaches a maximum around 400 keV, where Compton scattering dominates and adjacent detectors become scatter-analyzer pairs. For intense bursts, our simulations indicate that TTTS is sensitive to 10–50% linear polarization; their detection rate should be about five per year.

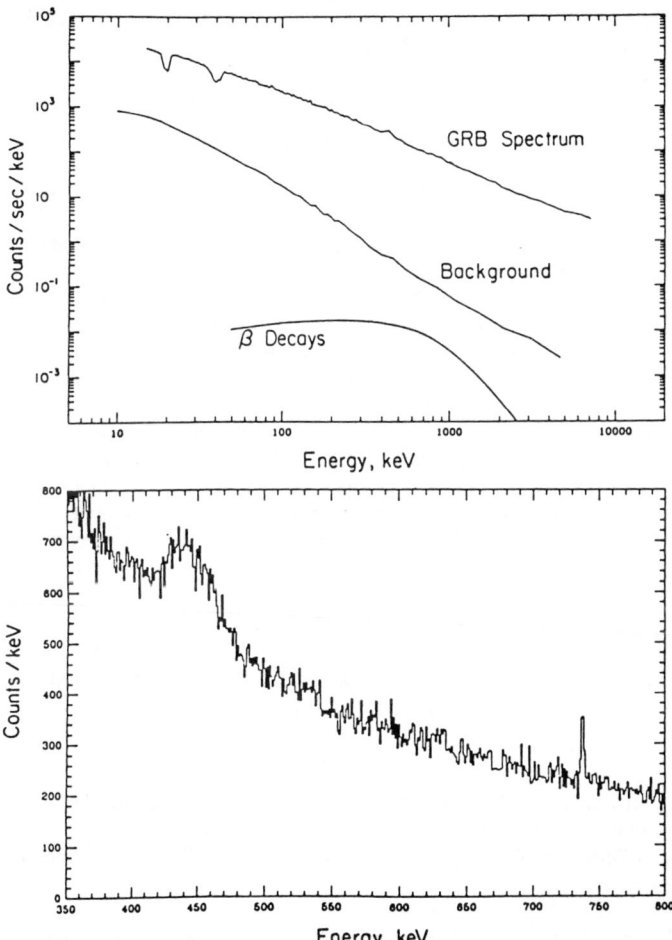

Figure 2. *(a)* Simulated GRB energy spectrum detected by the 12-detector array, plotted with low energy resolution. Two cyclotron features are clearly observed at 20 and 40 keV. *(b)* High resolution plot of the same spectrum, revealing a broad annihilation feature around 450 keV and a redshifted Fe line around 740 keV. The continuum and the high energy lines are based on actual observations of the November 19, 1978 GRB; cyclotron features with intensities observed in other bursts have been superimposed.

DEVELOPMENT STATUS

TTTS is quite similar in design and concept to a 12-detector balloon instrument built by the UCSD, UCB, Toulouse, and Saclay groups (J. Matteson, P.I.) for cosmic (non-transient) sources. It was successfully flown from Australia in 1989 for the first time. It is also very similar to the HIREGS 12-detector balloon instrument being built at UCB (R. Lin, P.I.), with collaboration from the UCSD and Toulouse groups, for the Max-91 program. Some of the differences between TTTS and these instruments are (1) TTTS does not require pulse shape discrimination or tight collimation (a simplification), but (2) it must be cooled in space for at least a year (an added complexity, but one which is solved by using the commercially available Lockheed cooler). Thus a minimum development effort is required.

DISCUSSION AND CONCLUSION

One Ge spectrometer has been flown in the past for transient studies (ISEE), and several will be flown in the future (Wind, Mars Observer); TTTS will be over an order of magnitude larger than those, and should therefore make substantial progress in our understanding of neutron star physics. We envision it as an ideal Space Station experiment; it has no preferred duty cycle, it needs only to be pointed towards the zenith to take data, and would provide a large quantity of excellent data in one year of operation. Its secondary objectives would include the fine spectroscopy of solar gamma radiation (several hundred flares per year are detectable even during solar minimum), monitoring X- and gamma-ray pulsars, surveying the sky for slow transients, and observing the galactic center 511 keV radiation. Although it is ready to be built now, it seems unlikely to fly before the late 90's. It is difficult to predict what missions might be operating simultaneously, but even in the unlikely event that no other transient experiments are operating at that time, the modest localization capability of the BGO shield and Ge array will assure that the instrument can function in a stand-alone mode. Finally, the experiment is recyclable; once retrieved and brought back to earth, it can be refurbished and reflown, for example as a focal plane instrument on the Pinhole-Occulter facility.

ACKNOWLEDGEMENTS

The TTTS concept was developed as a proposal in response to AO No. OSSA 3-88. The co-experimenters were T. Cline, J. Higdon, R. Lin, J. Ling, P. Luke, W. Mahoney, N. Madden, J. Matteson, R. Pehl, M. Pelling, L. Varnell, and W. Wheaton.

REFERENCES

Burns, M. L., Harding, A. K., and Ramaty, R. 1983, Positron-Electron Pairs in Astrophysics, AIP Conference Proceedings No. 101, AIP – New York.
Liang, E., and Petrosian, V. 1986, Gamma-Ray Bursts, AIP Conference Proceedings No. 141, AIP – New York.
Nast, T. 1986, *Adv. in Cryogenic Engr.*, **31**, 835.
Woosley, S. 1984, High Energy Transients in Astrophysics, AIP Conference Proceedings No. 115, AIP – New York.

DISCUSSION

Hugh Hudson

For the longer-range future, especially in imaging applications such as the Pinhole/Occulter Facility, it would be nice to think of larger Ge arrays than the ones currently being considered or flown. Why not make the TTTS larger? What would be the fractional incremental cost, say, to double the array size?

Hurley

Definitely, more germanium is better; there are three ways to get it. The first is to use larger detectors; the second is to use a larger array size; the third is to use more than one array. TTTS was based on the largest commercially available detectors (6.5 × 6.5 cm) at the time. Today it might be based on 7 cm detectors, and in the future, even larger ones, but these increases are still small. TTTS was also based on the largest array size with which we have experience, namely, 12 detectors. Increasing this number might add greatly to the complexity, although when the teething problems associated with this size array have been solved, we might feel comfortable with a larger one. This leaves the question of using more than one array. Under the Space Station AO, two complete flight models were required, and the equipment costs alone came to roughly $12 M for the two. So a very approximate estimate might be $10 M per array, when all costs are included. But note from the figure that it would not be simple to mount these units side by side.

Paul Gorenstein

Do you have any angular resolution or position determination capability?

Hurley

We could determine positions of bursts to $\sim 5 - 10°$ by comparing the responses of different portions of the BGO shield. In the unlikely event that no other transient experiments were in orbit simultaneously with the TTTS, this would allow us to deconvolve our energy spectra accurately. In the interest of simplicity, we chose not to build in a more accurate localization capability, although this is clearly an important objective to continue to pursue.

ASTROGAM: A MAGNETIC RIGIDITY SPECTROMETER FOR GAMMA RAY ASTRONOMY

J.H. Adams, Jr.[1], S.P. Ahlen[2], L.M. Barbier[3], J.J. Beatty[2], P.Carlson[4],
H.J. Crawford[5], R.L. Golden[6], K.E. Krombel[7], R.C. Lamb[8],
J. Lloyd-Evans[9], A.A. Marin[2], J.F. Ormes[3], M.E. Ozel[10], G.F. Smoot[5],
R.E. Streitmatter[3], A.J. Tylka[7], T. C. Weekes[11], and B. Zhou[2]

[1]E.O. Hulburt Center for Space Research, Naval Research Laboratory,
Washington, DC 20375-5000
[2]Boston University, Boston, MA 02215
[3]NASA Goddard Space Flight Center, Greenbelt MD 20771
[4]Manne Siegbahn Institute of Physics, Stockholm, Sweden
[5]University of California at Berkeley, Berkeley, CA 94720
[6]New Mexico State University, Las Cruces, NM 88003
[7]Universities Space Research Association, Columbia MD 21044
[8]Iowa State University, Ames, IA 50011
[9]University of Southampton, Southampton S09 5NH UK
[10]Cukurova University, Adana 01330 Turkey
[11]Harvard-Smithsonian Center for Astrophysics, Amado, AZ

Abstract

We present a new concept for a high energy gamma ray telescope with an extended energy range and excellent energy and angular resolution. Astrogam is a pair production telescope which was designed to use the magnetic field of the Space Station Freedom Astromag facility to separate the e^+-e^- pair and accurately measure the momentum of each particle (Eichler and Adams 1987). Astrogam will extend spectral measurements to higher energies, thus closing the gap between satellite and ground-based observations.

Introduction The spectrum of cosmic gamma rays in the 1 GeV to 1 TeV Energy (hereafter GTE) range is largely unexplored. Satellite-based gamma-ray telescopes have covered the high energy (HE) range from a few MeV to a few GeV, while ground-based experiments detecting extensive air showers cover very high (VHE) to ultra-high energies (UHE) above a few hundred GeV. The proposed gamma-ray telescope, Astrogam, will fill this gap in our knowledge of the astrophysical electromagnetic spectrum.

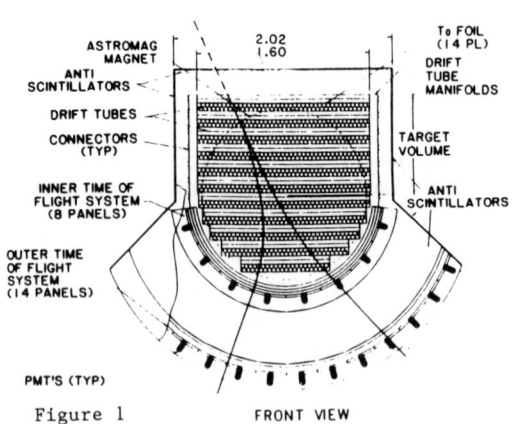

Figure 1   FRONT VIEW

The Instrument Design Astrogam is a wide field camera which makes an all sky survey. It observes each source as it passes through the field of view, collecting more photons per source than a pointed instrument of the same area and detection efficiency which divides its time equally among 13 sources per year. Figure 1 shows a cross section of Astrogam parallel to the plane of the Astromag coils. Gamma rays are detected by pair conversion in the target volume, primarily in the Ta foils. The magnetic field separates the pair whose trajectories are measured by drift tubes, allowing

---

Astrogam Parameters

Energy Range: 100 MeV to 1 TeV
Source Location Precision: 5-10 arcsec (bright sources)
Single Photon Angular Resolution: 20 arcmin (at 2 GeV)
Energy Resolution: 1% (1 to 100 GeV)
Geometry Factor - Efficiency Product: 7000 cm^2-sr
Equivalent Area[1] for a Point Source: 600 cm^2 (at 40° declination)
Field of View: 70° FWHM
Arrival Time Precision: 0.1 ms

[1] Astrogam is a scanning instrument which surveys the entire celestial sphere. It observes a particular source only part of the time, with a collecting area which depends upon the declination of the source. The "equivalent" area is that of an instrument which would collect as many photons as Astrogam by always pointing exactly at the source and detecting every incident photon.

---

the momenta of the pair to be measured. The magnetic field is strong enough to separate the pairs up to a photon energy of 1 TeV. The drift tubes, which use dimethyl ether, have a resolution of 34 μm. Precise mechanical alignment of the spectrometer is unnecessary because it will be aligned in flight using cosmic rays. The TDC on each drift tube is self-triggering and is read out if there is an event trigger. The trigger criteria are outbound particles in the time of flight (TOF) systems and no signal in the anti-scintillators. The trigger logic generates a common stop signal which is used to measure the drift times. Experience gained in balloon flights has shown that the common stop is accurate to better than a nanosecond, so it does not affect the drift tube resolution. Many sources of instrumental background have been investigated and means have been found to virtually eliminate all of them either in flight or during data analysis. The dimensions in fig. 1

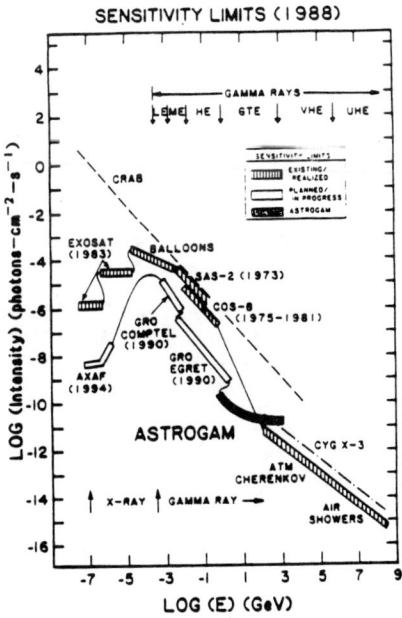

Figure 2

are in meters. Overall, the instrument is 2.86 m by 3.2 m. Its unique and significant features are:

1. **A wide energy range.** Astrogam will cover four orders of magnitude in energy, thus bridging the gap between satellite and ground-based observations. Figure 2 highlights this point and compares the sensitivity limit of Astrogam to that of other instruments. Table 1 shows the numbers of events we estimate Astrogam will collect from several sources in its first year of operation.

2. **Excellent source location precision.** Figure 3 shows Astrogam's single photon angular resolution as a function of energy. Multiple scattering error (medium dashes) dominates nuclear recoil (fine dashes) and measurement error (big dashes) below 20 GeV. Above this energy, each photon can be located to < 2 arcmin. For bright sources, Astrogam will locate the source centroid to within 5-10 arcsec. The data points are from a Monte Carlo

Table 1: Photon Statistics for Various Sources

Source	E > 1 GeV	E > 10 GeV	E > 100 GeV	E > 1 TeV
Galactic Center	125000	3800	67	1
Galactic Anti-center	25000	1400	62	2
Vela	23000	2200	210	14
Geminga	16000	2000	240	18
Crab	4300	270	13	0
2CG013+00	7700	4100	1800	500

simulation of Astrogam.

3. **Excellent energy resolution.** Astrogam will achieve 1% energy resolution because the magnetic rigidities of the pair are precisely measured in the Astromag field.

<u>Scientific Objectives</u> In the energy range from 100 MeV to about 10 GeV, Astrogam has 10 times the collecting power and energy resolution of EGRET on GRO. Astrogam can extend the observations that will soon be made by EGRET.

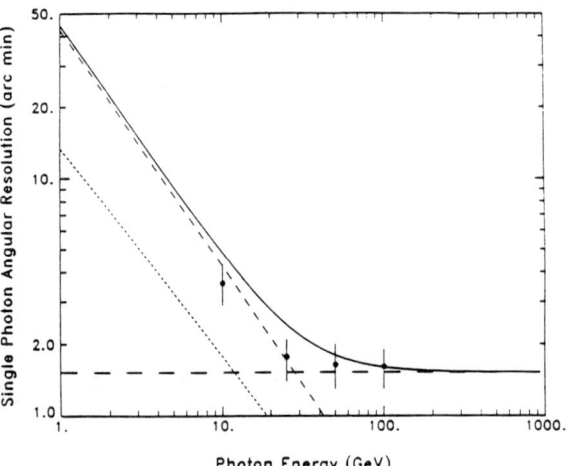

Figure 3

One example of such an extension has been suggested Adams and Fiedler (1990). They have pointed out that high velocity clouds which are located in the halo of our galaxy would have a reduced gamma ray emission above 100 MeV if they are located outside the galactic cosmic ray "leaky box" confinement volume. Table 2 gives their estimate of the observability of some of these clouds with a two year mission if they are within the "leaky box".

Astrogam would make possible observations in the GTE energy range. We could measure and map the diffuse gamma ray emission and complete our knowledge of

Table 2: Observability of High Velocity Clouds

Location	Diameter (degrees)	Column Density[a]	Signal[b]	Background[b]	Confidence Level (standard deviations)
104+9	5.2	2.6	4.4×10^{-6}	6.3×10^{-5}	4.4
113+12	4.8	6.5	1.1×10^{-5}	6.3×10^{-5}	9.1
100+36	3.5	3.0	5.1×10^{-6}	2.3×10^{-5}	5.1
98+26	5.0	3.6	6.1×10^{-6}	3.3×10^{-5}	7.4
South Pole	48.3	0.2	3.4×10^{-7}	2.3×10^{-5}	7.8

[a] in units of 10^{20} atoms/cm^2; [b] in units of cm^{-2}s^{-1}sr^{-1}

the spectra of many astrophysical objects. These possibilities are too new for theoretical investigations to have suggested what discoveries might be made, but some possibilities are suggested below.

Astrogam could resolve complex source regions like the Crab pulsar/Nebula system, the galactic center and the Geminga region and discover the true nature of the sources. Figure 4 shows a map of the Geminga region adapted from Spoelstra and Hermsen (1984), showing radio and X-ray sources. Also shown are typical error circles for COS-B, EGRET and Astrogam.

Astrogam will perform exploratory spectroscopy above 100 MeV. It could detect the narrow GeV gamma ray lines expected from the annihilation of weakly interacting massive particles (WIMP's), a candidate for the dark matter of the Galactic halo (Primack et al., 1988) provided that WIMP's account for a relatively small fraction of the closure density (Bouquet et al., 1989).

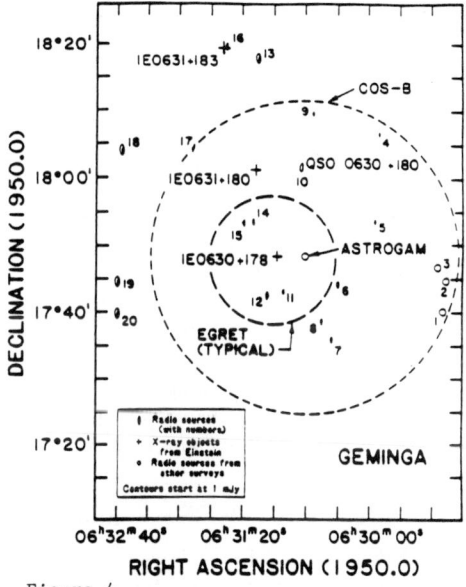

Figure 4

References
Adams, J.H., Jr. and Fiedler, R. 1990 (submitted to the 21th Intl. Cosmic ray Conf.)
Bouquet, Alian, Salati, Pierre, and Silk, Joseph 1989 Phys. Rev. D, 40, 3168
Eichler, D. and Adams, J.H., Jr. 1987 Ap. J., 317, 551-554.
Primack, J. R. et al. 1988, Ann. Rev. Nucl. Part. Sci., 38, 751.
Spoelstra, T.A.T. and Hermsen, W. 1984 Astron. & Astrop., 135, 135.

DISCUSSION

Richard Epstein

How could the operation of "Astrogam" complement ground-based Cherenkov telescopes?

Adams

Astrogam will overlap the ground-based observations of bright sources (ground-based observations extend down to 300 GeV). This will allow energy and collecting-power calibration of ground-based observatories. Also, Astrogam can scan the sky for new γ-ray sources which could appear or be strong at high energies (à lá Geminga). If new sources are found, their spectra could then be extedned to higher energies by ground-based observatories. Note that these ground-based observatories cannot scan for new sources. They must examine previously known source locations for evidence of high-energy gamma-rays.

Mark Leventhal

How will your instrument interfere with the function of Astromag as a cosmic-ray facility?

Adams

Astromag has two identical berths for instruments in the magnetic field. It is a facility, and these instruments will be changed out every 2 years. The initial operating complement will be two cosmic ray instruments. They will be resolution-limited after two years' operation. I expect several pairs of instruments to be deployed on the Astromag facility. Astrogam could be one of these.

Stanley Hunter

(1) Polarization of the γ-rays may be lost due to the magnetic field. Pair production will appear to occur preferentially perpendicular to the magnetic field. Please comment.
(2) Ta foils of 0.01 rad. length may be too thick; the electrons may be scattered before they can be measured. EGRET has 0.02 rad. length foils, which are too thick to observe polarization. Again, please comment.

Adams

(1) The track detectors will accurately measure the e^+ and e^- trajectories in the magnetic field. The plane of the pair can be reconstructed regardless of its orientation. It is possible that the resolution of this reconstruction may depend on the azimuthal angle of the pair plane, but it's not obvious to me that it would. We have not, however, investigated this in our Monte Carlo simulations. I do not believe that pair production will appear to occur perpendicular to the field. Monte Carlo simulation will reveal any such bias. (2) Scattering in the Ta foils is a limiting factor for both polarization measurements and angular resolution. Because the foils in Astrogam are thinner and many more photons are detected we can detect polarization, at least in the case of strongly polarized bright sources. Because scattering becomes smaller at higher energies, polarization measurements can be extended to higher energies.

SUPERCONDUCTING TRANSITION DETECTORS FOR LOW-ENERGY GAMMA-RAY ASTROPHYSICS

J.D. Kurfess, W.N. Johnson, G.G. Fritz,
M.S. Strickman, R.L. Kinzer, G. Jung
Naval Research Laboratory, Washington DC

A.K. Drukier, M. Chmielowski
Applied Research Corporation, Landover, MD

ABSTRACT

Low-energy gamma-ray detectors for the early 21st century will require much improved sensitivity, energy resolution and angular resolution. Superconducting devices offer some potential for meeting these requirements. In particular, we are investigating Superconducting Transition Detectors (STD's), in which the energy loss and location of a particle interaction are derived by sensing the transitions between the normal and superconducting states in a uniform array of small spherical granules. Such devices could prove especially advantageous for Compton telescopes and coded-aperture detectors.

INTRODUCTION

Progress in low-energy gamma ray astrophysics has moved steadily forward since the first observations were made over 20 years ago. The SIGMA experiment and GRO instruments will make significant advances over the next 5-10 years. These may be followed by a high spectral resolution mission now in the planning stage. The following generation of instruments will require angular resolutions of < 1 arc-min, energy resolutions comparable to or better than germanium detectors, and a line γ-ray sensitivity of < 10^{-6} γ/cm^2-s. Although concepts are being explored to meet some of these requirements with existing detector technology (scintillators, solid state detectors) alternative technologies, including high-pressure gas time projection chambers, liquid/solid noble gas detectors, and superconducting devices may offer considerable advantages. In this paper we discuss a program to investigate superconducting devices for these applications.

Superconducting detectors are sensitive to the breaking of bound pairs of electrons (Cooper pairs). The typical binding energy is: $\epsilon_{gap} \simeq \epsilon_o (1-T^2)$ with $T = T_o/T_c$ and $\epsilon_o \simeq kT_c$ where T_o and T_c are the operating and critical temperature respectively. We are considering low-temperature superconductors with a T_c typically several °K or lower. Materials such as Sn, In, Pb and Nb have transition temperatures accessible in liquid helium cryostats and have ϵ_o's of several millivolts. In other materials (Be, Ir, Hf and W) ϵ_o can be as small as several microvolts. The projected energy resolution of superconducting detectors, based on the statistics of secondary quanta, can be 0.1% or better at 100 keV.

Significant progress has been made in the development of cryogenic devices for X-ray detectors. The Wisconsin/NASA collaboration (Mosely et al. 1984, McCammon et al. 1989) has obtained $\Delta E=10$ eV for $E=6$ keV using a silicon bolometer with an implanted thermistor. Superconducting tunnel junctions have also been developed as X-ray detectors with an energy resolution of about 40 eV at 6 keV (Twerenbold et al. 1987, Zehnder et al. 1988, Krauss et al. 1989).

SUPERCONDUCTING TRANSITION DETECTORS

For higher energy applications, we are investigating techniques with potential for scaling to large detectors, while also providing excellent energy and positional resolution. Here, we discuss a class of devices, Superconducting Transition Detectors (STD's), utilizing a uniform array of spherical granules, tens of microns in diameter. (Drukier et al. 1972, Gonzalez-Mestres et al. 1988, de Silva et al. 1988, Turrell et al. 1989). To enable each granule to serve as a radiation detector, they are brought to a temperature just below the superconducting transition temperature in a cryostat to which a uniform magnetic field is applied. The superconducting transition temperature, T_H, in a magnetic field, H, is:

$$T_H = T_c \left[1 - \frac{3H}{2H_o} \right]^{1/2}$$

where T_c is the transition temperature in the absence of a magnetic field and H_o is the "critical field" at T=0, i.e. the magnetic field above which the material is not superconducting at T=0. This relationship is shown in Figure 1.

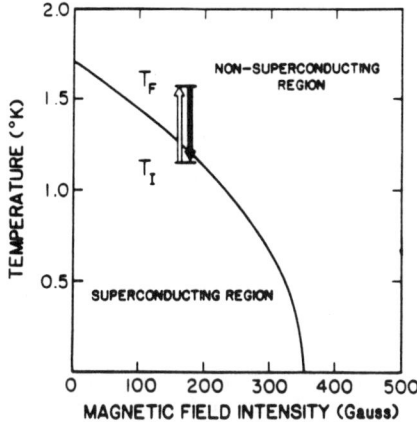

Figure 1. Typical temperature-magnetic field phase diagram for low-temperature superconductor.

Energy deposited by particle interactions will result in a rise in the granule temperature, causing the granule transition to the normal (non-superconducting) state. The transition will be accompanied by the penetration into the granule of the magnetic field which had been expelled by the superconducting granule (Meissner Effect). The disappearance of the dipole can be detected by loop antennae connected to a fast, high-sensitivity pre-amp. Figure 2 shows a typical signal from the transition of a 32-micron diameter Sn granule placed in a millimeter-diameter loop using GaAs FET's operating at liquid helium temperatures (Chmielowski et al. 1990). The rise time is ~ 40 nsec, but a time resolution of a few nanoseconds should be achievable on the leading edge.

In order to achieve energy resolution in a granular detector a fundamental requirement for low-energy gamma-ray applications is that the granule must contain the secondary electrons generated. For the following we assume that this is satisfied; superconducting Pb granules of diameter 20 microns and 85 microns would fully absorb 50% of 100 keV and 250 keV electrons, respectively.

A measure of the energy deposited in a granule can be obtained by measuring the time interval between the transitions to/from the normal state. The temperature history of a granule is shown in Figure 3. Prior to the interaction, the temperature of the granule, T_I, is biased just below the superconducting transition temperature, T_H, in the ambient magnetic field H. The energy deposited by the secondary electron increases the temperature to T_F. The granules are thermally decoupled from the dielectric substrate due to a mismatch of the phonon spectrum at the granule/dielectric surface; thus the time of heat escape is long, typically a few microseconds. For simplicity, we assume that the specific heat, C_v, is linear in the temperature (this approximation is only valid at very low temperatures) so that $T_F = T_I + \Delta T$, where $\Delta T = \Delta E/(C_v V)$. Following the initial rise, the temperature of the granule decays (approximately) exponentially. In this scheme the uncertainty in the energy loss is primarily due to the uncertainty in the time of the normal to superconducting transition.

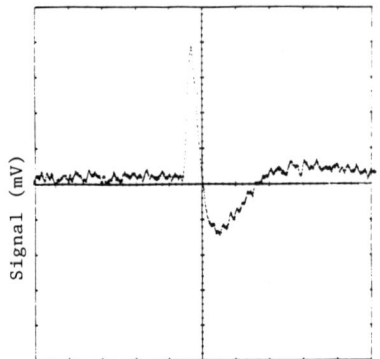

Figure 2. Signal produced by the superconducting-normal transition in a 32 micron diameter Sn granule. S/N is ~ 10.

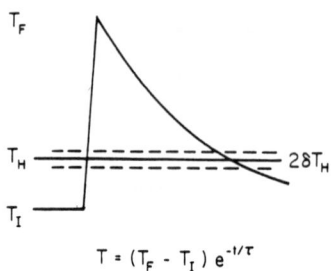

Figure 3. Temperature history of a STD granule following heating by an ionizing particle.

Maintenance of a uniform response in a large array of granules requires; 1) stability and uniformity of temperature and magnetic field; and 2) uniformity of the granule size and of the granules' defect density. We expect that energy resolution will be dominated by fluctuations in granule size, thermal conductivities, and the ambient temperature of the dielectric bath. These effects can be summarized as an effective uncertainty in the transition temperature δT_H. We have modeled the energy resolution to be related to this uncertainty by $(\delta E/E) \simeq 5(\delta T_H/\Delta T)$, where ΔT is the temperature rise induced in the granule.

Turrell et al. (1989) have developed a new method which permits the production of an ordered array of uniform granules. Spherical In granules of ~ 25 micron diameter were produced from a thin film deposited on a mylar foil using photolithographic techniques. The superior quality of granules produced in this manner was confirmed by a study of the hysteresis curves of the collective array. In the magnetic field of earth (H = ~0.5G) all granules in a 30x20 array underwent transition within a temperature range, δT, of < 10 mK, i.e. comparable to the temperature stability of the cryostat. This corresponds to $\Delta T/T$ of $< 3 \times 10^{-3}$. Based on these results we expect that $\Delta T_H/T_H \sim (2-5) \times 10^{-3}$. Thus, a typical energy resolution of 0.5% appears achievable.

The position of the interaction can be obtained by measuring the relative signal strengths at 3 or more loop antennae. The use of two orthogonal planes of readout loops, similar in concept to that employed in a multiwire proportional chamber, could provide two-dimensional localization. Using this approach, the Orsay group has developed a hard X-ray detector having 256 x 256 pixel resolution in a 5-cm x 5-cm detector (G. Waysand, private communication).

We expect that the most realistic implementation of the fast electronics for the position sensing will be the use of fast cryogenic GaAs front-end electronics mentioned previously. Alternatively, use of dc SQUID's would provide better signal/noise and related positional resolution; however their use may be limited by count rate considerations in large detectors.

APPLICATIONS OF STD'S TO GAMMA-RAY ASTROPHYSICS

The proposed R&D program will be applicable to gamma ray imaging applications based on high resolution Compton telescopes (Johnson et al. 1989) or coded-aperture telescopes. Johnson et al. (1989) estimate that a line sensitivity approaching 10^{-7} γ/cm^2-s is possible with a COMPTEL-sized high resolution Compton telescope. This represents a factor of 100 improvement over the sensitivities that will be realized by GRO, SIGMA or an NAE type instrument, and a factor of 10^3 - 10^4 in comparison to the few γ-ray line observations that have been reported to date.

With this expected improvement in sensitivity of a Compton telescope, scientific objectives that would be achieved include; 1) a sky survey detection of ~ 1000 AGN's observed in the 0.15 - 5.0 MeV region where the luminosity of AGN's peak; 2) detection of 20-30 Type 1 supernova/year in the lines of ^{56}Co; 3) detection and localization of hundreds of gamma ray burst sources to < 1 arc min^2 including a search for line emission with 100x improved sensitivity; 4) detection of red-shifted nuclear lines from neutron stars providing unique information on the equation of state of neutron star matter; and 5) high spectral resolution observations of the 0.15 - 10.0 MeV diffuse background.

REFERENCES

Chmielowski, M., et al., "Fast Low Noise Electronics for Superconducting Granular Detectors", to be submitted to J. Superconductivity.
da Silva, A., et al., Proc. 2nd Workshop on Low Temperature Detectors..., Annency-Le-Vieus, 1988, ed. L. Gonzales-Mestres et al., publisher Editious Frontiers.
Drukier A.D., et al., NIM 105, 285 (1972).
Gonzales-Mestres L., et al., Proc. 2nd Workshop on Low Temperature Detectors..., Annency-Le-Vieus, 1988, ed. L. Gonzales-Mestres et al., published Editious Frontiers.
Johnson, W.N., et al., these proceedings.
Krauss H., et al., Proc. 3rd Workshop on Low Temperature Detectors, Gran Sasso, Sept. 1989, ed. E. Fiorini, to be published.
MaCammon, D., et al., Proc. 3rd Workshop on Low Temperature Detectors, Gran Sasso, Sept. 1989 ed. E. Fiorini, to be published.
Moseley, S.H., et al., J. Appl. Phys., 56, 1257 (1984).
Turrell B., et al., Proc. 3rd Workshop on Low Temperature Detectors..., Gran Sasso, Sept. 1989. ed. E. Fiorini, to be published.
Twerenbold, D., et al., J. Appl. Phys., 61, 1 (1987).
van Feilitzsch, F., et al., "Workshop on the Superconducting Particle/Radiation Detectors", Torino, November (1987).
Waysand, G. 1989 (private communication).
Zehnder, A., et al., "Workshop on the Superconducting Particle/Radiation Detectors", Torino, November (1987), ed. A. Browne, World Scientific.

DISCUSSION

Alan Bunner

(1) Is the effective area of this detector just the area of the granules? (2) What filling factor of granules do you expect to achieve?

Kurfess

(1) The active detector volume is just the total volume of all of the granules. Practical hard X-ray and low-energy γ-ray detectors would be achieved by multiple layers of granules, so the effective area will be the cross-sectional area times the effective interaction probability in the detector thickness. For line γ-ray observations, the efficiency will also be limited by the requirement to totally absorb the secondary electron in a single granule. Practical limits on granule size for different applications remain to be determined. (2) The limitation on granule filling factor for a high-energy resolution instrument will probably be limited by requirements to achieve a uniform magnetic field at each granule. Filling factors above 10 − 20% are unlikely.

Richard Epstein

What is the maximum energy range of this type of detector?

Kurfess

A limitation of a detector consisting of many small granules, where one seeks good energy resolution, is the requirement for the secondary electron to be totally absorbed in a single granule. Absorption of 100 keV and 250 keV secondary electrons requires ~ 20 micron and 80 micron diameter tungsten granules, respectively. A hard X-ray (< 200 keV) detector is relatively straightforward using an appropriate thickness of layered arrays of granules. A compton telescope implementation using these devices, where multiple compton scatters are detected, could operate from 0.2 MeV up to several MeV.

Jonathan Grindlay

I have a question related to Alan's: How will you read out such a detector in 3-D? Surely the magnetic loops must then be in different planes, or in arrays. Will the loops then interfere? For high quantum efficiency the detector (with small filling factor) would have to be quite deep.

Kurfess

Three-dimensional readout will require multiple orthogonal readout loops. This clearly becomes a complication for any detector implementation. We expect that loop widths and separations of ~ 1 cm can result in ~ 1 mm positional resolution in one dimension. Interactions between adjacent loops will be a concern, but has not been considered at this time.

The answer to your second question is yes, the detectors we would forsee being implemented in hard X-rays and low-energy γ-rays would be rather thick.

SEGMENTED Ge DETECTORS AND MECHANICAL COOLERS
FOR FUTURE GAMMA-RAY ASTRONOMY INSTRUMENTS

Larry S. Varnell

Jet Propulsion Laboratory, 169-327
Pasadena, CA 91109

ABSTRACT

Future gamma-ray astronomy instruments will achieve better sensitivity by using segmented detectors to reduce radioactive background produced by energetic particle interactions in the detector and by using mechanical coolers to obtain extended mission lifetimes. To demonstrate the effectiveness of a segmented Ge detector in rejecting background events due to the beta decay of internal radioactivity, a laboratory experiment has been carried out in which radioactivity was produced in the detector by neutron irradiation. A ^{252}Cf source of neutrons was used to produce, by neutron capture on ^{74}Ge (36.5% of natural Ge) in the detector itself, ^{75}Ge ($t_{1/2}$ = 82.78 min), which decays by beta emission with a maximum energy of 1188 keV. By requiring that an event deposit energy in two or more of the five segments of the detector, the beta particles, which have a range of about one millimeter, are rejected, while most gamma rays incident on the detector are counted. Analysis of the experiment indicates that over 85% of the beta events are rejected, in good agreement with predictions. In the area of mechanical cooler development, Stirling cycle coolers currently being manufactured can meet the cooling requirements of Ge spectrometers, but questions of lifetime, reliability, and compatibility must be answered.

I. INTRODUCTION

To achieve better sensitivity to weak cosmic sources, future Ge spectrometers will incorporate segmented detectors to reduce the radioactive background produced in space by energetic particle reactions. Previous studies (Varnell et al. 1984) have estimated that 90% of the radioactive beta events can be rejected by using an externally segmented detector inside a thick active shield and requiring that only events which deposit energy in two or more segments be selected. Beta particles, which have a range of the order of a millimeter, will be rejected, while gamma rays from 0.15 to 8 MeV, which typically interact by Compton interactions in two or more segments, will be selected. Although the externally segmented detector was introduced several years ago, there has been no balloon or space flight to demonstrate its effectiveness. An alternative is to use particle beams from accelerators or neutrons from radioactive sources to irradiate the detector and produce internal radioactivity, simulating the energetic particle environment of space. Although these laboratory experiments cannot produce the complete range of background experienced in a balloon flight, they can measure the effectiveness of the segmented detector in rejecting internal background. We have carried out such an experiment by producing ^{75}Ge by neutron capture reactions with the detector material (36.5% ^{74}Ge). Simultaneous spectra are then taken of the activity in the detector under two conditions; a free spectrum in which all events in the detector are accumulated, and a gated spectrum in which events are accumulated only if they deposit energy in two or more segments. By comparing the spectra, the effectiveness of the detector in rejecting beta events can be measured.

Since Ge detectors must be cooled to temperatures near 80 K to operate, spaceflight qualified coolers must be developed which are compatible with the detectors. Mechanical, solid subliming cryogen, and passive radiative coolers have all been flown in the past. For future missions with requirements of low weight and extended lifetimes of the order of several years under different thermal environments, mechanical coolers, in particular the Oxford-type Stirling cycle cooler, are currently the technology of choice. There are still many questions of reliability, lifetime, vibration, and electromagnetic interference which must be answered, and a development program to study problems specific to cooling Ge detectors is being pursued at JPL.

II. SEGMENTED DETECTOR IRRADIATION EXPERIMENT

The five segment coaxial Ge detector used in this experiment has been described previously (Varnell et al. 1988). It is a reverse electrode detector, 5.5 cm in diameter and 5.5 cm long, divided electronically by segmenting the outer electrode. The total energy signal is taken from the center electrode. Each segment has its own signal chain of preamplifier, amplifier and lower level discriminator taken from the external electrode. There is an output from the discriminator whenever energy larger than the threshold is deposited in that segment. The discriminator level was set at 45 keV for these experiments to avoid possible pile-up problems due to high counting rates during the neutron irradiations. The five discriminator outputs are fed to a coincidence unit, which can be set to give an output when 1 or more, 2 or more, etc, discriminator pulses are present simultaneously. During the measurements, two spectra are recorded using the total energy signal from the center electrode. One spectrum is the free spectrum of all events. The second spectrum is gated so that only events which deposit energy above the threshold in 2 or more segments are recorded. These spectra can then be compared to determine the effectiveness of the segmented detector at rejecting events due to beta decay from internal radioactivity.

The neutron source for the irradiation was a sealed ^{252}Cf source emitting 1.4×10^6 neutrons/sec. The spectrum of fission neutrons from ^{252}Cf peaks at 1 MeV with an average energy of about 2 MeV. For the first experiment, the fast neutrons from the source were used to produce the inelastic scattering peaks from the Ge isotopes and other prompt activity. Then the source was removed and the radioactivity produced in the detector was measured. Spectra were recorded simultaneously for the free and gated modes of the detector during the irradiation and after the source was removed. The inelastic scattering lines in Ge and many lines from fission fragments were observed in the spectra obtained during irradiation, but any ^{75}Ge radioactivity produced in the detector was too weak to be observed above room background following the irradiation.

For the second experiment, paraffin wax with a thickness of 13 cm was placed between the source and the detector to thermalize the neutrons, and a shield of lead bricks was built around the detector after the irradiation to lower the background so that weak radioactivity could be observed. The detector was irradiated for 3.5 hours, then placed inside the lead shield. Spectra were recorded in intervals of 15 to 90 minutes for 6.3 hours, then longer intervals were used to observe the longer lived activity. Figure 1 shows the gated spectrum taken in an interval of 30 minutes beginning 75 minutes after the end of the irradiation. Gamma rays from ^{116}In ($t_{1/2}$ = 54.15 min) are labeled. An estimated 0.1 gram of indium was used in mounting the detector, but the large cross section for neutron activation results in the strong gamma-ray lines. The intensity of these lines demonstrates the importance of strict control over the materials used to build gamma-ray detectors for use in space. Figure 2 shows the free spectrum with the same ordinate scale, showing the beta spectrum from ^{75}Ge ($t_{1/2}$ = 82.78 min) with the indium lines. The energy scales are not quite equal because different multichannel analyzers with different conversion gains were used.

From the accumulated spectra, the beta continuum was obtained by subtracting the counts from the gamma-ray lines and from the background measured in the shield before irradiation of the detector. Over the energy range of the beta particles, 85% of the beta activity is rejected. This is in

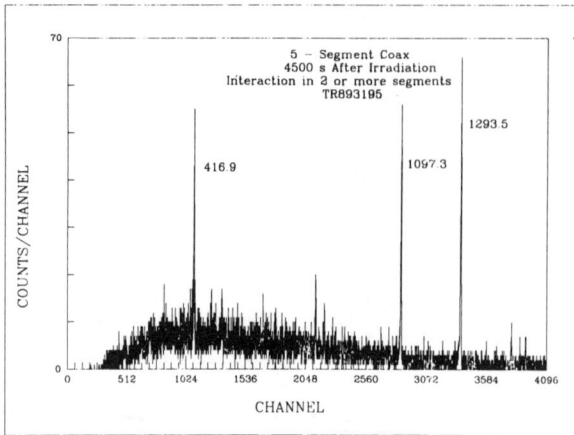

Fig. 1. - The gated spectrum requiring interactions in two or more segments taken 75 min after the irradiation. Gamma rays from ^{116}In ($t_{1/2}$ = 54.15 min) are labeled.

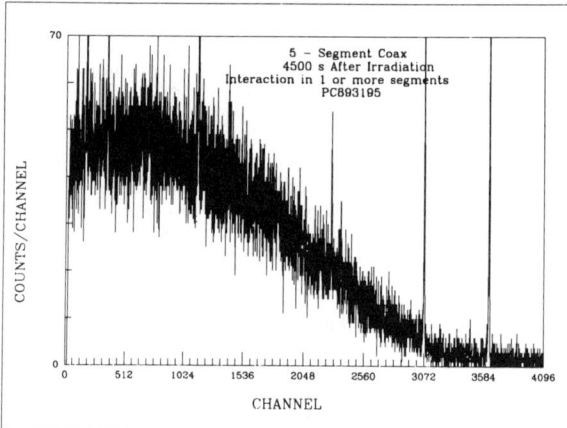

Fig. 2. - The free spectrum taken simultaneously with Fig. 1. The beta continuum is from the decay of ^{75}Ge ($t_{1/2}$ = 82.78 min), with an endpoint of 1188 keV. The 1097 keV line is at channel 3075, the 1293 keV line at channel 3631.

good agreement with the expected value. In 87.1% of ^{75}Ge decays, the transition is by beta decay to the ground state of ^{75}As, and calculations show that 95% of these betas will be rejected. In 11.5% of the decays, the transition is to the 264.6 keV level of ^{75}As, followed by a gamma ray of 264.6 keV. Only 20% of these transitions will be rejected by the segmented detector, but an additional 70% would be rejected by an active shield. A slight correction will be made for continuum counts due to Compton scattering of the In gamma rays when a measurement with an irradiated foil or a Monte Carlo calculation is completed. The gamma-ray efficiency of the segmented detector in the mode requiring interactions greater than 45 keV in two or more segments varies from 0.40 at 417 keV to 0.65 at 1293 keV relative to an ordinary detector, in good agreement with our previous measurement (Varnell et al. 1988). For discriminator settings of 10 keV, the relative efficiency in this mode would increase to 0.55 and 0.85, respectively, for the two lines. Taking into account the background reduction, the segmented detector in this mode has a factor greater than two better sensitivity than an ordinary detector. At energies below 400 keV, the efficiency of the two or more segment interactions mode decreases significantly. In this energy range, the front segment only mode of the segmented detector is used to improve sensitivity. In this mode, events are counted if there is an interaction only in the front segment. For gamma rays of energy up to a few hundred keV, there is a high probability of complete absorption in the front segment (nearly 100% from 15 to 100 keV). The front segment only mode for low energy gamma rays gives a high efficiency with a background for a five-segment detector only 1/5 that of an ordinary detector. Again, the sensitivity will be improved by a factor greater than two compared to an ordinary detector. Preparations are being made for a balloon flight in the Fall of 1990 to measure the detector performance in a space environment.

III. MECHANICAL COOLER DEVELOPMENT

There are now two commercial suppliers developing Stirling cycle coolers for space instruments; British Aerospace (BAe), represented in the United States by TRW in Redondo Beach, California, and Lucas Aerospace, represented by Lockheed in Palo Alto, California. Both coolers are derived from the Oxford University cooler used on the Improved Stratospheric and Mesospheric Sounder (ISAMS) instrument on the NASA Upper Atmosphere Research Satellite (UARS) to be launched in late 1991. BAe has worked under contract from the European Space Agency to evaluate and then make available commercially the Oxford-type cooler. The cooler provides 0.8 W of cooling at 80 K for an input power of 35 W. The weight of the cooler and electronics is less than 9 kg. The cooler operates at 40 Hz with an unbalanced force for the compressor of 32.9 N and for the displacer of 2.8 N, zero-to-peak. It is expected that this can be reduced by a factor of 10 - 100 by operating two units back-to-back. The Lucas cooler has similar specifications, with the significant difference that Lucas claims to reduce the heat load of a non-operating unit from 500 mW to 150 mW by reducing the displacer cold finger diameter from 10 mm to 7 mm. Concerns for space use of the coolers include detector microphonic noise from vibration, large current ripple fed back to the power supply bus, and expected operating lifetime. Tests are planned at JPL to measure the vibration of the BAe cooler in single and back-to-back configurations with a specially designed dynamometer. Modal vibration tests will be carried out on a Ge detector in a rugged mount to measure microphonic noise, and a vibration test of the detector coupled to operating coolers is planned. An Oxford cooler has been running at 80 K for over 2.5 years with no degradation, but there are no long life data on the BAe coolers, since the first production coolers are just now being delivered in December, 1989.

The segmented detector and electronics used in this experiment were built at Lawrence Berkeley Laboratory under the direction of R.H. Pehl. The assistance of members of the High Energy Astrophysics group at JPL is gratefully acknowledged. The research described in this paper was carried out by the Jet Propulsion Laboratory, California Institute of Technology, under contract with the National Aeronautics and Space Administration.

REFERENCES

Varnell, L.S., Ling, J.C., Mahoney, W.A., Jacobson, A.S., Pehl, R.H., Goulding, F.S., Landis, D.A., Luke, P.L., and Madden, N.W. 1984, IEEE Trans. Nucl. Sci. **31**, 300.

Varnell, L.S., Ling, J.C., Mahoney, W.A., Pehl, R.H., Cork, C.P., Landis, D.A., Luke, P.L., Madden, N.W. and Malone, D.F. 1988, in Nuclear Spectroscopy of Astrophysical Sources, ed. N. Gehrels and G. Share(New York:AIP), p. 490.

DISCUSSION

Robert Novick

What happens to the desired γ-ray detection efficiency when you use segmentation to reduce β-ray and other background signals?

Varnell

The detector efficiency depends on the lower level discriminator set on the segment signals. We have measured the efficiency for an LLD setting of 20 keV, and the efficiency varies from 0.3 at 300 keV to 0.85 at 2600 keV compared to an ordinary detector. (The present limit on the lower discriminator is cross-talk at the few keV level, probably due to the layout of our first cryostat. A new cryostat has been built and is being tested, which may allow LLD settings as low as 5 keV.) For low energy, 20 – 300 keV, the front segment is used in the mode that requires a front segment interaction and no interaction in any other segment. This reduces the background in the front segment mode by $1/N$, where N is the number of segments.

Kevin Hurley

This multiple segment detector is presently rather fragile. What are the prospects for mechanically hardening it to survive, say, a Shuttle launch?

Varnell

A rugged detector mount has been developed and built for us by the detector group at LBL. It uses the same design planned for the Ge Detector on the penetrator for the Comet Rendezvous mission, where the detector is held on the outside at both ends. Since the design is being used for several missions, vibration tests to the Shuttle levels and more will certainly be performed.

MERCURIC IODIDE ROOM TEMPERATURE GAMMA-RAY DETECTORS[*]

Bradley E. Patt, Jeffrey M. Markakis, and Vernon M. Gerrish
EG&G Energy Measurements Inc.
130 Robin Hill Rd., Goleta, CA 93117

Robert C. Haymes
National Aeronautics & Space Administration (NASA)
Astrophysics Div., Code EZ, 600 Independence Ave, SW Washington, DC 60546

and

Jacob I. Trombka
National Aeronautics & Space Administration (NASA)
Goddard Space Flight Center, Code 682, Greenbelt, MD 20852

ABSTRACT

High resolution mercuric iodide room temperature gamma-ray detectors have excellent potential as an essential component of space instruments to be used for high energy astrophysics. Mercuric iodide detectors are being developed both as photodetectors used in combination with scintillation crystals to detect gamma-rays, and as direct gamma-ray detectors. These detectors are highly radiation damage resistant. The list of applications includes gamma-ray burst detection, gamma-ray line astronomy, solar flare studies, and elemental analysis.

A. MERCURIC IODIDE PHOTODETECTORS[1]

Mercuric iodide (HgI_2) has been developed to a considerable extent, though not perfected, as a photodetector (PD) used in combination with a scintillation crystal to detect gamma-rays. Because of its large bandgap the material offers room temperature operation with superior electronic signal-to-noise ratio when compared to other semiconductor PDs. In addition, HgI_2 PDs offer significant advantages over the conventionally used photomultiplier tube (PMT). HgI_2 PDs coupled to CsI(Tl) scintillation crystals have yielded energy resolution superior to any other scintillator/photodetector combination.

The characteristics of HgI_2 photocells are as follows:

1. **Size.**

    There is a significant overall instrument size reduction afforded by substituting PMTs with HgI_2 PDs. PMTs are inherently large because of the electron optics and the required interdynode spacing. The volume for a 1 inch diameter active area HgI_2 PD is about one twentieth of the volume of a typical PMT with the identical active area. In instruments that might employ many detectors in arrays such as the flies eye array for directional detection, this advantage is multiplied by the number of detectors employed.

© 1990 American Institute of Physics

Feasibility for deploying large arrays of multiple detectors will be enhanced due to these size and weight advantages.

2. Active area.

Single crystal photocells with active areas ranging from 0.125 inches to 1.5 inches in diameter have been developed and tested. The detector active area is limited only by the size of currently available crystals. Larger active areas can be achieved by using a mosaic of HgI_2 crystals. The technique for fabricating crystal mosaics is in the development stage. HgI_2 PDs can be fabricated in shapes other than the conventional circular geometry to more fully cover the optical interface. A mosaic of approximately three elements will cover an 8 cm diameter scintillator such as the one used in the detector for the KONUS experiments[2].

3. Quantum efficiency.

The quantum efficiency (QE) of HgI_2 PDs is nearly unity in the visible range of the electromagnetic spectrum. PMTs on the other hand have QE's lower than 15% in this part of the spectrum. Because of this the HgI_2 PD is able to collect more of the scintillation light, and it will therefore have better photon statistics. Above 200 KeV, this results in superior energy resolution for the HgI_2 PD based spectrometer versus the PMT based spectrometer. Below 200 KeV however, the QE advantage is compromised by the dominant electronic noise in the preamplifier. We are currently working on improving the low energy spectral response.

4. Detector Performance.

The spectral performance of the HgI_2 PD based spectrometer is outstanding when compared to all other scintillation-photodetector type spectrometers: A 1 inch diameter HgI_2 PD coupled to a 1 inch diameter by 1 inch thick CsI(Tl) scintillator gave a [137] Cs resolution of 5.0% FWHM, and a 1.5 inch diameter PD coupled to a 1.5 inch diameter by 4 inch thick CsI(Tl) scintillator gave a resolution of 5.7% FWHM for the same source energy. For comparison, note that the best [137] Cs resolution ever reported using a PMT based spectrometer was 5.6% FWHM.[3]

5. Detector speed.

Detector risetimes are typically about 1 microsecond. A shaping amplifier time constant of at least four times that value is required to optimize electronic signal-to-noise. For most detectors shaping time constants of between 10 and 30 microseconds are used. This sets the limit on the count rate capability in an optimized HgI_2 PD based spectrometer to about 30 Kcps. This timing scenario also sets limits on the event time discrimination that can be used in gamma-ray burst direction determination.
Gamma-ray burst risetimes have been reported to be in the range 10-1000 ms, with durations in the range of a few tens of milliseconds to several hundreds of seconds.

6. Detector lifetime.

Some earlier PDs exhibited lifetimes of up to 1.5 years. Lifetimes are expected that far surpass this figure for more recently developed samples with improved entrance electrodes and that employ an encapsulating layer. A reasonable goal for a satellite based mission of two to three years of operating life is conceivable.

7. **Power supply requirements.**

Operating currents range from 30 pA to 1 nA for a bias of 500 volts. The DC stability of the power supply is not critical, and variations of up to ± 10% can be tolerated without affecting the PD gain. This is a significant advantage over PMTs whose DC precision is critical, as the PMT gain is highly sensitive to any variation in the DC level. The complexity of the power supply design for scintillation spectrometer instrumentation can therefore be significantly reduced when HgI_2 PDs replace PMTs.

8. **Areas of current research and future development.**

Emphasis is placed on areas which will improve important aspects of detector performance. These areas are:
a. Improvement in electronic signal-to-noise for heightened spectral response at the low energy end of the spectrum. There are two approaches that need to be investigated which have potential. The first is the development of assymetrically contacted PDs with reduced area back contacts will greatly reduce the noise to to PD capacitance. The second approach is the development of optimized pulsed or direct light feedback preamplifier electronics.
b. Techniques for increasing the detector active area in order to improve sensitivity. The approach here is the development of detector mosaics, and optimization of the geometaries and electronics associated with them.
c. Studies of new materials suitable as entrance electrodes that produce good performance increased longevity PDs. Photoluminescence studies are being conducted to determine the interfacial regions between candidate materials and HgI_2. Hydrogel contacts are being investigated.
d. The temperature dependance of HgI_2 photoresponse needs to be investigated.
e. Plastic - HgI_2 PD phoswitch anticoincidence systems need to be investigated.

B. VOLUMETRIC DETECTORS FOR DIRECT DETECTION OF GAMMA-RAYS

Mercuric Iodide is of interest as a solid-state gamma-ray spectrometer because of its large bandgap and high atomic number. These properties allow for room temperature operation and afford high gamma-ray cross section respectively. In addition, the material exhibits a much higher resistance to radiation damage than semiconductor detectors such as HpGe. Radiation damage experiments with HgI_2 detectors exposed to 1.4 GeV protons at an accumulated dose of 10^9 protons/cm^2 (equivalent to one year of bombardment by cosmic rays in space) showed no degradation in spectral response.

The characteristics of HgI_2 gamma-ray detectors are as follows:

1. **Size, efficiency, and spectroscopy considerations.**

In HgI$_2$, photoelectric absorption dominates other interaction processes to about 400 keV. At higher energies (up to several MeV) Compton scattering dominates. Detectors of thicknesses from 1.0 to 10.0 mm are used for increased detection efficiency and to minimize Compton escape events at high energies. For detectors of 2 to 4 mm thickness energy resolutions in the range 3% to 10% FWHM with peak to valley ratios of about 4.0 for the [137] Cs 662 keV photopeak are routinely obtained. In thicker detectors incomplete charge collection (hole trapping) limits the energy resolution. Gamma-ray counters of about 1 cm thickness have resolution limited to > 10%. Some work has been initiated to develop a spectral simulation model that will identify the extent of improvement required in the hole transport in HgI$_2$ in order to improve the resolution of thick detectors.

The mean free path for gamma-rays in the 400 keV to 10 MeV region in HgI$_2$ is between 1 and 4 cm, so that intrinsic detection efficiencies are between 5 % and 33 % for the 2 to 4 mm thick detectors. At 662 keV the linear attenuation coefficient in HgI$_2$ is about 0.58 cm^{-1}. This means that about 13% of the 662 keV gamma-rays are being stopped in a gamma-ray detector with typical thickness of 2.5 mm. Where more efficiency is required, it is necessary to go to thicker detectors at the expense of energy resolution.

In order to circumvent this trade-off between energy resolution and efficiency various schemes have been adopted. An approach being developed is to stack detectors to retain optimal charge collection and simultaneously fulfill the volume requirement for high efficiency. One such configuration used in a hand held delta rate meter employs eight 1 inch by 1 inch by 2 mm thick detectors where the detectors are mounted in such a way that the effective volume is 4 inches by 4 inches by 4 mm thick. In addition, pulse processing methods[4] which compensate for the incomplete hole collection have demonstrated significant improvements in energy resolution for detectors of 1 to 5 mm thickness.

2. **Detector lifetime.**

Recent advances in material processing at EG&G have allowed yields of 30% to be achieved in producing good quality HgI$_2$ gamma-ray spectrometers (3% to 11% FWHM at [137] Cs) of moderate thickness (2 to 4 mm). Yields of about 10 % were common previously. In addition, the development of new encapsulation methods have improved the long-term stability of HgI$_2$ detectors. Most recently, a set of detectors under constant test for the past six months have shown no degradation in spectral response.

SUMMARY

Mercuric Iodide detectors may be used in gamma-ray spectroscopy either directly or as a photodetector in combination with scintillation crystals. Mercuric Iodide photodetectors offer the following advantages over photomultiplier tubes: reduction in size and weight, greater response uniformity, higher quantum efficiency, improved spectral resolution, simplified biasing requirements, low supply currents, and insensitivity to magnetic fields. Mercuric Iodide semiconductor detectors for direct gamma-ray spectroscopy offer room temperature operation with good energy resolution and high efficiency. Both types of HgI_2 detectors are highly resistant to radiation damage.

A significant research and development effort directed towards applications in high energy astrophysics is still needed to bring this technology to the point it may be used with confidence in future space missions.

REFERENCES

[1] Markakis, J.M. *IEEE Trans. Nucl. Sci.*, Vol. NS-35, No. 1, Feb. 1988, 356.

[2] Mazets, E.P. & Golenitski, S.V. *Astrophys. & Space Sci.*, 75 (1981) 47.

[3] Persyk, D.E. & Moi, T.E. *IEEE Trans. Nucl. Sci.*, Vol. NS-25, No. 1, Feb. 1978, 615.

[4] Gerrish, V.M. & Beyerle, A.G. *Nucl. Inst. & Meth. Phys. Res.*, A283 (1989) 220.

This work was performed under the auspices of the US Department of Energy under Contract no. DE-AC08-88NV10617.

IV. NASA Capabilities in the 21ST Century

High Energy Astrophysics 21st Century Workshop
"Space Capabilities in the 21st Century"

Robert C. Rhome, P.E.
Assistant Associate Administrator for Space Science and Applications
National Aeronautics and Space Administration
December 13, 1989

I was asked to present to you today an overview of the infrastructure and enabling technology that NASA plans to have in place in the 21st century. To provide a clearer picture of this future infrastructure, this paper will be presented as though the 21st century has already begun, as though this were New Year's day of the year 2001. It will address the accomplishments of the 20th century, and the plans and aspirations of the next.

We've come a long way in the forty years we've been in space, and have made more leaps in technology and innovation than most of us dreamed possible. The new century spans before us with almost limitless possibilities for exploration and space science research. It is important to remember, however, that the potential for the 21st century would be less exciting without the achievements of the 20th century, particularly its last decade.

20th Century Science Accomplishments

The last ten years of the 20th century brought us more information about the universe in which we live than the total mankind has collected in the thousands of years which have preceded. Theories have been altered, textbooks changed. Basic questions have been answered, and many new ones have emerged. The last part of the twentieth century was the beginning of the momentum in space science research that is only now starting to build. The fits and starts related to budget and program uncertainties took their toll. But by faithfully implementing our Strategic Plan, we carved out an impressive science legacy (see chart 1).

The nearly complete Great Observatories program is perhaps OSSA's finest accomplishment. The Hubble Space Telescope, launched in March of 1990, continues to operate well, returning information about the universe on the visible spectrum at a rate ranging from 32 kilobytes to 1 megabyte per second. Our foresight in designing this instrument to be serviceable on-orbit has accommodated servicing missions that has and will continue to keep the $1.8 billion instrument functioning optimally for at least another five years. Although the Gamma Ray Observatory, launched in 1990, has reached the end of its intended mission life, the immense volumes of data it produced will not be fully analyzed for perhaps decades.

The Advanced X-Ray Astronomy Facility (AXAF), launched just a few years ago, is expected to function for at least ten to twelve more years. Scientists have only begun to sift through the data, since AXAF provides a 100-fold increase in sensitivity and 1,000 times more capability for spectroscopy over any previous x-ray observer. The Space Infrared Telescope Facility (SIRTF), after much debate over "who" and "where" and "how much" and "how long", has just recently been launched into a high Earth orbit. It will probe the distant and ancient universe with as much as a three order of magnitude increase in sensitivity over the Infrared Astronomical Satellite, flown more than 18 years ago in 1983. Serviceable but currently beyond the reach of our servicing technology, SIRTF is expected to remain operable for a total of at least five years, and will carry the Great Observatories program well into its second decade.

The orbiting telescope of SIRTF is paired with the somewhat delayed airborne Stratospheric Observatory for Infrared Astronomy (SOFIA) in the culmination of the Great Observatories program. Each of these two telescopes has unique and complementary capabilities for infrared studies. Coordinated observations combining the sensitivity of SIRTF with the larger aperture of SOFIA have already provided us with unprecedented observations in the infrared.

The Great Observatories are only a part -- albeit a very large part -- of the great increase in astrophysics data we've collected in the last decade. The Explorer program, both the medium-class program as well as the smaller Scout-class program, continues to provide us with frequent flight opportunities to achieve first class science objectives. The Cosmic Background Explorer (COBE), successfully launched and deployed on November 19, 1989, was the first in our medium-class Explorer program. COBE's relatively short two-year life span brought us much closer to understanding the origins of the universe.

After COBE, we launched the Explorer Platform in 1991 to provide a permanent spacecraft on which to base various missions of two to three year lengths. Its first payload, the Extreme Ultraviolet Explorer, carried out our first deep, all-sky survey in the extreme ultraviolet. Its second payload, the X-Ray Timing Explorer, carried out timing studies of compact objects, both galactic and extragalactic. The Far Ultraviolet Spectroscopic Explorer, whose Phase A study was completed in 1989, fundamentally advanced our understanding of galaxies, stars, and planetary systems on the basis of far UV studies.

Consistent with our Strategic Plan which called for moderate missions to support major science themes, OSSA initiated in the late '80's a new type of Explorer program: the Small Explorers (SMEX) Program. The objective of the SMEX program was to provide frequent, small, quick turn-around space missions. So far we've launched an average of one to two payloads per year on Scout-class rockets, depending on mission cost and availability of funds and launch vehicles. The initial missions chosen in this program were the Solar, Anomalous, and Magnetospheric Particle Explorer (SAMPEX), the Submillimeter Wave Astronomy Satellite (SWAS), and the Fast Auroral Snapshot Explorer (FAST). This was a good decision on our part, and we plan to continue these types of quick-turnaround, low cost missions well into this next century.

The activation of the Space Station provided us with a new vantage point from space to support further astrophysics investigations. Space Station attached payloads now include the Large Area Modular Array of Reflectors (LAMAR), which produces high-resolution spectroscopy of cosmic x-ray sources; and the X-Ray Background Survey Spectrometer (XBSS), which measures the energy spectrum of diffuse soft x-ray emitting regions of the interstellar medium.

This science legacy was made possible because we had developed a Strategic Plan for our science programs, and assembled innovative, inventive, and inquisitive science teams to implement it. We also exploited the evolving space transportation, communications, and technology infrastructure and capabilities to carry out these missions.

Evolution of Space Infrastructure

The space science of the future remains dependent on a robust infrastructure. While attaining assured and reliable access to space was the main concern of the last century, exploration and space research are the dominant themes of the 21st century. Without a strong infrastructure in the form of transportation -- to orbit, the Moon and beyond -- communications, space technology, and of course the Space Station, we cannot expect to continue with the incredible leaps of understanding that we started in the 1960's, some four decades ago.

Transportation in the 1990's

Because Shuttle technology was more than 20 years old by the early 1990's, it had to be vastly upgraded to improve efficiency and reliability. These upgrades included improvements to the main engines and the replacement of the old solid rocket motors with Advanced Solid Rocket Motors. The ASRM's added up to 10,000 pounds of upmass capability. An Extended Duration Orbiter capability was also added so that Columbia could support experiments with durations up to 16 days, and Endeavor could support experiments up to 28 days. Shuttle flight rates increased from nine in 1990 to an almost steady 13-14 per year to keep pace with the demand for science payloads as well as Space Station assembly payloads to be put into orbit.

We learned many lessons from the Challenger accident in 1986, and in the ensuing 14 years NASA adopted and implemented a mixed fleet strategy to assure constant access to space (see chart 2). Titan, Atlas, and Delta unmanned launch vehicles developed in the 1960's were upgraded in the '90's to augment Space Shuttle capabilities in delivering payloads to low-Earth orbit. These expendable launch vehicles (ELV's), using upper stages such as the Centaur and the Transfer Orbit Stage, were also used to launch robotic missions, such as the Mars Observer and Lunar Observer missions. The trade-off between existing technology and desired spacecraft characteristics, however, has remained the science program manager's dilemma. Building to the existing launch capability or striving to drive the evolution of that capability has never been an easy task, especially when budgets are concerned.

After the release of the President's 1988 National Space Policy, NASA was required to procure all ELV launch services commercially or through the U.S. Air Force (see chart 3). Pegasus and Taurus were among the first ELV's to be developed commercially. Pegasus is a winged vehicle that is carried aloft by a transport-class aircraft. After ignition of its first stage motor, it can reach suborbital or orbital trajectory, and is capable of placing up to 900 pounds in low-Earth orbit. Taurus, first launched in 1991 using a Pegasus-derived configuration, can place up to 3,000 pounds into polar low-Earth orbit and over 800 pounds into geosynchronous orbit.

In the late 1980's, a number of small commercial ELV providers came forward with conceptual designs for ELV's capable of launching up to 2,000 pounds to low-Earth orbit. Many of these concepts were specifically designed to evolve into larger lift capabilities. Some designs considered at the time were the American Rocket Company (AMROC), Conestoga, Liberty, and E Prime Aerospace Corporation (EPAC) concepts. Not every idea got off the drawing board -- some never got beyond our imaginations. But those concepts which most closely captured the "demand" of the space science and exploration community, flourished. Others, quite simply, did not.

With the construction of the Space Station and its growth toward becoming a Lunar transportation node, it became clear that we would need heavy lift vehicles capable of transporting far larger payloads to low-Earth orbit. One such newcomer was the Shuttle-C, developed in the mid-1990's, as a shuttle-derived cargo vehicle in which the Shuttle orbiter is replaced by a cargo carrier. The Shuttle-C's design is therefore not new, but one that expands upon the current Space Shuttle program infrastructure. It uses existing and modified Shuttle-qualified systems, including main engines, solid rocket boosters, and a slightly modified external tank with enhanced structural interfaces. Shuttle-C increased our lift capability from 17.3 metric tons with the Shuttle to 60 to 70 metric tons. While the

Shuttle-C is still in use in 2001, it is being augmented with the Advanced Launch System, discussed below.

Communications

The return to flight in the late 1980's began an unbelievable increase in demand for data tracking, acquisition and processing. The demand for near continuous coverage created the need for vastly updated earth-to-orbit communications links. The old, reliable, but antiquated network of ground stations around the globe, used since the Apollo era, were gradually replaced in the '80's and early '90's with the Tracking and Data Relay Satellite System (TDRSS). Operating with four satellites plus an orbiting spare, the system provided round-the-globe coverage for Earth-orbiting payloads. The TDRSS maintained Earth-to-orbit communications more quickly and with better reliability, especially with the addition of the second ground terminal at White Sands in early 1993. This space-based network provided about 90 percent coverage for low Earth missions such as the Space Station.

But even before the TDRSS was completed in the mid-1990's, we recognized that it could not handle nearly enough data, at nearly the needed rate. The increased spacecraft capacity of the late 1990's required an enhanced replacement just to keep even with the demand. The science community has a voracious appetite for information, and we have nearly always struggled to keep up with it. The Advanced TDRS System (ATDRSS), which we began to implement in 1997 to 1998, was designed to improve operational capability, meet the performance needs of future users, and reduce the overall life cycle costs for the system. Current projections indicate that the system will be sufficient to satisfy various communications requirements for the next ten to fifteen years.

In terms of ground networks, the Deep Space Network (DSN) continues to be used as a world-wide telecommunications network that provides tracking and data acquisition support to a variety of interplanetary and Earth-orbiting spacecraft (see chart 4). The DSN has the most sensitive capability in the world for receiving very weak signals from distant spacecraft, as well as performing observations in radio astronomy. The primary elements of the DSN are three tracking complexes located at approximately 120 degree intervals around the Earth. These complexes provide nearly continuous tracking of deep space spacecraft, Earth-orbiting spacecraft, and the Shuttle.

Quite clearly, communications networks are vital to operating space vehicles and observatories, and to acquiring and disseminating information gathered from space. The NASA Communications Network, or NASCOM, is a global network interconnecting tracking and data acquisition facilities with NASA centers that has been operating for a number of years now. NASCOM provides operational support to NASA missions in carrying telemetry, command, voice, and data information. The NASCOM has evolved over the years into a totally digital network using communications satellites optical fiber circuits, to support rapidly increasing data rates and volumes. Matching user needs with systems capability on a time scale, consistent with hardware development schedules and available budget resources, however, remains a hurdle to this day.

Space Station Freedom

The Space Station Freedom was NASA's "infrastructural leap" in the last decade of the 20th century. It provided a permanent manned base from which to study the Earth, the universe, and man's ability to adapt to space. Today, at the beginning of the year 2001, the Assembly Complete phase of the Space Station Freedom has been finished for nearly two years. It took a lot of work to get to this point -- 29 Shuttle trips over 5 years, starting in

1995, to assemble, outfit, and logistically support the manned base. In addition, OSSA provided Science Utilization Flights to exploit the early Station opportunities in support of science investigations. Assembly of the Station in orbit was a challenge of enormous proportions. But the task was one which blazed the trail for ambitious missions in the future which will require on-orbit assembly, test, checkout and operations. There were some capability casualties along the way, but we managed to slowly "buy back" essential elements lost in the budget throes of the early 1990's.

The baseline Station includes the U.S. elements, international components, and two unmanned platforms. The heart of Freedom's manned based is a horizontal boom structure, 154 meters, or 508 feet long. Four special purpose modules -- two U.S., one European, and one Japanese -- are attached to the boom at its midpoint. Each module has an atmosphere nearly identical to Earth's: 80 percent nitrogen and 20 percent oxygen kept at sea-level pressure. In these modules, eight men and women perform experiments, monitor the space complex, maintain equipment, and handle repairs.

The Station supports research in a wide variety of disciplines, including microgravity, material processing, medicine, Earth observation, life science, astronomy, and space physics. The Station is also used to develop and advance a broad range of space technologies such as automation and robotics, advanced structures and materials, power generation, space electronics, and communications.

Because of the extensive work done in and around the Station, a number of remote and robotic tools were needed. In the mid-1990's the Orbital Maneuvering Vehicle (OMV) was tested and completed to be used as a small, reusable free-flying spacecraft, a "space tug" to service orbiting spacecraft. Because of budget constraints, the OMV was initially Shuttle-based and therefore not available for use at or near Station when the Orbiter was not present. The OMV is now used mostly in conjunction with Station operations, capable of delivering payloads to higher altitudes; refueling, reboosting, or otherwise repositioning spacecraft; and, as needed, deorbiting spacecraft in a controlled fashion. It will continue to be used to service and upgrade the Station, as well as to service low-orbit payloads such as the Hubble Space Telescope and AXAF.

Robotic servicing has been a critical element in the construction, maintenance, and repair of space infrastructure and space experimentation. Robots like the Flight Telerobotic Servicer (FTS) perform tasks which would otherwise result in an always risky Extravehicular Activity (EVA). Satellite repair and spacecraft refueling are just two examples of robotic servicing that could be performed under the control of Freedom's crew members. Space robots ultimately will have the work capability of a space-suited astronaut, helping in the on-orbit assembly and maintenance of the manned base, and minimizing the amount of EVA and therefore the risk involved. The concept of marrying the FTS with the OMV to further enhance satellite servicing has been under study and may well come to fruition in the early part of the 21st century.

A Canadian Mobile Servicing System (MSS), equipped with a manipulator arm, also has been installed to assist in Freedom's assembly. The MSS is based on the Canadian expertise acquired in creating the Shuttle-based Canadarm, used extensively in the '80's and '90's. The MSS's robot arm also further reduces the need for EVA's, and helps deploy, dock and redeploy a visiting Orbiter. The MSS is mounted so that it is able to not only translate along Freedom's truss structure, but also able to change the plane of the truss on which it operates. It can therefore assemble, retrieve, and transport payloads around the Station with relative ease.

21st Century Infrastructure

From the vantage point of the year 2001, we have infinite possibilities for where to go next. Although we have relied in part on our collective imaginations to describe the future of space exploration, we are guided to a large degree by early work performed in 1989 related to human exploration of the Moon and Mars. This is what preliminary glimpses of the future, as we may envision it, will look like (see chart 5). While NASA is placing equal planning emphasis on exploration of Mars as on exploration of the Moon, most of the comments here will center on Lunar exploration, since Lunar exploration and habitation appears more immediately conducive to astronomical studies.

The 21st Century Space Station

To maintain tight control over costs, NASA took an evolutionary approach to building Space Station Freedom. Station's ability to evolve is critical to any subsequent role as an exploration support base, since the thrust in the early part of the 21st century will be to first more fully explore the Moon, and then Mars and beyond. In addition to the continued support of its science role, Freedom will perform two essential functions in these new exploration initiatives. First, Freedom will serve as an on-orbit laboratory for conducting research and developing technology required for exploration missions. Freedom is the ideal location for such research because no terrestrial laboratory can adequately simulate the characteristics of the space environment.

Second, it will serve as a transportation node for assembling, testing, processing, servicing, and recovering Lunar and Mars vehicles. As such, it can also supply crew support, data management and communications systems, and logistics services for these activities.

Four major milestones may very well mark Station's evolution as a research and technology test-bed and as a transportation node. The first post-Assembly Complete configuration will support the Lunar transfer vehicle verification flight. The truss structure will be augmented to include lower keels and a lower boom, and the power will be increased from 75 kilowatts to 125 kilowatts. The crew will be increased from 8 to 10, and a service track assembly for the Lunar transfer vehicle will also be added.

The second configuration will support expendable Lunar transfer vehicle operations. A second habitation module will be added to accommodate the transient Lunar mission crew of four. A Lunar transfer vehicle hangar also will be added to the service track assembly to protect the vehicle from orbital debris damage during construction.

The third configuration will support reusable Lunar transfer vehicle operations. Two permanent crew members will be added to support increased life sciences research, Lunar transfer vehicle servicing operations, and maintenance of Freedom. Additional solar dynamic power units will increase on-board power to 175 kilowatts, and a second remote manipulator arm will be added.

The fourth configuration, and at this point the last in the preliminary exploration sequence, will be designed to support Lunar and Mars operations. This "sporty" version of the Station boasts upper keels and booms as well as a support structure for on-orbit assembly and checkout of the vehicle elements, in addition to a mars vehicle assembly facility.

At this stage, Freedom will have 12 permanent crew members, 4 transient Lunar or Mars crew, enclosed Lunar/Mars vehicle hangars, and two remote manipulators and mobile transporters.

The Lunar Base

The next step in the exploration strategy will most likely be the development of a permanent Lunar outpost -- the timeframe for which is unclear even now -- but clearly the initial steps will take place within the first decade of the 21st century. The first two missions to the Moon will be unmanned cargo missions delivering the equipment necessary to begin the operation. The cargo missions will most likely be followed by a piloted mission with a crew of four, which will stay on the surface for up to 30 days. The crew will begin to set up the habitation module and support systems, as well using the rover for location studies. Within the next decade, conceivably as early as the year 2012, the Lunar outpost will be permanently manned, with a crew of 12 living there for up to one year at a time. But we can expect several budget battles between here and there.

Nonetheless, when the pressurized laboratory module is in place, indepth scientific experiments will begin. Within a few years, a manned pressurized rover will allow us to support scientific exploration activities at a distance from the outpost, and excursion vehicles will allow us to tend experiments on the farside of the Moon.

Transportation

An effort of this magnitude will require a vastly increased lift capability, particularly for Mars exploration. In the late 1990's, we began designing the Advanced Launch System (ALS), an unmanned family of launch vehicles designed to deliver to orbit broad ranges of cargo size and mass. Like the Shuttle-C, the ALS will probably be shuttle-derived, and its primary design objectives will be low cost per flight, high reliability, and high operability. These vehicles will deliver up to 140 metric tons to orbit, carrying payloads of up to 12.5 meters in diameter and 30 meters in length. For Lunar payloads, one concept is to use two-booster vehicles with a 98 metric ton lift capacity. For Mars payloads, a three-booster vehicle with up to 140 metric ton lift capacities may best suit our needs.

The plans for Lunar/Mars exploration create a demand for personnel transport and support of the Space Station that extends well beyond the currently projected life of the existing Shuttle fleet. The Shuttle design is now almost 30 years old. Not only do we now need a more efficient and reliable way to put crew in orbit, we also need an assured crew return capability in case of emergency. Analyses of options for an assured crew return capability have now been underway since the early 1990's, and we are also currently studying an Advanced Manned Launch System to more reliably and quickly put men into orbit.

The Lunar/Mars exploration plans also call for improved methods of transporting cargo and personnel to the respective bases. We envision multiple reuse interplanetary transfer vehicles to transport people and cargo from low-Earth orbit to low-Lunar or Mars orbit, and for planetary excursion vehicles for transportation between planetary orbit and the surface. Of course, much more work needs to be done in this area.

A typical Earth to Lunar Base scenario might look like this. At the Station, the crew, payloads, and propellants are loaded onto the Lunar transfer vehicle that will take them to low-Lunar orbit. The Lunar transfer vehicle meets with an excursion vehicle in Lunar orbit, which will either be parked in Lunar orbit or will ascend from the Lunar surface. Crew and payload are transferred, the excursion vehicle descends to the Lunar surface, and the transfer vehicle will return to the Station. The transfer vehicles will be serviced and maintained at the Station, and the excursion vehicles will be serviced and maintained from the outpost.

Telecommunications, Navigation, and Information Management

As with the other elements of space infrastructure, these exploration initiatives require an evolutionary development in communications and data management support capabilities (see chart 6). Being able to tell others of new insights, or being informed by someone or some device of new scientific data, is crucial to the "businees" of science. It is no surprise, then, that the increased demand for communications capabilities that began with support of robotic missions, now continues with the Lunar outpost missions, and will continue with the human exploration of Mars. The proposed telecommunications, navigation, and information management architectures for these exploration initiatives require a veritable plethora of:

- o Earth-based acquisition and control networks
- o Lunar and Mars telecommunications relay networks
- o Navigation support elements
- o Data acquisition and control network for robotic missions
- o Human mission operations support

Data required for the Lunar missions will obviously include high- and low-rate video, voice, science and engineering telemetry, and commands for transmission on the link between the Moon and Earth. It has been estimated that we will need a preliminary data rate of 10 megabytes per second to handle compressed, high-rate video transmissions. Mission interfaces for Lunar and Mars exploration will exist at four different locations: in transportation vehicles, on the Space Station, at Lunar and Mars surface terminals, and on Earth.

Low-Earth orbit service will be provided by the ADTRSS, as mentioned before, and will also support Earth launches and landings to and from the Station. Beyond low-Earth orbit, service will be provided by an expanded Earth-based Deep Space Network. A centralized, dedicated support facility will control and monitor the Earth-based tracking and data acquisition network. The capability will exist to support several separate processing data strings at once. Unattended operations techniques are currently being developed using expert systems. These systems may reduce the number of operators required for operations and data quality assessment, permit rapid detection of data path failures, and provide automatic rerouting to available channels.

At the Lunar outpost, users will communicate with Earth through two paths: to the outpost terminal by a hardwire and then to Earth, or through a direct link to Earth. Lunar transit vehicles will also be able to communicate directly to Earth. These Moon-to-Earth links will be supported by up to two 34-meter antennas installed at each complex of the Deep Space Network.

At the Lunar farside, experiments will be linked to the surface terminal by hardwire, then via a relay satellite to Earth. Missions beyond reach of the surface terminal will use radio links through the relay satellite to Earth. A Lunar telecommunications relay satellite will be positioned in orbit behind the farside to provide coverage for orbiting vehicles, farside surface terminals, and the orbit insertion of piloted vehicles.

Space Technology

A number of critical technological developments have taken place to make Lunar habitation a near-reality. Getting there and staying there meant that we needed certain enabling technology, much of which is still being developed. The Pathfinder Program, begun in 1989, has provided us with most of this technology.

Mobile robotic systems were developed for automated and human exploration of extensive areas of the Moon and Mars. Mobility, guidance and avoidance of hazards, and lightweight, compact power technology had to be developed for these planetary rovers. The Sensor and Guidance systems for the Autonomous Lander were developed to provide safe landings at potentially hazardous sites, as well as autonomous resupply operations near the Lunar outpost. Power is a critical element not only for support of the outpost, but support of various Lunar experiments as well. High-performance, low-mass, high-reliability surface power had to be developed, including both nuclear and solar power systems. Optical communications technology enabled us to use lasers to transmit high rate data from distant spacecraft with highly efficient small-size optical transceivers. The development of advanced vehicle propulsion systems helped us get there faster and safer, and use of technology such as aerobraking meant we could get faster reentry more efficiently. All of these developments were driven by some operational requirement.

21st Century Space Science

Once we have the space infrastructure in place we can get down to the business of what we're there for: science. The early studies published in 1989 gave us our first glimpse into the future of space science as it relates to human space exploration in the 21st century. These assessments offered a reference set of instruments such as the Lunar Transit Telescope, the Optical Imaging Interferometer, the Submillimeter Imaging Interferometer, and the Large Optical Telescope. Farside studies such as these will provide us with yet more information about the universe from a stable, dark platform. Arenas for discussion like the High Energy Astrophysics 21st Century Workshop help to make concepts like these and many others a bit closer to becoming a reality. When the 21st century infrastructure and enabling technology I have talked about today is in place, we will be able to attain leaps in understanding that make those in the 20th century pale by comparison.

This overview was presented from a 21st century perspective, just to give a good idea of how we think things will look, from where we stand now. But we didn't start in the 21st century, we started in the 20th century -- and continue today, December 13, 1989. Clearly, the space infrastructure we find in the 21st century will be a direct by-product of the vision we define now for a robust Astrophysics science program (see charts 7-13).

There is no doubt that with every new technological development and every new scientific discovery between now and the year 2001, the picture will change. But the basic goals we have now will not change: to seek to better understand the universe in which we live. There is still, and will always be, so much to be done. Despite the grand scale of our achievements now, they will no doubt only prove to have been a beginning.

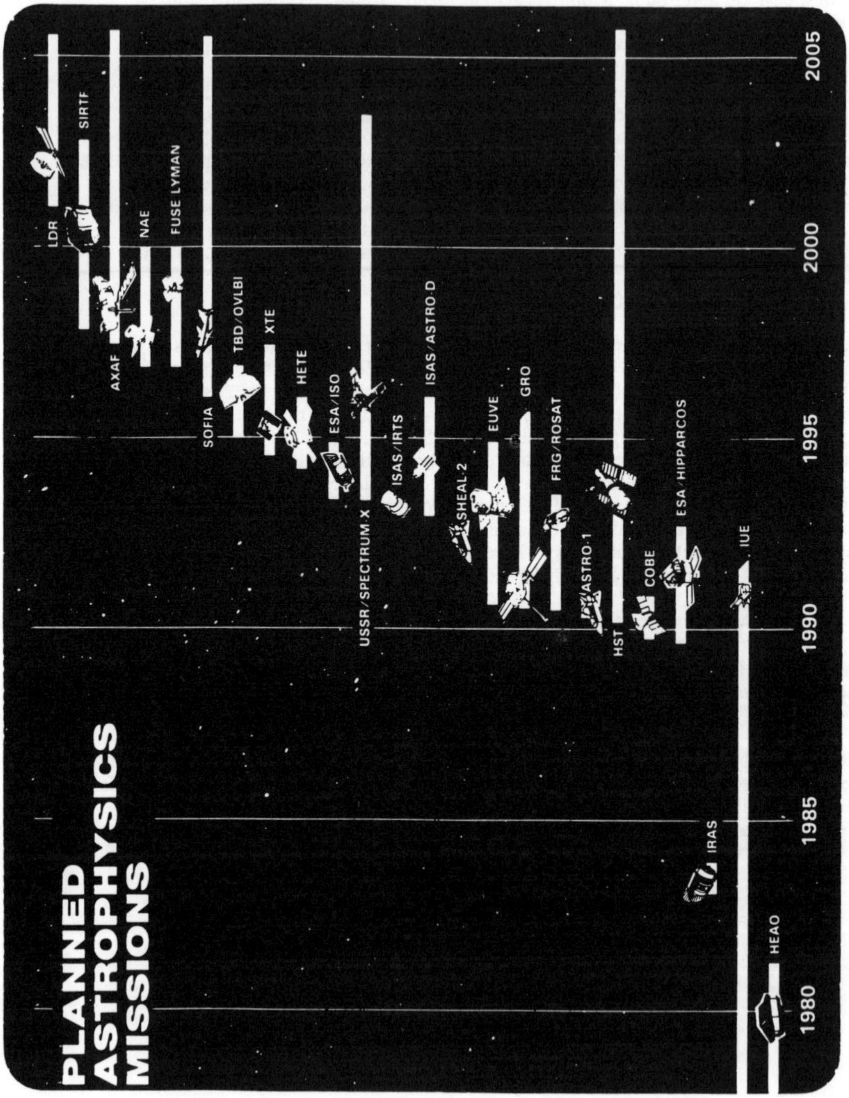

274 Space Capabilities in the 21st Century

Chart 1

U.S. EXPENDABLE LAUNCH VEHICLES

VEHICLE	AVAILABILITY	PERFORMANCE (I = 28°) LBS			PAYLOAD FAIRING DIAMETER	ESTIMATED VEHICLE LAUNCH SERVICES COSTS 1989 $
		LEO	GTO	GSO		
SCOUT - ETR - WTR	NOW	570 460			2.9 AND 3.5 FT	$10-20M
DELTA II MODEL 6925 MODEL 7925	FEBRUARY 1989 JUNE 1990	8,780 11,110	3,190 4,010	1,600 2,000	8 AND 9.5 FT 10 FT (FEBRUARY 1990)	$40-50M
ATLAS I	NOW	13,000 12,550	5,150 4,950		11 FT 14 FT	$65-70M
ATLAS II*	EARLY 1991	14,950 14,500	6,100 5,900		11 FT 14 FT	$70-80M
ATLAS IIA	1992	15,700 15,250	6,400 6,200		11 FT 14 FT	$80-90M
ATLAS IIAS	1992	19,000 18,500	8,000 7,700		11 FT 14 FT	$100-110M
TITAN II * (WTR)	SEPTEMBER 1988	4,200			10 FT	$30-35M
TITAN III WITH SRMU	JULY 1989 LATE 1991	30,500 38,000	11,000		13.1 FT	$130-145M
TITAN III / TOS	JULY 1989		13,000		13.1 FT	$145-200M
TITAN III / IUS WITH SRMU	JULY 1989 LATE 1991			4,200 5,000	10 FT	$145-200M
TITAN IV * / NUS WITH SRMU	EARLY 1989 MID 1991	39,000 49,000			16.7 FT	$175-230M
TITAN IV * / IUS WITH SRMU	EARLY 1989 MID 1991	49,000	15,000	5,200 6,600	16.7 FT	$230-300M
TITAN IV * / CENTAUR WITH SRMU	MAY 1990 MID 1991			10,200 13,500**	16.7 FT	$230-300M

* NOT COMMERCIALLY AVAILABLE
** CURRENT CENTAUR IS STRUCTURALLY LIMITED TO 11,500 LBS

Chart 2

SMALL ELV COMMERCIAL LAUNCH COSTS

COST PER POUND TO ORBIT ($89)

	QUOTED PRICE ($M)	PAYLOAD TO POLAR ORBIT (LBS)	COST / POUND TO ORBIT ($)
SCOUT	10 - 12	460	21,739 - 26,090
SCOUT II	15	920	16,304
ATLAS E	N/A	1,800	N/A
TITAN II	N/A	4,200	N/A
PEGASUS	6.3	840	7,500
TAURUS	15.0	3,000	5,000
ILV-S	7.5	590	12,712
ILV-I	12.0	3,000	4,000
CONESTOGA II thru V	10 - 20	900 -2,200	9,090 - 11,110
EPAC S-I	25 - 35	1,200	20,830 - 29,170
S-II	20 - 40	4,500	4,440 - 8,890
S-III	50 - 60	9,040	5,530 - 6,640
S-IV	70 - 80	12,520	5,600 - 6,400
POSEIDON C3	TBD	850	TBD
LIBERTY	5.0*	400 - 1,530	12,500

* BUY OF THREE

Chart 3

ML-89-9-37
11/03/89

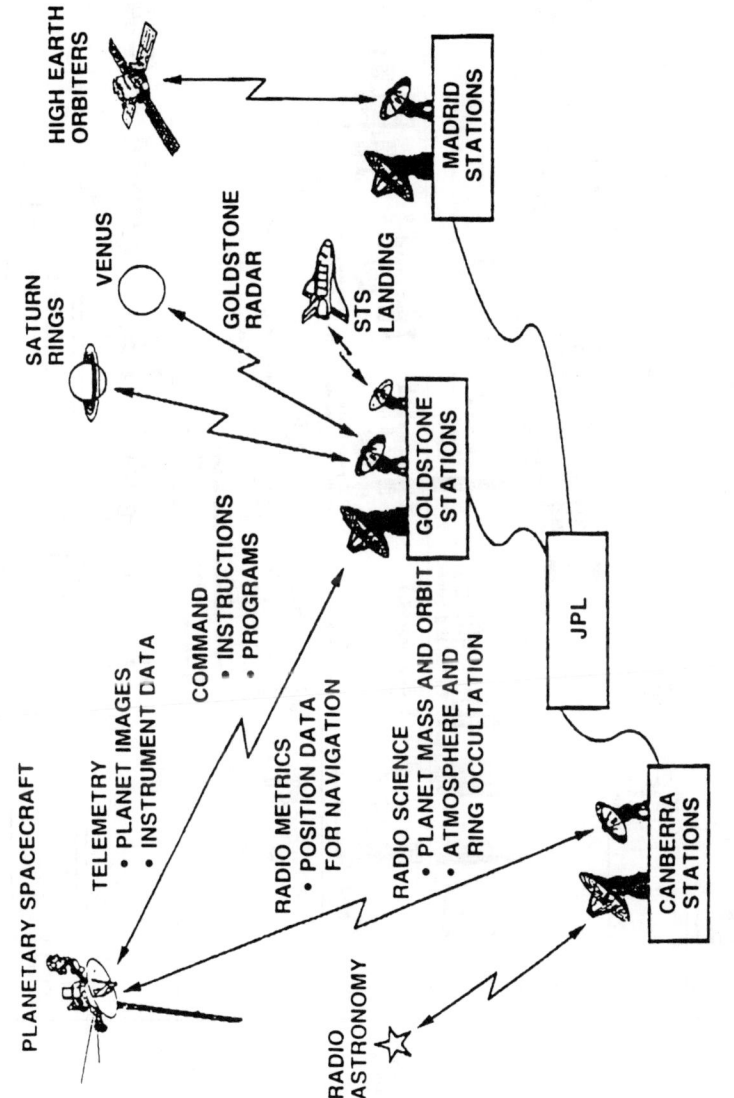

Chart 4

278 Space Capabilities in the 21st Century

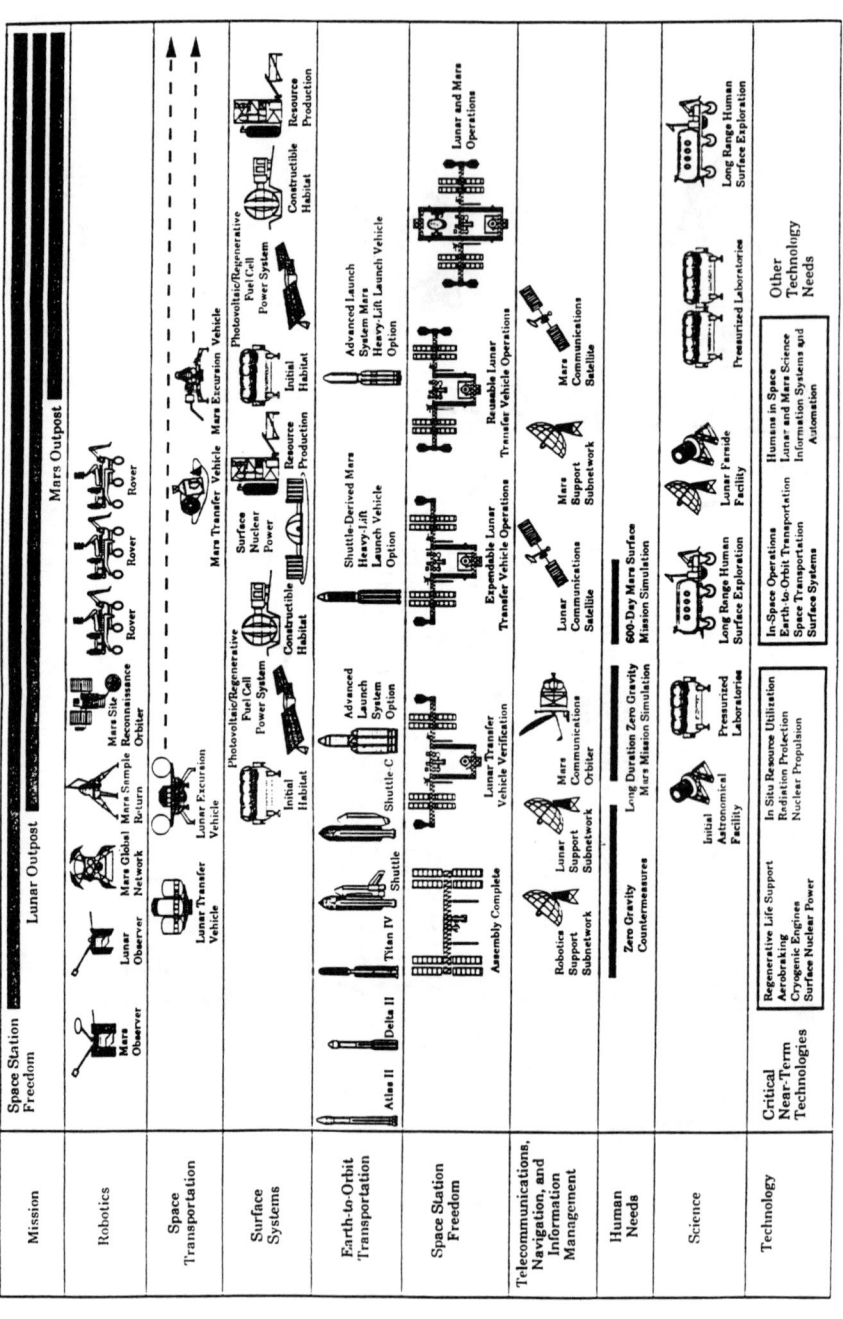

Chart 5

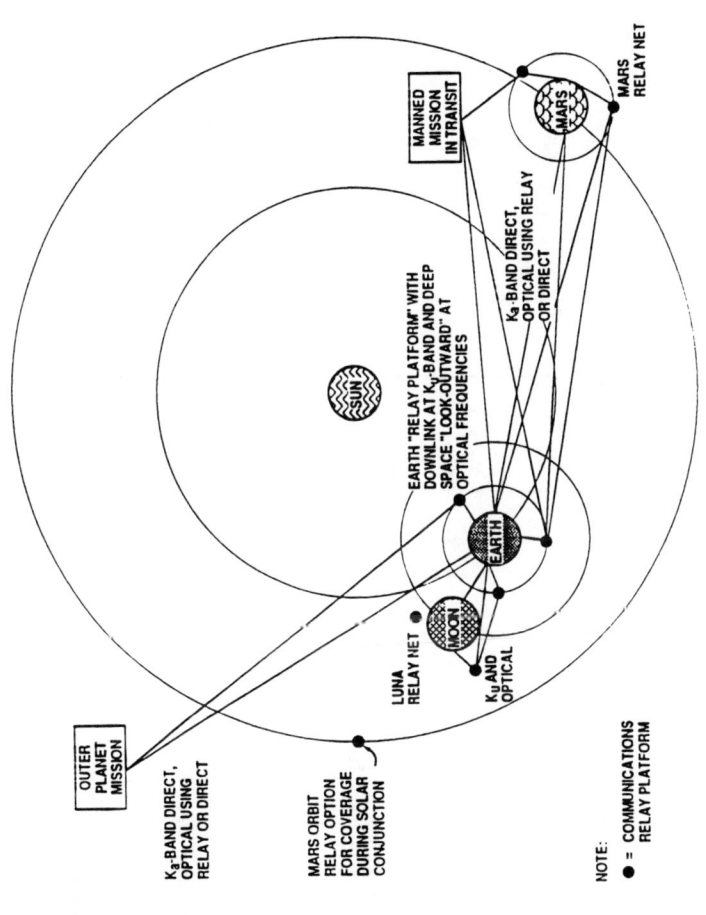

Chart 6

OSSA INTEGRATED MISSION LIST

Date: 12/22/89
Page: 1

SUB-DISCIPLINE	MISSION NAME	ACRONYM	ACTIVITY	LAUNCH DATE	LAUNCH VEHICLE	COMMENTS
ASTROPHYSICS	Advanced Very Long Baseline Interferometry	AVLBI	DEL	2005	ELV	Evolves from VLBI Joint USSR/JAPAN/OSSA Project; requires DSN
	Advanced X-Ray Astrophysics Facility	AXAF	DEL	1997	STS	
	Advanced X-Ray Astrophysics Facility	AXAF	SVG	2001, 2004, 2007, 2010	STS	SVG on 3 yr centers; last SVG mission retrieves AXAF in 2012
	Astronomical Obs./Broad-Band X-Ray Telescope	ASTRO/BBXRT	LVA	1990	STS	ASTRO 1
	Cosmic Ray Isotope Experiment	CRIE	DEL	1990	ELV	Co-manifested with CRRES on an Atlas I/June 1990
	Diffuse X-Ray Spectrometer	DXS	DEL	1991	STS	

Chart 7

OSSA INTEGRATED MISSION LIST

Date: 12/22/89
Page: 2

SUB-DISCIPLINE	MISSION NAME	ACRONYM	ACTIVITY	LAUNCH DATE	LAUNCH VEHICLE	COMMENTS
ASTROPHYSICS	Energetic Heavy Ion Composition	EHIC	DEL	1991	ELV	Co-manifested with NOAA-I on an Atlas E/May 1991
	EXPENDABLE EXPLORER PROGRAM	EEP				EEP intends to use Delta-class ELVs to launch larger P/Ls than the SMEX program
	Nuclear Astrophysics Explorer	NAE	DEL	1999	ELV	
	Follow-On Payloads TBD	EEP TBD	DEL	2001, 2002, 2003, 2004, 2005, 2006, 2007, ...	ELV	18 month launch windows
	EXPLORER PROGRAM					
	Extreme Ultraviolet Explorer	EUVE	DEL	1991	ELV	
	Extreme Ultraviolet Explorer	EUVE	SVC	1994, 1996, 1998, 2000, 2002, 2004, 2006, ...	STS	First SVC carries XTE; next SVC will be FUSE; subsequent SVCs TBD. Launches could occur one/year if international joint ventures are selected.

Chart 8

OSSA INTEGRATED MISSION LIST

Date: 12/22/89
Page: 3

SUB-DISCIPLINE	MISSION NAME	ACRONYM	ACTIVITY	LAUNCH DATE	LAUNCH VEHICLE	COMMENTS
ASTROPHYSICS	Advanced Composition Explorer	ACE	DEL	1997	ELV	ELV (Delta-Class) required due to interplanetary orbit. Space Physics (Code ES)
	Gamma Ray Observatory	GRO	DEL	1990	STS	
	Gravity Probe-B	GP-B	DEL	1996	STS	
	High Energy Transient Experiment	HETE	DEL	1994	STS	Joint project with Argentina
	Hubble Space Telescope	HST	DEL	1990	STS	
	Hubble Space Telescope	HST	SVC	1993, 1995, 1996, 1998, 2000, 2002, 2005	STS	Last SVC mission retrieves HST

Chart 9

OSSA INTEGRATED MISSION LIST

Date: 12/22/89
Page: 4

SUB-DISCIPLINE	MISSION NAME	ACRONYM	ACTIVITY	LAUNCH DATE	LAUNCH VEHICLE	COMMENTS
ASTROPHYSICS	Large Deployable Reflector	LDR	DEL	2001	STS	
	Large Deployable Reflector	LDR	SVG	2003, 2005, 2007, 2009	STS	SVG on 2 yr centers
	Roentgensatellite	ROSAT	DEL	1990	ELV	Joint project with West Germany
	Shttl Pallet Sat-Orb & Retrvbl Far & Extr UV Spect	SPAS-ORFEUS	LVA	1992	STS	ORFEUS is the first flight of the ASTRO-SPAS carrier. Retrieval on same mission as delivery.
	Shuttle Relativity Explorer	STORE	LVA	1993	STS	Mission under review
	SMALL EXPLORER PROGRAM	SMEX				
	Fast Auroreal Snapshot Explorer	FAST	DEL	1993	ELV	Scout-class; Space Physics (Code ES)

Chart 10

OSSA INTEGRATED MISSION LIST

Date: 12/22/89
Page: 5

SUB-DISCIPLINE	MISSION NAME	ACRONYM	ACTIVITY	LAUNCH DATE	LAUNCH VEHICLE	COMMENTS
ASTROPHYSICS	Solar Anomalous Magnetospheric Particle Explorer	SAMPEX	DEL	1992	ELV	Scout-Class; Space Physics (Code ES)
	Submillimeter Wave Astronomy Satellite	SWAS	DEL	1992	ELV	Scout-Class; Astrophysics (Code EZ)
	Follow-On Payloads TBD (Small Explorer Program)	SMEX TBD	DEL	1994, 1995, 1996, 1997, 1998, 1999, 2000, ...	ELV	Approximately two flights/yr from 1994 on
	SPACE STATION ATTACHED PAYLOADS	SSAP	DEL	2002, 2003, 2004, 2005, 2006, 2007, 2008, ...	STS	SVC on approximately 1 year centers for instrument changeout
	Large Area Modular Array Of Reflectors	LAMAR	DEL	1998	STS	
	X-Ray Background Survey Spectrometer	XBSS	DEL	1997	STS	Retrieval after one year

Chart 11

OSSA INTEGRATED MISSION LIST

Date: 12/22/89
Page: 6

SUB-DISCIPLINE	MISSION NAME	ACRONYM	ACTIVITY	LAUNCH DATE	LAUNCH VEHICLE	COMMENTS
ASTROPHYSICS	SSAP CONCEPT STUDIES:					
	Energetic X-Ray Observatory for the Space Station	EXOSS	DEL		ELV	
	High Res. Imaging Spectroscopy at Terahertz Freq.		DEL		STS	
	Orbiting Stellar Interferometer	OSI	DEL		STS	
	Soft X-Ray Telescope	SXT	DEL	1991	ELV	NASA instrument on Solar-A launched aboard a Japanese ELV
	Space Infrared Telescope Facility	SIRTF	DEL	1998	ELV	SVC mission eliminated
	Spectrum-X	MOXE/SXRP	DEL	1993	ELV	NASA instrument on USSR flight

chart 12

OSSA INTEGRATED MISSION LIST

Date: 12/22/89
Page: 7

SUB-DISCIPLINE	MISSION NAME	ACRONYM	ACTIVITY	LAUNCH DATE	LAUNCH VEHICLE	COMMENTS
ASTROPHYSICS	X-Ray Multi-Mirror Mission	XMM	DEL	1998	ELV	NASA instrument on ESA flight

DISCUSSION

Giuseppina Fabbiano

You said that C. Pellerin has science missions planned for 2000-2010. What are these?

Rhome

The Astrophysics Division has formulated a strawman plan for astrophysics that can be accomplished from a lunar base. This scenario includes optical, infrared, and submillimeter interferometers. In addition, it is likely that continuing use will be made of the Space Station Freedom, as well as low- and high-Earth orbit platforms for astrophysics in the 21st century.

For a list of currently planned astrophysics missions, please see the OSSA Integrated Mission List — Astrophysics, included in my presentation.

Steven Kahn

In you discussion of the Astrotech 21 program and of the relation to OAST, you made a distinction between "enabling technology" and "technology enhancement." How does NASA define "enabling technology"? If the tolerance on a given device needs to be a factor of a few better, is that enabling technology or technology enhancement?

Rhome

"Enabling technology" is technology that must exist before a specific project or mission can proceed. "Technology enhancement" includes improvements to an existing technology's quality, performance, and/or science return. An example of enabling technology is that of the closed loop life support system, the development of which is essential before manned exploration of Mars is possible. Examples of technology enhancement are imporvements to optical communications technology needed in Space Station evolution and improvements to current launch vehicle capabilities to increase mass to orbit.

Gerald Share

Many of the high-energy experiments would benefit from a low-inclination orbit. A near equatorial orbit would reduce background considerably. Are there any opportunities for launching U.S. vehicles into low-inclination orbits?

Rhome

There are a number of opportunities for launching spacecraft into low-inclination orbit. The determining factor in inclination is the size of the spacecraft (i.e., payload). Currently, NASA is using Scout-class expendable launch vehicles, and which launch site that is used to obtain a certain inclination is very mission specific and must be determined on a case-by-case basis.

The standard orbit for the Shuttle is 28°, and larger Shuttle-attached payloads can routinely be flown at this inclination for periods up to 16 days (when an Extended Duration Orbiter kit is manifested with the payload). The Space Station also will operate at 28° for those attached payloads requiring durations well beyond 16 days.

(Unknown)

What is the true cost of Space Shuttle launches?

Rhome

The marginal cost of the last Shuttle launch in an annual series of launches is about $90 million.

(Unknown)

Is it possible that the scientific capabilities of Space Station Freedom might be enhanced through negotiation with our international partners?

Rhome

Yes; for example, the U.S. budget for hardware used to support science instruments is finite. It is likely that we may be able to negotiate with our international science partners such that they would fund, design, build, test, and provide accommodation hardware (pointers, etc.) in exchange for a U.S. offer to European scientists to use U.S.-allocated space/resources on the Space Station truss for a specific period of time.

V. Mission Concepts

Kevin Hurley and Richard Mushotzky, Chairmen

X-RAY INTERFEROMETRY FOR SUB-MILLIARCSECOND AND SUB-MICROARCSECOND IMAGING

Christopher Martin

Columbia Astrophysics Laboratory
and Department of Physics
Columbia University

Abstract

X-ray interferometry has the potential for observations of unprecedented angular resolution and far-reaching astrophysical consequences. I discuss two x-ray interferometer observatory concepts: a sub-milliarcsecond interferometer which can resolve nearby stellar coronae and jets in AGN's, and a sub-microarcsecond resolution design which could resolve and measure the diameters and inclinations of accretion disks in AGNs and determine the orbital parameters of x-ray binaries. The designs exploit the large angle reflection capabilities of multilayer mirrors, and use proven optical interferometric techniques being developed for gravity wave beam detectors to provide the necessary metrology. Observations of sources fainter than 1% of 3C273 and Capella should be possible. For the sub-microarcsecond concept, a lunar platform is essential for providing a stable, kilometer-length baseline.

I. The Motivation

The angular size of a variety of x-ray emitters is given in Figure 1.

Active Star Coronae. Imaging stellar coronae would provide critical information on the emission volume, distribution, magnetic field configuration, relationship to companions in binary systems, and the time variability of these conditions.

Active Galactic Nuclei. The prospect of actually imaging an accretion disk around a massive black hole stands as unprecedented experimental challenge with a fantastic payoff. Resolution of an accretion disk would stand as definitive proof of the massive black hole/accretion disk paradigm for AGNs. The aspect ratio would give the disk thickness and inclination, the position angle the orientation on the sky and relative to radio and optical features such as jets and the galactic disk. We could search for X-ray jets, and determine their contribution to the total x-ray flux.

X-ray Binaries. With astrometric capability that can be incorporated in an interfero-

292 X-Ray Interferometry

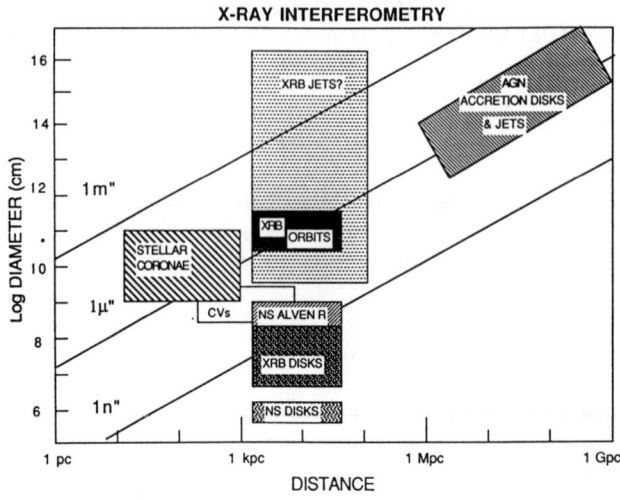

Figure 1: Angular diameter of various x-ray sources.

metric observatory, it would be possible to trace the orbit of the x-ray source in a binary system, and make an unambiguous determination of the orbital elements and stellar masses. We could also search for extended emission in x-ray binaries produced by jet-like structures that have been observed in the radio at larger distances.

II. Design Guidelines

To exploit the significant potential of x-ray interferometry, we require angular resolution < 0.1-1 milliarcsecond and eventually < 0.1 microarcsecond, sensitivity < 0.1 F(3C273) and sufficient to image ∼100 nearby stellar coronae, as well as good position angle and baseline coverage. These requirements can be met with a two-beam Michelson interferometer. Two samples of the wavefront, separated by a baseline B are recombined coherently to produce interference. The amplitude of the fringe contrast, obtained by varying the relative path lengths over one wavelength, is called the *visibility amplitude*. This gives considerable information about the spatial extension of the source, and exact information when the source distribution is known (as it is for visible stellar disks). To reconstruct true images, the *visibility phase* must be recovered as well. An x-ray version of the Michelson two-beam interferometer can in principle be constructed. A proposed 20 m baseline interferometer is illustrated in Figure 2. However, a number of considerable technical challenges must be surmounted.

1. Path length equalization: Coherent interference between two widely separated wavefront samples occurs only when the path lengths are equal to within a coherence length

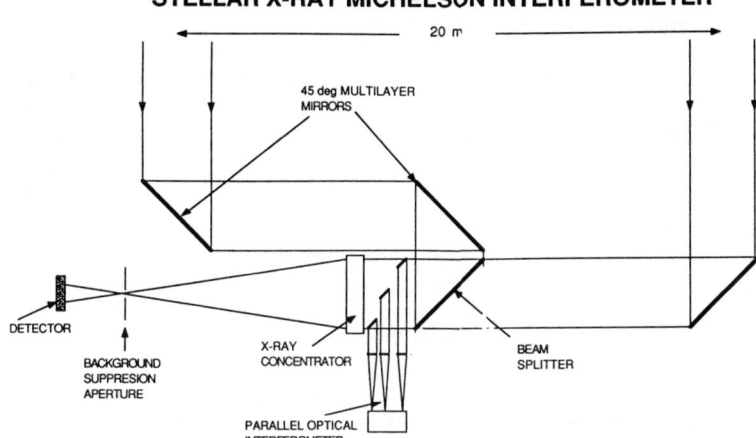

Figure 2. X-ray Michelson interferometer for 0.1 milliarcsecond resolution at $\lambda \simeq 50$Å.

$l_c \simeq \lambda(E/\Delta E)$. In the x-ray band, reflections must be minimized for reasons of efficiency, and normal incidence reflections cannot be exploited to permit path-length compensation. Symmetry of the two optical paths is essential, and the use of sidereostats limited. Active control of the optical paths must be used during the observation to maintain the equality to within $l_c \simeq 50$Å $\times 100 = 0.5$ μm. This can be accomplished using a parallel optical interferometric metrology system, and piezo-electric actuators.

2. Full-aperture path length knowledge to $0.1\lambda_x$: The most formidable challenge in the construction of an x-ray interferometer is maintaining wavefronts in the two beams to within a fraction of a wavelength, over the full baseline and aperture. As long as the path lengths are kept equal to within the coherence length, the relative path lengths only need to be monitored, not controlled to $0.1\lambda_x$. At an x-ray wavelength of 4 nm, optical path metrology is required to $\sim$2 Å, or 1 part in 10^{11} of a 20 m baseline. The development of interferometric techniques for gravity wave "beam detectors" has rendered metrology at this level comparatively routine. Note that the ultimate goal of beam detectors is to measure a gravity wave induced strain of $< 10^{-20}$ over tens of kilometers, a task that is a ten million times more demanding than required for XRI. Using a multi-pass Michelson or Fabry-Perot interferometer, with a 10W laser and 50 passes, the minimum detectable x-ray phase change in 20 nsec is 6×10^{-6} rad!

3. Beam combination: The beams must be made to overlap, coherently and in such a way that the resulting interference fringes are detectable. The former condition would be difficult to obtain in grazing incidence systems with large baseline, since the interferometer length $L_{int} \sim 15B/\gamma N_r$, where γ is the graze angle in degrees and N_r is the number of reflections prior to combination. A lunar based interferometer requires sidereostats, which cannot be made at grazing incidence. Figure 2 shows a design using multilayer mirrors operating at 45°. In this configuration, the instrument length is equal to the baseline. The multilayer automatically provides a reduced bandpass and corresponding increased coherence length. The penalty is a factor of 100-200 bandwidth reduction. Altitude sidereostats can be constructed from 45° mirrors for lunar basing of a long-baseline interferometer (see below).

4. Beam interference and fringe detection: One can imagine several methods for producing detectable interference fringes. In one, the two beams each pass through a beam splitter, as in Figure 2, resulting in interference of nominally parallel plane wavefronts. A beam splitter can in principle be constructed from multilayers, with good reflection and transmission efficiencies. For example, a 45 layer W/C mirror can reflect 8% of s-polarized radiation and transmit 35% (Lee 1982). At grazing incidence, a broad-band beam splitter could be made by coating a thin diamond film with a layer of gold or nickel with a thickness chosen to make the absorption and reflection efficiencies comparable (although these would vary with wavelength). Beams can and have been succesfully combined using transmission gratings (Rocket 1985). With an unresolved source, constructive interference occurs when the two optical paths are equal modulo a wavelength. Variation of the optical path of one beam by $\lambda/2$ produces destructive interference. By varying the relative path lengths over one wavelength, the amplitude of the complex visibility function can be determined from the variation in received intensity in the fringe detector at one baseline. (cf. Shao *et al.* 1986 for a description of this technique as applied to optical interferometry). Visibility is most easily measured in the pupil plane. Path length errors produced by a variety of influences result in fringes in the pupil plane. As the relative optical paths are varied over a wavelength, the fringes will become alternately light and dark (with an amplitude contrast determined by the visibility). Imaging these fringes in the pupil plane allows increased tolerance to these path length errors. To eliminate diffuse x-ray background, an x-ray concentrator is used after beam combination, followed by a field aperture in the focal plane.

5. Tracking: Source tracking is the most critical problem for maintaining phasing. Stability is required to $\Delta\theta_s < l_c/B$ with knowledge to $\Delta\theta_k < 0.1\lambda/B$ radians. For a stellar inteferometer, this corresponds to $\Delta\theta_s < 5$ m$''$ and $\Delta\theta_k < 5$ μ''. For sub-microarcsecond interferometry, $\Delta\theta_s < 20$ μ'' and $\Delta\theta_k < 20$ n$''$ must be achieved, making lunar basing essential. I present here only an outline of some possible solutions. A parallel optical interferometer, sharing the x-ray interferometer optical path, can be used to track the optical image. This could be implemented on a free-flyer or a lunar based device. This requires a sufficiently bright unresolved optical source. Alternately, an independent optical interferometer can use bright stars to establish an inertial reference. The XRI is then tied in with multi-pass interferometers. An independent x-ray interferometer could also be use bright x-rays sources to establish an inertial reference. Again, the main XRI is tied in with multi-pass interferometers. In each of these cases, lunar basing dramatically reduces the bandwidth of pointing errors. Errors in the lunar ephemeris will be *low frequency*, so that high bandwidth inertial stabilization will not be required. Accurate inertial reference updates obtained periodically can be interpolated to obtain the instantaneous orientation for each x-ray photon arrival. Seismological disturbances can be tracked with multi-pass optical interferometers.

III. Design Concepts

Stellar Intereferometer: Figure 2 illustrates a stellar x-ray interferometer designed to operate at $\lambda \simeq 40 - 200$Å, with λ chosen to coincide with a bright coronal lines such as C V 40.3Å, Si IX 55.3Å, Fe XIV 60.0Å, Fe XIII 76.0Å, or the Fe complex at 170Å. With a baseline of 20 m, the ultimate resolution would be 0.05-0.25 milliarcseconds, depending on λ. The observing candidates are all optically bright stars, and would furnish large fluxes for tracking and metrology with a parallel optical interferometer. If the x-ray and optical interferometers share the same optical paths, then there is the potential for measuring fringe visibility phase as well as fringe visibility amplitude. Since the optical stellar distribution is known, this can be done by comparing the relative positions of the optical and x-ray fringe. Phase recovery would permit true image reconstruction of the x-ray coronae. A rough calculation of the throughput with 2 m mirrors shows that Capella would give 100 ct/s, permitting a 1% visibility measurement in 10^2 s at one baseline. One thousand baselines could be measured in 10^5 s. A star with 0.1 IPC ct/s would give 10^4 cts in 10^4 s. One hundred baselines could be measured in 10^6 s.

Accretion Disk Interferometer: A long-baseline x-ray interferometer requires lunar basing. The x-ray wavefront is sampled with a 2-4 km baseline by two 2.4 m diameter 45° flats. These are equipped to track the source in altitude with a simple rotation around the axis between the mirrors and the central fringe detection turret. Azimuth tracking is performed by rotation of the two altitude mirrors around the central turret. The two beams are combined in the central turret using a multilayer beam splitter. The baseline is fixed during the source transit, as long as the altitude mirrors remain at the same distance from the central turret. Multiple baselines can be sampled successively by either moving the altitude mirrors radially, or by constructing several sets of altitude mirrors. The QSO 3C273 produces about 1 ct/s, providing a 1% visibility measurement over 10 position angles × 10 baselines in 10^6 s.

References

Lee, P. 1982, *Opt. Commun.*, **43**, 237.

Rocket, P. 1985, Proposal to the National Science Foundation.

Shao, M., *et al.* . 1988, *Astr. Ap.*, **193**, 357.

DISCUSSION

Webster Cash

I think you have been a little harsh on the potential science at one milli-arcsecond. Clearly this is a question we need (and can) answer. But consider that α Cen shows a 7 milliarcsecond stellar disk. Then there are binaries, and don't forget the promise of extragalactic observations.

Martin

A three order of magnitude increase in angular resolution will undoubtedly lead to unexpected new science.

Claude Canizares

As you know, the LIGO (Laser Interferometer Gravity Wave Observatory) will measure of strains of $\sim 10^{-22}$ or 10^8 Å over 10 km! So your requirements are quire modest!

HIGH THROUGHPUT X-RAY ASTRONOMY
Paul Gorenstein
Harvard-Smithsonian Center for Astrophysics

ABSTRACT

Many of the scientific objectives of x-ray astronomy for the 1990's and the twenty-first century require high throughput. Three phases of high throughput telescopes are discussed: 1m^2, 10m^2, and 100m^2 effective areas. The era of the 1m^2 telescope will commence in the late 1990's with the launches of XMM and LAMAR. A 10m^2 facility could be built with existing technology and launched as a free-flyer on a rocket like the Soviet Energia. With its relevance to x-ray astronomy's scientific objectives the 10m^2 facility should be given high priority as the first major initiative of the twenty-first century. A 100m^2 facility could occur in the middle of the twenty-first century. Although it is rather speculative, an approriate site for the 100m^2 telescope is a lunar base where x-ray optics can be manufactured in situ.

I. INTRODUCTION

In the future, there will be increasing emphasis upon deriving unambiguous results directly from x-ray observations. They should be as quantitative and as definitive as those from optical telescopes. Hence, x-ray measurements must be carried out with much more precision and for many times more objects than they were on the first generation of satellites.

Table 1 is based upon various papers presented in the initial sessions of this workshop. Their common theme is a requirement for a level of throughput that is much larger than any missions currently scheduled. The throughput ranges from a few square meters for surveys to about 10 m^2 for studies of large scale structures. Several speakers described a need for systems as large as 100 m^2 to study very distant galaxies, the origins of the x-ray background, and for ultra-fast timing. A very high throughput x-ray telescope was previously recommended by the Task Group on Astronomy and Astrophysics of the Space Science Board in a study conducted by the National Academy of Sciences for the period 1995-2015 (Nat. Acad. Press., Wash. DC 1988). For nearly all of these objectives good angular resolution is needed to image, resolve sources from background, and avoid confusion. In addition, it is essential to have good spectral resolution to resolve lines from continuum. Above 2 kev, the energy resolution should be at least as good as that of a CCD, i.e. about 100 ev, for Fe line measurements. Below 2 keV the spectral resolution should be much better, $E/\Delta E > 100$, for resolving lines in the rich complexes emitted by plasmas with temperatures below 20 million degrees.

To illustrate the throughput requirements of a twenty-first century facility, consider the study of an object detected in a deep survey measurement of the *Einstein* Observatory IPC carried out in 4×10^4 seconds of net observing time. This object has a flux of 7×10^{-14} ergs/cm^2-sec in the 0.8-3.5 keV band. The number of counts detected above background is about one-hundred, which is enough only for existence, position, and crude spectral information. On the basis of optical identifications, the typical object is a quasar at z = 0.5. Future goals would be to measure its redshift, and obtain information concerning physical conditions in the accretion disk around the central black hole, directly from x-ray observations. It is generally accepted that quasars are driven by the same mechanism responsible for x-ray emission in Seyfert galaxies. Matsuoka et al, 1989 and Nandra et al, 1989 have detected features at about 6 to 7 keV in the spectra of several Seyferts observed by *Ginga* which appear to be iron lines with equivalent widths of about 100 eV. George et al, 1989 describe a model in which the iron lines arise from the fluorescence of clouds or filaments near the central engine emitting the continuum radiation.

Table 1

Scientific Objectives For X-ray Astronomy in the Twenty-first Century and the Instrument Requirements

Speaker Scientific Problem	keV	Eff. Area	Ang. Res.	Spec. Res. $E/\Delta E$	Time Res.	Sensitivity and Other Requirements
R. Rosner Stellar Deep Survey	0.1-5	0.3m²	3"	10³		Large Sample, 10^{-15} ergs/cm²s
D. Lamb X-ray Timing, cont. presence in Space	0.5-10				0.1 msec	Multiple Modest Missions
G. Fabbiano High Sensitivity Obsv. of Normal Galaxies	0.1-10	10-100m²	1"			Large Samples Large Distance
R. Mushotzky AGN Observations	1-10	$\geq$ 2m²	~15"	40		Large Distance 5×10^{-16} erg/cm²s Need Fe Sensitivity
M. Geller Large Scale Structure	1-10	3m²	10"			Large Cluster Sample Large area of sky
K. Wood Fast Timing Measurements	1-20	100m²			10^{-6} sec	Post-XTE Mission Need Associated Imager
E. Boldt All Sky Background Isotropy	2-10	> 1m²				All Sky Coverage 2×10^{-14} ergs/cm²s
R. Griffiths Origin of X-ray Background	0.5-10	10m²	$\leq$ 1"	10³		10^{-19} ergs/cm²s

According to their model the iron lines are broadened to a width of about 1 keV by a combination of doppler shifts from the high velocities of clouds in an accretion disk and gravitational effects from the black hole. The line profile reflects the velocity distribution of the clouds. In order to measure the red shift, width, and structure of an iron line it should be detected with a significance of at least 5 σ. With an equivalent width of only 100 eV and an actual width of 1 keV, at least 250 counts from the line above the continuum are required. The requirement of 250 counts detected in the line (redshifted from z = 0.5) translates into 45,000 total counts between 0.8 and 3.5 kev or 450 times as many counts as were obtained in the *Einstein* deep survey. Since many objects should be observed, the measurement must be accomplished in 10^4 seconds or about one-quarter the time of an *Einstein* deep survey. Hence, an instrument with 1800 times more throughput than the *Einstein* Observatory is required, or 3×10^5 cm² of collecting area. Furthermore, the collecting area must be maintained to at least 5 keV to include the red shifted line. (The *Einstein* Observatory had very little area at 5 keV.)

Consider another objective described at this workshop, the study of the large scale structure of the universe up to $z = 0.5$ by mapping the distribution of clusters of galaxies in three dimensions. The CfA red shift survey has revealed a structure of surprisingly large scale (Geller and Huchra, 1989). It is based upon measuring individual galaxies and can only be carried out to $z \simeq 0.1$. Detecting clusters of galaxies through their x-ray emission and measuring their redshift from the Fe line seems to be the best (and perhaps the only) method of extending the red shift survey to larger distance. The criterion is that redshift of an Fe line from a typical cluster ($L_x = 10^{44}$ ergs/sec, $T = 5 \times 10^7 K$, Fe/H = 0.3 cosmic) at $z = 0.5$ be measured to a precision of 600 km/sec or 1 part in 500. The equivalent width of the line is 300 eV and its energy should be measured to ~ 10 eV. Assuming the energy resolution is 100 eV, 100 photons are required. At $z = 0.5$, the flux in an Fe line from a typical cluster is 2×10^{-7} photons/cm²-sec. Hence, with an instrument of effective area A(cm²) observing for time, T(sec) we require that:

$$AT (2 \times 10^{-7}) > 100,$$

The instrument field of view is limited to 0.5° at 5 keV by the reflectivity cut-off angle. In order to cover 1500 square degrees 3000 pointings are needed. In one year, the net time per measurement, T, (assuming 50% efficiency) is limited to 5000 sec. Therefore, A = 10^5 cm².

Another indicator of large scale structure is the distribution of AGN's as described by E. Boldt at this meeting. Their structure may differ from that of clusters of galaxies. A study of their distribution can be carried out simultaneously with that of the clusters. The angular resolution of the instrument should be good enough ($< 20''$) to distinguish between the point-like AGN's and the extended clusters.

II. FUTURE HIGH THROUGHPUT MISSIONS

a) Introduction

We consider three phases of future high throughput missions, characterized by successively larger instruments with effective areas of 1 m², 10 m², and 100 m² (1 keV). The phases are:

- The late 1990's   1m²
  Actual Programs:
  XMM (0.5 m²)
  LAMAR (1 m²)
- 2010   10m²
  Single Launch Into Earth Orbit
- 2040   100m²
  Lunar Base
  In Situ Manufacturing of Optics.

The first phase consists of two 1 m² class missions that are scheduled for the end of the century, XMM and LAMAR. The second phase, which will start about the year 2010 is the subject of this workshop, "High Energy Astrophysics in the Twenty-First Century". The instrument has a throughput of 10 m² and is responsive to the needs of most of the scientific studies listed in table 1. A modest extrapolation from present technology indicates that it should be possible to place this facility into Earth as a free-flying observatory orbit with a single launch. The third phase of high throughput is largely speculative, with respect to its scientific justification as well as the technology needed to develop it. Based on existing technology, a 100 m² will be too large and too massive to be launched as a single entity. Given that NASA intends to concentrate its resources on establishing a lunar base, the most likely and possibly the only site for a 100 m² facility is the lunar surface. The transport of massive x-ray optics to the moon will probably prove to be more expensive than constructing the facility in situ from lunar materials.

b) Programs of the 1990's

All x-ray satellites with imaging telescopes currently scheduled or being planned during the next decade are listed in Table 2. The next significant increase in throughput will be Japan's ASTRO-D mission with four telescope modules to be launched in 1993. Its capabilities will be foreshadowed by flights of the two module BBXRT experiment aboard the Space Shuttle starting in 1990. The next important step in throughput will be the two large telescopes being built by the Danish Space Research Institute for the Soviet Union's Spectrum-X-Gamma mission. Two high throughput missions with substantially larger collecting area, improved angular resolution, and much better spectral resolution below 2 keV than prior missions are being planned for the end of the century. They are ESA's XMM and the LAMAR Space Station experiment. These two facilities will mark the first era of high throughput x-ray astronomy.

Table 2
Future X-ray Telescope Missions of Long Duration

Mission	Country or Agency	Year of Launch	Effective Area of Mirror at 1 keV	Angular Resolution*
(Einstein Observatory	USA	1978	200 cm²	10")
ROSAT	W. Germany	1990	430	4"
ASTRO-D	Japan	1993	1000	2.5'
SAX	Italy	1993	175	1'
SPECTRUM-X-GAMMA	USSR	1994	2500(a)	2.5'
			400(b)	30"
SPEKTROSAT	Germany	1995	430	2"
AXAF	USA	1997	1000	0.5"
XMM	ESA	1998	5000	20"
LAMAR	USA	1998	10000	30"

(a) **XPECT**, (b) **JET-X**   *HPW

c) High Throughput Telescope Techniques

A twenty-first century high throughput facility (HTF) with 10 m^2 of collecting area can be discussed within the context of current technology. Several techniques for fabricating high throughput telescopes are listed in table 3 and discussed below.

Table 3
High Throughput Telescope Technologies

Techniques	Missions	Advantages	Comments
Epoxy Replication on CFRE Substrate	XMM-Wolter 1 Optics	Good Ang. Res.	Mandrels expensive Needs Development
Electroforming	SAX Jet-X (Spectr-X-γ) Northwestern Rocket Salyut	Good Ang. Res.	High Weight
Foil Cones	BBXRT Astro-D SODART (Spectr-X-γ)	Low Cost Light weight	Poor Ang. Res.
Bent Flats Float Glass (or Ep. Rep. CFRE)	LAMAR	No Mandrels or Simple Mandrels	K-B Optics Adaptable to Larger size

Epoxy replication

ESA's XMM Observatory has three telescope modules (Jensen, Ellwood, and Peacock, 1989). XMM's x-ray mirrors are Wolter type 1 optics with the parabola and the hyperbola integral on the same substrate. Fabrication is a two step process. Substrate shells made of carbon fiber embedded in a epoxy matrix (carbon fiber reinforced epoxy or CFRE) are laid up upon steel mandrels which are made to a shape tolerance of 10 microns and a surface roughness below 0.1 micron. A precision mandrel, slightly smaller in diameter than the CFRE substrate shells, is made of glass. Each glass mandrel is polished to an excellent surface finish and then coated with gold. Epoxy is injected into a 100 micron space between the substrate shell and the glass mandrel. The gold layer plays a dual role. It acts as a release agent for the replication process and as the reflecting material after it is transferred from the mandrel to the substrate.

The XMM method of producing mirrors requires considerable development. Although epoxy replication (upon beryllium substrates) was successful for the two small mirrors of EXOSAT (30 cm diam.), the requirements of the three large XMM mirrors (70 cm) are much greater. There are 58 nested shells in each XMM module compared to two for EXOSAT. The XMM mirrors are required to have high reflection efficiency at 7 keV. Hence, their surface smoothness must be much better than the EXOSAT mirrors which were not required to reflect efficiently above 2 keV. To date results have been encouraging (Egle et al, 1989). Each of the 58 shells requires its own highly polished mandrel. As the mandrels require a great deal of time and effort, the XMM method of mirror fabrication is expensive.

Electroforming

Two mirror systems scheduled for flight aboard satellites in the 1990's are being "electroformed", i.e. fabricated by electroplating. The first is two small modules that will be aboard SAX, an Italian national program (Butler, 1990). These mirrors constitute only a relatively small part of the SAX payload which is devoted primarily to hard x-rays. The second program is Jet-X (Wells et al, 1990), one of the two telescope experiments scheduled to be aboard the Soviet Union's *Spectrum-X-Gamma*. Jet-X will have much less throughput than the XSPECT/SODART telescopes that are also aboard *Spectrum-X-Gamma*, but is expected to have much better angular resolution.

Electroforming of x-ray telescopes is being pursued at the Czech Academy of Sciences who provided solar x-ray telescopes for the *Salyut* and *Phobos* spacecrafts, and Northwestern University (Ulmer et al, 1988) where small grazing angle telescopes optimized for short wavelength are being developed.

The SAX group has used aluminum mandrels which are machined to a double cone. The surface of the mandrel is smoothed by an acrylic dip polishing process as in the case of foil cones (see below). Similar to epoxy replication, a gold layer is evaporated over the mandrel to function both as a release agent, and as the x-ray reflecting material after it is transferred to the replica. Separation is accomplished by cooling the plated mandrel. Differences in the coefficient of thermal expansion between the aluminum mandrel and the nickel replica cause the two to separate. Performance of the SAX mirrors indicates that this process will produce mirrors with angular resolution better than an arcminute, at least in small mirrors (Citterio et al, 1988). The disadvantages of electroforming are the requirements for mandrels and high weight because of the high density of nickel. To reduce weight, compound materials consisting of thin electroformed skins with a glass sphere/epoxy filler are being developed.

Foil cones with Acrylic Lacquer

X-ray telescopes made of thin foils were first made at the Goddard Space Flight Center for the BBXRT Space Shuttle experiment (Serlemitsos, 1981) which is scheduled for flight in 1990. Aluminum foils, about 0.013 cm thick, are rolled into cones that approximate the parabolas and hyperbolas of a Wolter 1 telescope. The surface is made smooth on a scale of a few angstroms by a lacquer dip coating. The alignment and stability of the foils limit the resolution to about 2 arcminutes. Foil telescopes are low cost, relatively easy to manufacture, and light weight. They offer excellent throughput over a broad band.

Foil optics will be used in two instruments that will be launched in the mid-1990's, *ASTRO-D*, and the *XSPECT/SODART* telescope of the Soviet mission, *Spectrum-X-Gamma*. Foil mirrors offer the best ratio of weight to effective area of all mirror technologies that have been developed so far. It remains to be seen whether the angular resolution can be better than an arcminute. That would justify its use on systems that are much larger than *Spectrum-X-Gamma*.

The technique of acrylic dip polishing that has been developed to smooth the surface of aluminum foils can be used with other materials and other types of telescopes. This increases the number of options for the development of a high throughput facility for the twenty-first century.

Bent Flats, Kirkpatrick-Baez Mirrors

One telescope approach differs radically from the others. The Kirkpatrick-Baez (K-B) X-ray mirror is composed of flats that are curved slightly into parabolic cylinders. An imaging telescope consists of two successive sets of nested plates that are orthogonal to each other. Such a telescope with an angular resolution of a few arcminutes was used in a successful series of rocket flights some ten years ago. More recently, a 20 cm x 30 cm telescope has achieved an angular resolution of 31 arcseconds half power width projected along an axis, (Fabricant, Cohen, and Gorenstein, 1988). The principal advantage of the K-B geometry is simplicity. All the reflectors are identical flats prior to bending and every reflector functions independently of every other. The problem of producing a good mirror consisted of two tasks: mass producing a thin, lightweight, stiff flat with a smooth surface, and imparting the correct parabolic curve appropriate to the plate's distance from the optic axis. The latter has been accomplished under computer control (Fabricant et al, 1986). The flat does not have to be monolithic. It can be be segmented along its length and height as long as the relative alignments of the segments are maintained. A disadvantage of the K-B mirror is that for the systems commonly used the effective area above 6 keV is only half that of a Wolter 1 mirror (or its conical approximation) of the same focal length. Alternatively, a focal length that is 1.5 times larger is required to obtain equal efficiency. However, below 4 keV the K-B and Wolter mirrors are equally efficient.

Float glass is an inexpensive commercial material and has excellent surface smoothness on a scale below a micron. Its limitations are relatively high weight and deviations from flatness which will limit the angular resolution to no better than 25 arcseconds. Composite materials, for example the method of epoxy replication upon CFRE substrates being developed for XMM reflectors, offer a possible lighter weight alternative to float glass. In the K-B geometry the number of mandrels is small and they are relatively inexpensive flats. The major difficulty of composite materials is that the replication process leaves internal stresses which make it more difficult to produce high fidelity flat reflectors than cylindrical ones.

The LAMAR space station experiment is based upon K-B optics. K-B optics are potentially important for the much larger twenty-first century high throughput facility. In the K-B geometry, the mandrel are very simple, and it is possible to divide large reflectors which are nearly flat into smaller segments. Thus, the K-B geometry does allow the production of much larger mirror systems, perhaps more easily than the Wolter 1 configuration.

III. LAMAR SPACE STATION EXPERIMENT

a) Experiment Description

The Large Area Modular Array of Reflectors (LAMAR) experiment was selected as an attached payload for the Space Station. The institutions involved in the scientific development are the Smithsonian Astrophysical Observatory with overall responsibility, the Marshall Space Flight Center, the University of California (Berkeley), who have responsibility for the gratings, and the University of Chicago. The University of Arizona and the University of Texas (Arlington) will collaborate in the investigation of composites as an alternative reflector material. A sketch of LAMAR integrated into a pointing system and deployed on the Space Station is shown in figure 1. The instrument budget is very small compared to other instruments of comparable size but the cost of launch, deployment in space, and the pointing system are provided separately. Most of the observing time especially after the first year will be allocated to a NASA guest observer program.

304 High Throughput X-Ray Astronomy

Fig. 1. LAMAR Experiment within Pointing System on Space Station.

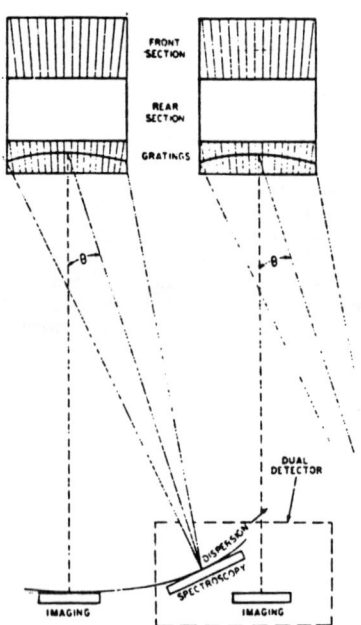

Fig. 2. Imaging and dispersive spectroscopy detectors of LAMAR.

The aperture of LAMAR is 2 m x 2 m providing a mirror effective area of about 10^4 cm^2 at 1 keV. The two dimensional modular array will be between 3×3 and 5×5 modules depending on how long a focal length the pointing system allows. The longer system (8m) is more desirable because fewer detectors are required. LAMAR performs imaging and dispersive spectroscopy simultaneously. Each module contains a K-B telescope, a stack of reflection gratings which intercepts about 40% of the rays, and a dual section position sensitive gas proportional counter. One section detects the direct image, the other, the dispersed spectrum which is displaced to one side. The arrangement is shown in figure 2. For two-thirds of the modules the blaze angle and line density of the gratings are optimized at 16 Å, for one-third, at 60 Å. LAMAR has no moving parts.

LAMAR's net effective area including the detector efficiency for imaging and spectroscopy are shown in table 4 and table 5. The dispersive spectroscopy is effective below 2 keV. A simulated spectral observation of Capella is shown in figure 3. A comparison of LAMAR's effective area and resolution with AXAF and XMM is shown in table 6.

b) Comparisons between LAMAR and XMM

Moderately High Resolution Spectroscopy

LAMAR and XMM are the highest throughput missions planned. They are also the only high throughput missions with angular resolution better than an arcminute and the only missions offering high throughput dispersive spectroscopy below 2 keV. The two are complementary. XMM's goal for angular resolution is 20" HPD as compared to LAMAR's goal of 30" HPW along an axis which was achieved by a smaller mirror. Although their effective area is the same at 7 keV, XMM has much better energy resolution above 2 keV thanks to CCD detectors whereas LAMAR, at least its initial configuration, has only imaging proportional counters. On the other hand, LAMAR has more effective area below 2 keV. LAMAR has large effective area in the 44-80 Å band for both imaging and dispersive spectroscopy whereas XMM does not extend longer than 30 Å. The significance of the greater spectroscopic bandwidth is illustrated in figure 3, the dispersed spectrum of Capella. LAMAR is sensitive to both bands, 8-30 Å (left) and 44-80 Å (right). XMM is not sensitive in the 44-80 Å band. Furthermore, LAMAR has three times as much effective area as XMM in the 8 - 30 Å band.

Figure 4 illustrates absorption edge features expected in many AGN's and quasars from material around the central engine (Kallman and Mushotzky, 1985). The most detectable feature is the oxygen edge where the quasar continuum flux is high. Indeed, such an oxygen edge feature was probably observed in one BL Lac object by the *Einstein* Observatory (Canizares and Kruper, 1984). The figure shows the bands of LAMAR (L-S) AND XMM (X-S) relative to the emitted spectrum for a quasar at $z = 0$, $z = 1$, and $z = 2$. At $z = 1$, the XMM acceptance band is becoming marginal whereas the red shifted oxygen absorption edge is still well within the LAMAR band. With three times as much spectroscopic effective area as XMM in the 6-30Å band, LAMAR can study absorption features of many more AGN's.

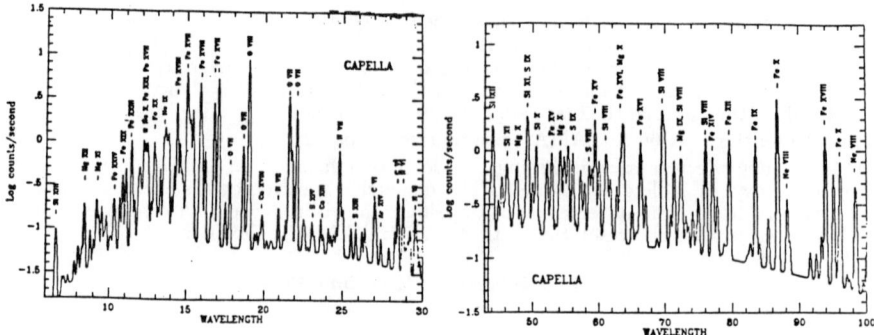

Fig. 3. Simulated spectrum of Capella in two bands, 44 - 80 Å, right panel and 8 - 30 Å, left panel as it would be observed in LAMAR. XMM is sensitive only to the shorter band.

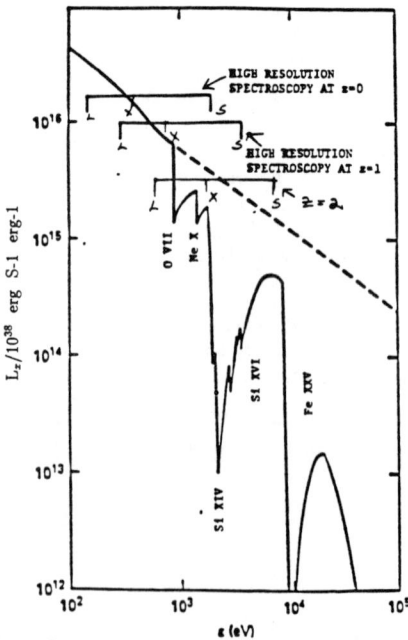

Fig. 4. Theoretical absorption line spectrum of a quasar (Kallman and Mushotsky, 1985). The lines between "L" and "S" denote the acceptance band of LAMAR in the rest frame of the quasar when it is at distances of z = 0, 1, and 2. The segment of the line between "X" and "S" is the acceptance band of XMM.

Table 4
Imaging Capabilities
On Axis Ang. Resolution: 30″, 50% Power width
Field of View: 40′ FWHM at 2 keV, 30′ FWHM at 6.2 keV

Energy	(W.L.)	Net Effective Area (cm^2)*
0.28 keV	(44Å)	5700
1.0	(12)	4000
4.0	(3)	2700
7.0	(1.8)	1600

Limiting sensitivity is 10^{-14} ergs/cm^2-sec (0.2-8keV). It is set by confusion limit of 1 source/30 detection cells and is reached in 3×10^3 sec.
*On Axis, including detector efficiency

Table 5

High Resolution Spectroscopy Capabilities.
Field of View: 6′ x 12′

Energy	(W.L.)	Net Effective Area* (cm^2)	Resolution (E/ΔE)
0.15 keV	(80Å)	305	450
0.28	(44)	433	260
0.50	(24)	414	320
0.78	(16)	736	230
1.0	(12)	734	180
2.0	(6)	158	80

*Including Detector Efficiency

Table 6

Comparison of Spectroscopy Capabilities at Longer Wavelength

W.L.	E	LAMAR		XMM		AXAF	
		E/ΔE	Area	E/ΔE	Area	E/ΔE	Area
16 Å	0.775 keV	230	720 cm^2	300	250 cm^2	600*	60* cm^2
24 Å	0.516	315	380	400	120	850*	25*
60 Å	0.207	350	400	-	-	850	20**

*Medium Energy + High Energy Gratings, **Low Energy Gratings

Surveys

Future x-ray surveys should not only be more sensitive than ROSAT, but should also extend beyond ROSAT's 2 keV cut-off to provide spectroscopic information for distinguishing between stars and extra-galactic objects. LAMAR is much better suited for surveys than XMM because LAMAR's two-dimensional field of view is four times larger and LAMAR has a larger bandpass and larger effective area. The combination of broader field of view and higher throughput make LAMAR nearly an order of magnitude more effective than XMM for surveys. In fact, surveys can be easily carried out on the Space Station by pointing towards the zenith or during the return slews of the pointing system from the setting of a source.

Upgrade Capability

Although the Space Station is not the ideal platform for astronomy, it is unique in providing access to the instrument for upgrade. It is likely that x-ray detector technology will be considerably more advanced in a decade. We can expect imaging detectors with better energy resolution, larger format, and better spatial resolution from improved CCD's, microcalorimeters, and superconducting devices. Hence, after several years of observing, it should be possible to replace LAMAR's imaging proportional counters with future state of the art detectors. The cost is very low compared to constructing an entire new observatory with LAMAR's throughput.

IV. HIGH THROUGHPUT FACILITY (HTF)

The first goal for the twenty-first century is to develop a high throughput x-ray facility with an effective area of 10 m^2. It is possible to estimate its weight and volume by scaling from current technology. Table 7 is a list of the area to weight ratio for several x-ray mirror types and of the complete system.

Table 7

Effective Area to Weight Ratios of Focussing X-ray Telescopes

Telescope Type	Res.	Spacecraft	Area/Weight cm^2/kg Mirror	System
Al Foil Cones	2'	ASTRO-D Spect-X-γ	30	2.5
Composites (CFRE)	20"	XMM	10	2.5
Electroforming	30" - 1'	SAX, Jet-X	3-10	
Float Glass	30"	LAMAR	4	2
High Resolution	1"	AXAF	0.5	0.03

The various techniques do differ in area to weight ratio of the mirror. Foil cones are the most efficient but their angular resolution is, at least at present, not adequate for a 10 m^2 instrument. When the total system weight, i.e. including detectors plus structure, is considered the differences among the various techniques diminish. If we take 2.5 cm^2/kg as the ratio, the mass of a 10 m^2 HTF is 40 metric tons. Table 8 shows the capacity and cost of several launch vehicles. The 10 m² facility can be launched as a free flying observatory by the Soviet Union's Energia rocket. The Energia's cost launch is relatively low at $60 M (as of 1987) and well within the cost constraints of major programs. The cost of the Energia may be artificially low because of a subsidy from the Soviet Union. This could be a Soviet contribution to an international program. Alternatively, a U.S. rocket known as the "Advanced Launch System" with capability similar to the Energia is planned for the beginning of the twenty-first century and could also be used to launch HTF.

The HTF could consist of ten modules, each with an aperture of 2 m and a focal length of 20 m. To reduce volume, the HTF can be folded up to half its length for launch as planned for both ASTRO-D and *Spectrum-X-Gamma*.

It is expected that improved dispersive and non-dispersive spectrometers will be available for the HTF. The goal is $E/\Delta E$ is 300 to 1000.

Table 8

Payload Capacity and Cost per Pound of Payload Delivered to Orbit*

	Shuttle	Titan 4	Delta	Proton	Energia
Launch Cost Per Pound	$6,800	$5,100	$3,275	$750	$300
Payload	24 tons	20 tons	5.5 tons	20 tons	100 tons
Cost of Maximum Payload	320 M	200 M	36 M	30 M	60 M

* Newsweek Estimates August 17, 1987

V. X-RAY ASTRONOMY FROM A LUNAR BASE

a) Introduction

At the present rate of progress the era of a 100 m² x-ray facility will occur in the middle of the twenty-first century. Its scientific justification and the methodology for its construction are both highly speculative at the present time, when the largest x-ray telescopes have been no larger than 0.1 m². An extrapolation over three orders of magnitude in throughput and a 50 year time span is likely to prove to be irrelevant. However, current scientific objectives as described by the speakers in the opening sessions of this workshop and current technology do provide some scientific justification and technical guidelines. The following scientific studies require a 100 m² facility.

- Large scale structure of the universe up to $z = 2$ through a survey in three dimensions of the distribution of clusters of galaxies and AGNs.

- Evolution of the accretion disk in AGNs up to $z = 2$ by measuring the redshift, width and structure of Fe fluoresence lines.
- X-ray observations of large samples of normal galaxies including ones at large distances.
- The x-ray background and the source content at large z.
- Very high time resolution studies of compact binaries.
- Very high resolution spectroscopy and polarization studies.

a) Construction of the X-ray Optics

Furthermore, for most of these studies the 100 m² facility should have much better angular resolution than the earlier instruments because it will be observing more distant objects of much smaller angular size. The goal is 1″ which apparently conflicts with the requirement for high throughput. By scaling current technology, the estimated mass of a 100 m² facility is 400 tons. This exceeds the launch capability of a single rocket. Currently, NASA intends to establish a base on the Moon. This goal is likely to preclude construction capability elsewhere in space. Consequently, the Moon would be the only possible site for the 100 m² facility. However, it may not be economically possible to transport 400 tons of cargo to the Moon for an x-ray observatory. The alternative is to construct the massive components, i.e. the optics, the optical bench and the pointing system in situ, from lunar materials such as 0, Na, Al, Si, Ti, Fe, and Ni. X-ray detectors, star sensors and electronics which account for much less mass and are too complex to manufacture on the Moon, would be transported from the Earth. A possible configuration for the 100 m² facility is an array of 20 modules each consisting of a telescope of 5m diameter and a focal length of 50 m. Each module could have its own pointing system. If large optics are constructed on the Moon, their weight is no longer the predominant constraint that it was for the free-flying satellite. Hence, there may be no need for exotic materials. In some respects, the task of constructing optics is less difficult on the Moon than on the Earth because of the lower gravity, the vacuum, and not having to contend with vibration and acceleration during launch.

The type of optics that seems to be best suited for lunar manufacture is the bent flats or K-B design used by the LAMAR Space Station experiment. The number of mandrels that need to be sent from the Earth is minimal. Large reflectors can be segmented into smaller units. The precise figure can be tuned with cosmic x-ray sources directly on the optical bench/pointing system using the actual detectors while the star trackers keep the system in a fixed celestial position.

The process for manufacturing flats may be one of the following three:

- Glass factory: Fabrication facilities may exist because glass would be made from lunar materials for solar collectors.
- Chemical vapor deposition in the lunar vacuum,
- Electroforming using nickel or iron extracted from lunar soil: Electroforming can be used to make the structural components of the mirror, the optical bench, as well as the reflector flats.

A tele-robotic capability is needed to extract nickel, electroform the components, and assemble them.

c) High Angular High Resolution

There remains the problem of improving the angular resolution from 20″ to 1″ by the time a 100 m² facility is developed. This collecting area is three orders of magnitude larger than AXAF's. Perhaps, this goal will be feasible in fifty years as manufacturing precision improves. An alternative method is to combine focussing with a two-grid rotating modulation collimator (RMC). The RMC produces a temporal modulation at each point in the focal plane that can be transformed into finer scale angular information. By this technique, the RMCs of SAS-3, Ariel 5 and Hakucho were able to refine the angular resolution of collimated proportional counters from 15 degrees to a few arcminutes, or more than two orders of magnitude. The task of improving the resolution from 20″ to 1″ is actually more modest than what was achieved by these instruments. A two grid RMC with a separation of 30 m and a wire spacing of 75 microns would provide a resolution of 1″. Diffraction would not be the limiting factor for wavelengths shorter than 5 Å. The disadvantage of the RMC are a factor of four loss in effective area, additional size and complexity, and the cost of transporting some number of 100 kg pairs of wire grids to the Moon. On the other hand, the RMC improves the resolution over the entire 1 degree field of the telescope, whereas the 1″ field of a pure focussing telescope is only about 15′. The larger field is important for surveys. The problem of distinguishing intrinsic temporal variations from those caused by the rotation of the collimator can be solved, as it was on Hakucho, with RMCs on different modules rotating synchronously in opposite directions. Hence, the summed count rates remain constant.

VI. REFERENCES

Butler, R. C., 1990, in *High Resolution X-ray Spectroscopy of Cosmic Plasmas*, P. Gorenstein and M. Zombeck eds. Cambridge University Press, Cambridge, UK (in press).
Canizares, C. R., and Kruper, J., 1984, *Ap.J. (Letters)*, **278**, L99.
Citterio, O. et al, 1988, *Applied Optics*, **27**, 1456.
Egle, W., Bulla, H., and Kaufman, P., 1989, *Proc. of the S.P.I.E.*, **1160**, 432.
Fabricant, D. G., Conroy, M., Cohen, L. M., and Gorenstein, P., 1986, in *Proc. of the S.P.I.E.*, **640**, 164.
Fabricant D. G., Cohen, L. M., and Gorenstein, P., 1988, *Applied Optics*, **27**, 1456.
Geller, M., and Huchra, J., 1989 *Science* **246**, 897.
George, I.M., Nandra, K., and Fabian A. C., 1990, *M.N.R.A.S.*, (in press).
Jensen, P. L., Ellwood, J. M., and Peacock, A., 1989, *Proc. of the S.P.I.E.*, **1160**, 525.
Kallman, T. R., and Mushotsky, R., 1985, *Ap.J.*, **292**, 49.
Matsuoka, M., Yamauchi, L., Piro, and and Murakami, T., 1989, *Ap.J.*, (in press).
Nandra, K. A., Pounds, K. A., Fabian A. C., and Rees, M. J., 1989, *M.N.R.A.S.*, **236**, 39.
Serlemitsos, P., 1981, in *X-ray Astronomy in the 1980s* NASA Tech. Mem 83848, 441.
Ulmer, M., Matsui, Y., Bedford, D. K., Simneff, G. M., and Takar, S., 1987, *Applied Optics*, **26**, 3852.
Wells, A., Lumb, D. H., Pounds, K. A., Stewart, G. C., Ashenbach, B., Brauniger, H., Hasinger, G., Trümper, J., Citterio, O., Scarsi, L., Peacock, A., and Taylor, B., 1990, in *High Resolution X-ray Spectroscopy of Cosmic Plasmas*, P. Gorenstein and M. Zombeck eds. Cambridge University Press, Cambridge, UK, (in press).

DISCUSSION

Claude Canizares

To me, one of LAMAR'S greatest strengths is its ability to do surveys. Have you considered spending half the observing time over several years to do a survey of stars, clusters, and AGNs?

Gorenstein

We certainly consider surveys to be one of LAMAR's important objectives. There are several options. One is performing surveys during the return slew between the rising and setting of a source. Another is to schedule a dedicated survey for a significant period of time as ROSAT plans for its first six months. This can be done by pointing the instrument in an anti-Earth direction and changing its inclination periodically. For better or worse, there is plenty of time to develop the optimum strategy.

Kent Wood

For reference, XLA at 100 m^2 weighs 25 tons or less with some allowance for improvement. What is the largest aperture that can fit into a Shuttle flight and on Energia?

Gorenstein

The LDEF which is due to be retrieved by Space Shuttle in January 1990 has an area of $14' \times 30'$. That aperture is the largest one I know of that can fit into the Space Shuttle. At 100 tons the weight-lifting capacity of the Soviet Energia is four times that of the Space Shuttle. I do not know if this translates into four times the aperture, because I am not familiar with the limitations on length and diameter of the payload.

Richard Mushotzky

Could NASA consider an "unmanned" modular spacecraft, e.g., use "lego" technology, to assemble modules spacecraft?

Gorenstein

This certainly would be important technology for scientific satellites and very appropriate for a modular high throughput facility. However, the current emphasis on establishing a Lunar base may not leave much room in the NASA program for developing a construction capability elsewhere in space.

Steven Kahn

Although you emphasized the point several times in your talk, I really want to comment again on the need for a facility with an array of telescopes that allows frequent access for instrument switch-out. In principle, if such a process could be made cost effective, it could solve a sociological problem now facing the field namely that it has become very difficult for a university-based research group to become involved in space experiments without involving a major national laboratory or industrial contractor. If LAMAR exists on the Station or the Moon, and if there are frequent flights, it might eventually become possible for university groups to build a science-specific experiment in its own laboratory, at reasonably low cost.

Gorenstein

What you suggest would be excellent for X-ray astronomy. However, at this time the prospects for frequent visits to an experiment on the Space Station (or a Lunar base) seem remote.

THE X-RAY LARGE ARRAY (XLA)

K.S. Wood
E.O. Hulburt Center for Space Research
Naval Research Laboratory
Washington, DC 20375

ABSTRACT

The X-ray Large Array (XLA) is a mission to advance scientifically rich fields of X-ray timing and compact object studies well beyond limits of current or planned timing instruments, such as Ginga or XTE. XLA allows variability at natural timescales of compact objects to be resolved both temporally and spectrally. The array also achieves angular resolution in the milliarcsecond range for the first time in X-ray astronomy. Focusing optics are not needed, although some imaging with a coded aperture is useful. Without optics, XLA is feasible in the near future at reasonable cost.

INTRODUCTION

XLA is a mission concept for an X-ray array distinguished by collecting aperture $\gg 10$ m^2, nominally 100 m^2. It will become achievable around the turn of the century under present scenarios for space development. Without breakthroughs in technology, it opens up several orders of magnitude in observational parameter space. Goals in X-ray astrophysics inaccessible to smaller instruments come within reach. XLA is best for observing tasks where the principal aim is to gather many photons rapidly from comparatively bright sources. It is complementary to AXAF, strongest where AXAF is inappropriate.

A large array can be built at the Space Station, but might also be appropriate for a LEO free flyer or a lunar base. The ideas presented here were developed in studies, papers and proposals over the past few years (see Dabbs 1987; Wood and Michelson, 1988) as a concept for a 100 m^2 array attached to the Space Station. <u>Large arrays will become possible somewhere</u>, and there will result parallel growth in the feasible size of space-based astronomical instruments. <u>X-ray timing work needs exactly this growth in area to achieve a major advance</u>. Payloads of order 1 m^2 have been flown for more than a decade. Other characteristics of the baseline XLA concept are that it would have a field of view of 1 square degree and cover energies from 0.25 to 100 keV. Pointing to a few tenths of a degree would suffice. XLA would detect the Crab in 1 μs and 3C273 in 1 ms. This discussion summarizes first the scientific role of such a facility, then turns to the technical approach.

SCIENTIFIC PROGRAM OF XLA

The XLA scientific program might be described as "photon rich X-ray astrophysics", the result of observing ~1000 bright X-ray sources with a very large X-ray collector, each source receiving substantial observing time. Much of the power of this approach comes from reaching very short timescales, which are the dynamical timescales of interest in compact objects. A large increase in photon throughput brings orders of magnitude of quantitative improvement in four kinds of measurements. These are (i) fast photometry, (ii) fast time-resolved spectroscopy, (iii) ultrafine angular resolution, and (iv) mapping of highly extended diffuse sources.

Photometry refers to accumulation of a series of counts as a function of time in some chosen energy band. The shortest integration time is set by source brightness and detector aperture. With a 100 m^2 array a source such as the Crab Nebula gives 4 million photons per second, enough in one microsecond to constitute a detection and in 20 microseconds to determine source brightness to < 15% uncertainty. Photometric measurements are applicable to periodic, aperiodic and quasiperiodic phenomena. They can be used to study millisecond pulsars, black holes, X-ray bursts, eclipses and many other phenomena. It is not necessarily true that the level of modulation will be anywhere near 100%. Sometimes a basic issue hinges on observing subtle modulations that are small fractions of the total flux. With suitable summations of data, a large array can work at the level of 10^{-4} or even 10^{-5} of the total flux from the source, <u>particularly at high frequencies</u>. This can be applied to quasiperiodic oscillations, searches for vibrations or pulsations in neutron stars, and accretion shock phenomena in AM Her stars. (See further discussion in the companion article in this volume (Wood 1990) and in Wood and Michelson (1988).

Fast time-resolved spectroscopy is the process of making many (millions) of determinations of the X-ray spectrum of a source in rapid succession, with a short integration time for each spectrum. Spectral resolution depends upon the collector. If we assume a proportional counter then the resolution is about 18% at 6 keV. One can detect line or edge features as they appear and vanish, or determine rise and fall times of X-ray spectral components.

A by-product of having very large area and fast timing is very fine angular resolution. The idea is simple in principle. The large aperture permits a source such as an active galaxy or a cluster of galaxies to be seen in milliseconds. If an occulting edge can be made to sweep slowly across the field, sources can be resolved to a limit which is the angle scanned in the limiting integration time. This can be better than any other current means of achieving high angular resolution in X-rays. With XLA in LEO, the occulting edge that one would use is the limb of the moon, which can move < 1arcsec s^{-1} in favorable circumstances. Combined with the ability of a large array to see, for example, bright active galaxies in 1 millisecond, this would translate into milliarcsecond angular resolution, although limitations of knowledge of lunar terrain would degrade it by about an order of magnitude. The lunar limb observed from LEO eventually gives access to about 10% of the sky for occultation purposes. This 10% happens to include many interesting sources including the Galactic Center, the Crab Nebula, and the brightest X-ray quasar, 3C273. Terrain and sky access limitations can be overcome by placing an orbiting artificial occulter in an orbit having the radius of the lunar orbit, but perpendicular to it. (Wood and Breakwell, 1987). This occulter could be steered to intersect the line of sight to targets of interest over the full sky. A similar technique is available from the lunar surface and is described by Novick (1990) in these proceedings. Angular resolution is inferior from the moon but full sky access is easier to achieve.

TECHNICAL APPROACHES AND ISSUES

XLA is optimized for tasks where area -- not the product of area with integration time -- is critical, such as study of very fast processes. For such investigations the performance gains realized from using focusing optics are marginal or nonexistent and there can even be penalties associated with the high energy cutoffs of mirrors. A mirror system trying to do the XLA science would have to achieve the same frontal area (100 m^2) and energy range (.25 - 100 keV). Mirrors now in use are very limited above 7 keV, and to reach higher energies requires lengthening the mirror, which then makes for greater volume and weight per unit frontal area. Costs deriving from the weight and volume of mirrors realistically would delay by many years the date when a large area would come on line as a facility. A 100 m^2 array without optics could be carried to orbit in one or two Shuttle flights, but an equivalent mirror system would require many Shuttle flights or a different vehicle. It is also not clear that a mirror array would ever be used for a significant fraction of its lifetime on very bright sources. These are an important aspect of the XLA science.

Preference for X-ray optical systems is sufficiently strong that it may help to belabor this point with signal-to-noise estimates for particular applications. Consider first the task of detecting aperiodic modulations in accretion flow onto a pulsar such as Hercules X-1. Suppose we try to see modulation of order 30% on a timescale of 0.1 ms. The background rate from 100 square meters of proportional counters with a background rate equal to that of Spartan-1 amounts to 6 photons per millisecond, or an expectation of 0.6 background events in 0.1 ms. The signal from Her X-1 would be about 40 photons in the same 0.1 ms interval. The main limitation is the Poisson fluctuation in the 40 signal photons, and the background is irrelevant. Variations in the background matter even less. As a second example, consider search for a millisecond pulsation in a source such as Sco X-1. The rate from the 100 m^2 array is of order 30-100 million photons per second, so that periodic modulations at the level of 10^{-4} become detectable in about 30 seconds. In this time the total number of events from all forms of background is < 2×10^4, so that background effects again are negligible.

For investigations involving variability at longer timescales and low flux levels the XLA facility includes a coded aperture having a nominal collecting area of 1 m^2. Consider an imaginary case in which we are examining a new kind of periodic signal that has a timescale of 1 second and corresponds to .1% of a 100 mCrab source, a task which is only within the reach of an aperture >> 1 m^2. Further imagine that we are worried about the hazard that this signal might actually come from 10% modulation of a 1mCrab source in the same field. The coded aperture images the field periodically during the observation. It will find sources down to < 0.1 mCrab. Variability of such sources on timescales > 100 s will be seen as image-to-image variations by the coded aperture. The case is highly artificial, because in a 1 square degree field there is an expectation value of <0.2 for having a 1 mCrab source, and the known sources at this flux level are mostly constant (clusters) or vary at timescales > 100 s (AGN). All of the scientific objectives proposed for XLA are of the sort just described: either the background and the faint sources in the field are completely irrelevant or else devoting ~1% of the total array to a coded aperture provides excellent monitoring for remote possibilities.

XLA does not try to duplicate science done by instruments such as AXAF but rather emphasizes the science such instruments cannot do. Thus for monitoring hour-to-hour variations in an AGN, AXAF would be the appropriate instrument, but for searching for variability on the order of a second or less in the same objects XLA would be the instrument of choice, because AXAF would be limited by its area.

XLA is large enough to require assembly in space. The Space Station provides a suitable assembly site. A pre Phase A study (Dabbs, 1988) found no insuperable obstacles for its accommodation at the Space Station and there are advantages in having it there for maintenance and data handling. There is some possibility that costs could be minimized by siting XLA on an unmanned platform. Siting alternatives as well as assembly options need to be studied competitively, but it is clear that Space Station suffices. Weight reduction, design simplicity, and mass replication of units are keys to cost reduction. The NASA pre Phase A study estimated XLA could be built for <$300M, scaling from HEAO-1 experience and using standard economies of scale. Spartan-1 detectors had a unit cost about a factor of ten lower than that of HEAO-1, so there is a possibility this number could be further reduced. The use of detectors of this design should be viewed as a strawman, and better designs would be sought. The performance requirements are really for minimal cost, weight, and volume per unit area, long detector lifetime, and suitably broad energy range. The cost of transporting the array to its intended site is influenced by the stowed configuration. Detector weight and effective depth (volume per unit area) govern how many modules can be packed into the Shuttle.

SUMMARY

A large X-ray array will bring about photon-rich X-ray astronomy. This would be a powerful instrument for extracting information about physical conditions and processes in bright X-ray sources. It would address a diversity of topics, and would not be a single-purpose instrument. The sources in question are known to exist and observing them at their natural dynamical timescales will bring rich rewards. There are several important items that require further study. A size versus science tradeoff study should explore what can be done with arrays of 10 - 100 m^2, as a function of area. The realistic range for fabrication cost per unit area, weight per unit area, and volume per unit area should be examined. Detailed simulations of the instrument performance for various observing programs should be carried out. This should include a study of what fraction of the array is optimally devoted to a coded aperture or Fourier Transform Telescope. The optimum site should receive attention. The ratio of the frontal area devoted to non-imaging and imaging needs to be optimized. Such studies will not proceed without community interest. Astrophysicists should realize the potential in timing beyond the reach of $1m^2$ instruments and assign it priority in future planning.

REFERENCES

Dabbs, J. et al., 1987, Report on Pre-Phase A Feasibility Study for XLA, NASA/MSFC.
Novick, R., 1990, these proceedings.
Wood, K.S., 1990, these proceedings.
Wood, K.S., and Michelson, P.F., 1988 in International Symposium on Experimental Gravitational Physics, eds. P.F. Michelson, Hu Enke, and G. Pizella (World Scientific Publishing: Singapore), p. 475.
Wood, K.S. and Breakwell, J.V. 1987, Acta Astronautica, 15, 9.

The Study of Large-Scale Structures with Wide-Field X-ray Optics

Richard Burg, Christopher J. Burrows and Riccardo Giacconi

Space Telescope Science Institute

Abstract

By utilizing Wide-Field X-ray Optics we can create samples of large numbers of high redshift clusters and groups suitable for studying both evolution and structures on the largest scales in the universe. We describe a Scout-class explorer mission capable of carrying out this survey.

1.0 The Study of Large-Scale Structures - Introduction

The major breakthroughs in the study of the structure and the constituents of the universe have come at moments when new survey instruments, with unique capabilities, became available to motivated observers. We need only recall what was achieved by Hubble with the Mount Wilson 100-inch, by Zwicky and Abell with the Palomar 48-inch, by the CfA redshift survey team with the availability of new electronic detectors for optical spectroscopy. We feel that a comparable advance in our understanding of evolutionary trends in the large scale structure of the universe is imminent in X-ray astronomy.

The formation and evolution of clusters of galaxies, and of large scale structure will only be fully understood when surveys at redshifts greater than one become available. We believe these surveys must be done in the X-rays, have high angular resolution, high sensitivity, and be done in contiguous regions.

A detailed study of the applicability of currently planned missions to cluster research and large scale structure has led us to believe that the presently planned instruments have limitations that are not related to collecting area but are caused by limited field of view. In a survey mission, aperture size and angular field of view are equivalent in their contribution to sensitivity and categorization of sources, but mission cost rapidly increases with collecting area. At high redshift, the only unambiguous identification of clusters is as extended objects in the X-ray; the optical images of the galaxies are too faint and foreground/background contamination is too high. Studies of clusters of galaxies (and large scale structure) at redshifts greater than unity are presently frustrated by the lack of high redshift samples. AXAF is an observatory uniquely capable of detailed studies of such phenomena, but in the absence of a high redshift cluster catalogue, it will not be able to solve definitively the problem of cluster formation and evolution. We have therefore studied X-ray observatories that could successfully carry out such surveys. We have designed optics that achieve high spatial resolution (better than 2.5 arc-seconds rms) over a large field of view (at least 1 degree). We intend to propose a Scout-class explorer, utilizing these optical designs, to carry out a program to study clusters and large-scale structures at high redshifts.

2.0 Clusters of Galaxies and Large Scale Structure

2.1 Clusters of Galaxies

X-ray observations reveal a wealth of information about the formation and dynamic evolution of structures consisting of gravitationally bound galaxies and gas. Several authors have pointed out that X-ray observations of such systems may offer important advantages with respect to studies in other wavelengths domains particularly at early epochs of the universe (Kaiser 1986, Shaeffer and Silk 1988). The fundamental properties of clusters that can be deduced from X-ray measurements are luminosity, mass, temperature, metallicity, and morphology. Collectively, they provide insight into the origin and evolution of clusters. The missions of the previous 25 years did not have the capabilities to make the necessary measurements. Although these properties can be measured in the optical domain, it is not practical to do so for distant clusters because of the intrinsic sparseness of bright galaxies.

Extensive observational and theoretical studies of the X-ray luminosity of Abell clusters by richness class show that the X-ray emission from a cluster depends not only on its mass but also on the initial conditions, the subsequent dynamical evolution of the system, and the mechanism and history of the gas injection (Burg et al. 1989, Cavaliere, Burg and Giacconi 1989). From these studies we conclude that the luminosity function at the current epoch is not sufficient to disentangle the various contributions of cosmology (the density fluctuation spectrum) and astrophysics (structural evolution and gas injection).

If we approximate, on the scale of clusters, the fluctuation spectrum with a power law, one can derive scaling laws which describe the redshift dependence of cluster parameters such as characteristic mass, luminosity, density, etc. (Efstathiou, Fall, and Hogan 1979, Kaiser 1986). These scaling laws are a consequence of the hierarchical scenario, where clusters and groups are part of a continuous population. This scenario predicts that there is substantial substructure, and that it is scale invariant. Optically selected high redshift clusters are not suitable for these studies since they are severely biased toward the richest systems which generally are fully collapsed, erasing most traces of their origins.

As shown by Cavaliere and Colafrancesco (1988), who combined scaling laws with gas injection mechanism, evolution can range from pure density to pure luminosity evolution, and the observed number density can grow or decrease. The interpretation of the evolution of the luminosity function is therefore difficult, and requires carefully planned observational strategies. If we study high redshift and low redshift clusters with the same mass and density, then we are studying objects that were formed at the same step in the hierarchy. What will be different is essentially the evolution of the internal structure and of the gas. With large enough samples we can probe the astrophysical evolution directly. We can also test for the validity of the scaling laws predicted from the hierarchical clustering scenario. If they do exist, we can measure the local index, which is related to the slope of the power spectrum of fluctuations. This determination would be independent from that obtained in correlation analysis.

Using the general concept of scaling we can predict that clusters will have a minimum size of between 5 and 10 arc-seconds (diameter). Thus, any survey appropriate for study of the evolution of clusters at high redshifts must have angular resolution of at least 5 arc-seconds.

2.2 Large-Scale Structures

Clusters are also known to be the tracers of the galaxy distribution over large scales (Gregory and Thompson 1979, Bahcall and Soneira 1983); this effect can be seen especially well in the CfA slice (de Lapparent *et al.* 1986). The CfA group catalog (Ramella *et al.* 1989) further indicates that the distribution of groups follows the large scale distribution of galaxies. Thus, an X-ray survey of clusters leads naturally to an elucidation of the large-scale structure of the universe. The study of structures on the largest scales can only be done with a survey over a contiguous area.

3.0 The Wide Field X-Ray Optics

We have recently finished a study of grazing incident telescopes (Werner 1977, Burrows, Burg, and Giacconi 1989) where we have incorporated our scientific goals directly into the design process. Simply stated, we wish to maximize the solid angle over which we can successfully discriminate clusters from point sources. This requirement was translated into an optimizing ray tracing code where the solid angle weighted point response function was minimized over the desired field of view.

The design starts with standard two mirror conics, to which polynomial corrections up to fifth order are added. This has the effect of introducing spherical aberration on axis but compensates for off-axis aberrations. We are left with an optical system which has slightly worse than the theoretical on-axis performance (but still comparable to the real performance imposed by manufacturing tolerances) but which is over an order of magnitude better at the edge of the field. Most importantly, the manufacturing tolerances of these telescope designs are no higher than the conventional Paraboloid-Hyperboloid configuration. The best surface of focus is still parabolic with a radius of curvature which is no larger than the comparable Paraboloid-Hyperboloid design.

4.0 The Wide Field X-Ray Telescope

We have designed a telescope (see table 1) which has an angular resolution of better than 2.5 arc-sec rms when averaged over a thirty arc-minute radius field of view (figure 1). The design is optically fast; 1.5 meter focal length and 30 cm diameter aperture. With a 10 per cent filling factor and four modules this system has a collecting area of approximately 300 cm^2. The efficiency for survey work can be expressed as the product of field of view and area (Omega-A). For the Wide-Field X-ray Telescope Omega-A is 1,080,000 sr-sq. cm. For AXAF Omega-A is 306,000 sr-sq cm. The Wide-Field X-ray Telescope is roughly three times faster than AXAF for survey work!

A mission to survey 100 square degrees at a uniform sensitivity of 8×10^{-15} erg s^{-1} cm^{-2} (and an additional 1000 square degrees at a sensitivity of 3×10^{-14} erg s^{-1} cm^{-2}) utilizing the Wide-Field X-Ray Telescope design can be completed in four years. We estimate that we will find greater than 3000 clusters (and groups) with redshifts greater than 0.5 (many hundreds of them greater than redshift 1). In addition this survey will discover approximately 30,000 AGN and 1,500 galaxies.

With a CCD detector, tiled to match the focal surface, we estimate ~5% of the clusters will have measuable redshifts. This will allow an immediate determination of the luminosity function and correlation function for a significantly large, high redshift, sample of clusters and groups.

The weight for this mission is estimated to be 280 kilograms. The mirror alone weighs 115 kilograms. The Wide-Field X-ray Telescope will fit in a Scout-II or Pegasus launch vehicle, either of which is capable of lifting it to a 550 km equatorial orbit.

We have initiated a collaboration with the Osservatorio di Brera, Milan (lead by G. Chincarini) to carry out this mission In particualar the mirror fabrication group in Milan, lead by O. Citterio, is now making a test mirror for this program.

The Wide-Field X-Ray Telescope survey will complement and extend our studies of clusters of galaxies and of large scale structure. It is a mission of the 1990's which will provide a sample of clusters that will be appropriate to study with AXAF and with new missions of the twenty-first century.

Acknowledgement: We would like to thank Pierre Bely, Guido Chincarini and Oberto Citterio for help in preparing this paper.

Table 1: Wide Field X-ray Telescope - Mission Summary

Weight	280 kg		
Length	1.5 m		
Area	300 sq. cm		
FOV	1 sq. degrees		
Orbit	550 km 3 degrees		
Launch Veh.	Scout-II or Pegasus		
Resolution	2.5" rms		
Energy Range	0.4 - 3.0 keV		
Survey	100	1,000	sq. degrees
Time	300,000	30,000	s/sq. degree
Smin	8 e-15	3 e-14	ergs/s/sq. cm
#Clusters	1,000	2,000	
#AGN	10,000	20,000	
#Galaxies	500	1,000	

Figure 1 - Point Spread Function vs. Off-Axis Angle for Wide-Field X-ray Telescope

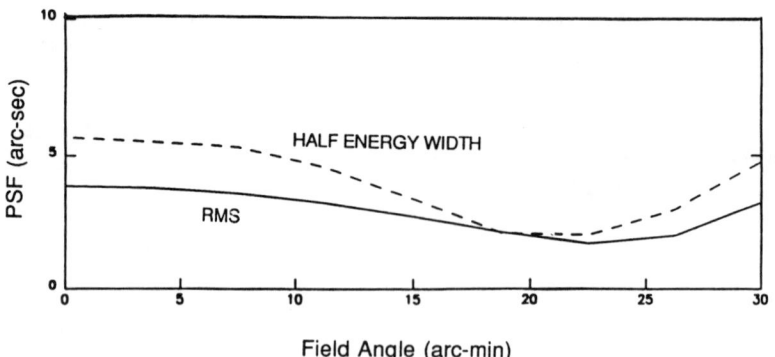

5.0 References

Burg, R., R. Giacconi, W. R. Forman, and C. Jones, 1989, in preparation.

Burrows, C., Burg, R., and Giacconi, R. 1989, in preparation

Cavaliere A., R. Burg and R. Giacconi, 1989 in preparation.

Cavaliere A. and S. Colfrancesco, 1988. Ap. J. 331, 660.

de Lapparent, V., M. J. Geller, and J. P. Huchra, 1986, Ap. J. Letters, 302, L1.

Efstathiou, G., S. M. Fall and C. Hogan, 1979. MNRAS 189, 203.

Gregory, S. A. and L. A. Thompson, 1978, Ap. J. 222, 784.

Kaiser, 1986. MNRAS 222, 323.

Shaeffer R. and J. Silk, 1988. Ap. J. 333, 509.

Werner, W., Applied Optics, 16, 764.

DISCUSSION

Martin Weisskopf

How sensitive are you to your assumption that 5" is the minimum cluster size?

Burg

Not very sensitive. Our estimates are based on larger sizes ($\sim 20''$) that we used for simulating ROSAT Deep Surveys.

George Ricker

Could you comment on what the properties of AXAF would be if its mirrors were figured in a similar manner to the WFXT? What would be the radius of the field of view with < 1 arc second images? What would be the best image quality within this field of view? Please compare with the current W-S design for AXAF.

Burg

We have looked into AXAF, and we feel this design approach is very promising. Essentially the PSF can be 1" or better over the field of view of the CCD detector. Best resolution will be limited by manufacturing errors.

Hugh Hudson

This design sounds like the one to be flown on Solar-A in 1991. It is a hyperboloid-hyperboloid based on a design by Nariai. A wide field of view is useful for solar observations too! See the details given in Catura's review in these proceedings.

Burg

Thank you very much. We didn't know about this work.

William Priedhorsky

How much of your too 270 kg is payload mass? Are the mirrors nested? What is your upper wavelength cutoff? What detector do you have in mind?

Burg

The mirrors are nested (there are 10) and we estimate the weight to be roughly 115 kg. We plan to use CCD detectors, and the energy range is approximately $.4 - 3.0$ keV.

The EXOSS Mission for Hard X-ray Astronomy

J. Grindlay (CfA), T. Prince (CIT), M. Weisskopf (MSFC) and G. Skinner (Birmingham)

ABSTRACT

We describe the basis for an Energetic X-ray Observatory on Space Station (**EXOSS**), which has been accepted by NASA as a Concept Study for the Space Station Attached Payloads program. With its high sensitivity, **EXOSS** will complement the capabilities of **AXAF** and **GRO** by covering the important 10-300 keV energy range (with extended response from $\sim$3 keV - 1 MeV). The principal scientific objectives of **EXOSS** will be to study in detail active galactic nuclei (AGN) and quasars (some 10,000 should be detectable) as well as compact galactic sources (accreting white dwarfs, neutron stars and black holes) to probe both non-thermal and high temperature thermal phenomena and the fundamental nature of these objects. High energy imaging studies of diffuse sources, both supernova remnants and galaxy clusters, as well as surveys will also be conducted. As such, **EXOSS** will achieve many of the objectives of the major Hard X-ray Imaging Facility (HXIF) for astrophysics proposed in the "Space Science in the 21st Century" report of the Space Science Board.

1 INTRODUCTION

The imaging and spectroscopy of astrophysical objects in the hard X-ray and soft gamma-ray band, from approximately 10-500 keV, is of key importance for high-energy astrophysics. In this energy range, a wide variety of important non-thermal and high-temperature thermal phenomena can be studied in detail. Many of these are now only poorly understood and many can be studied *only* in the hard X-ray and soft gamma-ray energy range. Imaging techniques and detector technology are now sufficiently advanced that a long-lived capability for high sensitivity observations in the hard X-ray and soft gamma-ray band is technically feasible and scientifically compelling. We have therefore proposed to develop a Concept Study for an Energetic X-ray Observatory on Space Station (**EXOSS**). Although implementation of this proposal, which was accepted, has been delayed, we describe here the rationale for such a mission which could also be developed as a free flyer.

The baseline concept for **EXOSS** includes two coded-aperture imaging telescopes which operate in the energy range from 3 keV to $\sim$1 MeV with optimized performance in the 10-300 keV range. The Imaging Gas Proportional Detector (IGPD) consists of a large-area array of Xe gas imaging proportional counters, which combine good sensitivity at energies below 60 keV with good energy resolution (5-10%) and excellent spatial resolution (0.5 mm). The Imaging Phoswich Scintillation Detector (IPSD) consists of a large-area array of NaI/CsI position sensitive phoswich detectors having high sensitivity and good energy resolution (10 - 15%) from 30 keV to 300 keV, with extended response up to 511 keV. The principal technical characteristics of the baseline instruments, which overlap in sensitivity in the $\sim$40 - 60 keV band, are given in the accompanying Table 1.

EXOSS has been conceived as a Guest Observer facility, available to the community for a wide range of investigations including survey observations. One of the first activities during the Concept Study will be formation of a Science Working Group to refine and develop an optimal instrument complement and scientific observation strategy.

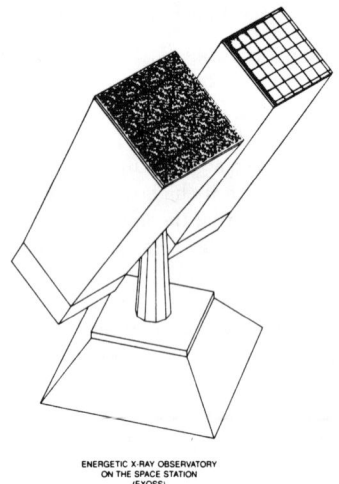

ENERGETIC X-RAY OBSERVATORY
ON THE SPACE STATION
(EXOSS)

Figure 1. Overall layout for EXOSS telescopes

	Table 1 Baseline Instruments for EXOSS	
	Imaging Gas Proportional Detector (IGPD)	Imaging Phoswich Scintillation Detector (IPSD)
Detector Elements	Xe filled MWPC (2 atm) 16 modules 32 cm x 32 cm area per module 870 cm² (with collimator) 10 cm active depth	NaI/CsI phoswich 4 modules 90 cm x 90 cm area per module 6885 cm² (with collimator) 2 cm NaI/2 cm CsI
Total Effective Area	1.4 m²	2.8 m²
Energy Range	3-60 keV (with response up to 100 keV)	20-300 keV (with response up to 1 MeV)
Sensitivity	1×10^{-6} photons/cm²-s-keV (3σ at 20 keV, 10^5 s)	3×10^{-7} photons/cm²-s-keV (3σ at 200 keV, 10^5 s)
Coded Aperture	7 m, mask/detector spacing 2 mm cell-size 157 x 157 element Galois URA 0.6 mm tungsten	7 m, mask/detector spacing 2.4 cm cell-size 37 x 37 element Galois URA 2 cm Pb-Sn-Cu
Pixel Size	1 arcmin	12 arcmin
Field of View	2.6° x 2.6° full width (collimator: 1.3° FWHM)	7.3° x 7.3° full width (collimator: 3.6° FWHM)

2 Hard X-ray Astronomy

The study of cosmic X-ray sources at energies above those accessible to grazing-incidence telescopes, approximately 10 keV, and extending up in energy into the gamma-ray range at energies in excess of 200-300 keV, remains one of the most scientifically promising and yet unexploited regions of the electromagnetic spectrum. Exploratory observations have been made at hard X-ray and soft gamma-ray energies, with the result that a limited number of sources have been detected and studied. These pioneering instruments, as well as currently planned detectors, have been limited to observing relatively bright sources, representing only a fraction of the interesting objects.

2.1 The General Case for a New Capability

Whereas the introduction of grazing-incidence imaging for X-ray astronomy (below 10 keV) has enabled spectacular gains in our understanding of many diverse classes of objects and astrophysical problems, most of these are fundamentally thermal in nature: from the hot coronae of stars and diffuse gas in clusters of galaxies to the thermal plasmas evident on the surfaces of neutron stars in X-ray bursts or (possibly) even the accretion disks around massive black holes in quasars. At higher energies, on the other hand, these same objects can be expected to reveal a diverse range of primarily non-thermal phenomena in which production of energetic particles is accompanied by radiation spectra extending out to significantly higher energies. It is likely that our entire view of many astrophysical objects and phenomena has been severely limited thus far by the "thermal bias" which dominates at lower energies. Evidence abounds that nature has not buried all under a thermal blanket, and that there is a rich variety of fundamental physical problems lurking in the non-thermal universe.

However, non-thermal phenomena themselves have their own low-energy bias: virtually all such processes are characterized by power-law spectra with spectral indices (for either photon or particle number spectra) significantly greater than one. As recognized long ago in "Cocconni's Law", such processes are best studied and sources are most likely to be detected at the lowest energies possible for the non-thermal phenomena of interest still to dominate. Enough dynamic range *must* be available (e.g. factor of ~ 30) to differentiate between competing processes and underlying thermal components of source spectra.

The thermal universe is also particularly interesting and not devoid of phenomena at hard X-ray photon energies of 20 keV and above. Accreting black holes such as Cyg X-1 have hard X-ray spectra extending out to well above 100 keV which may be a signature of a very hot Comptonizing plasma surrounding the inner accretion disk. Active galactic nuclei (AGNs) and QSOs may also be dominated by pair plasmas at ~ 500 keV. High-temperature Comptonized spectral components are also suspected for other classes of compact objects, such as the low-mass X-ray binaries (LMXBs) with low magnetic fields and low accretion rates. Thus, the scientific return is likely to be maximum for a hard X-ray observatory designed primarily for the 10-300 keV band, with significant response up to 1 MeV. Not only is this the optimum range for much of the science and the detectability of sources, it is also the optimum energy range for the design of low-background, high-sensitivity experiments, such as **EXOSS**

This is an opportune time to consider an advanced mission for hard X-ray imaging. A long-lived facility for observations in the 10-300 keV energy range is vital to complement the observing capabilities of **AXAF** and **GRO** which concentrate on energies below 10 keV and above 100 keV respectively. Although this important decade of the electromagnetic spectrum, where non-thermal and thermal phenomena often meet is not covered by any planned or proposed missions with the sensitivity and resolution proposed for **EXOSS** , the energy band *must* overlap with both **AXAF** and **GRO** for high sensitivity studies of objects in common.

2.2 Space Station Attached Payload Concept for EXOSS

The Energetic X-ray Observatory on the Space Station (**EXOSS**) has been approved for a concept study for the Attached Payloads Program for the Space Station. **EXOSS** would employ large-area modular scintillation detectors and gas proportional counters together with arrays of coded-aperture masks for two-dimensional hard X-ray imaging. An overall view of the **EXOSS** concept is shown in Figure 1, is summarized in Table 1 and is described in detail in Section 5. The two telescopes include a 2.8 m^2 (with collimator) modular array of position-sensitive phoswich scintillation detectors and a 1.4 m^2 (with collimator) modular array of Xe gas proportional counters. The scintillation detector array provides the required high-sensitivity with moderate angular resolution (12 arc minutes) in the important 20-300 keV energy range (with response to > 511 keV), while the gas counter array combines good response below 20 keV (down to ∼3 keV) with improved energy resolution and excellent angular resolution (∼ 1 arcmin) up to about 60 keV (with response to 100 keV). The relative areas of the scintillator and gas proportional detectors have been chosen to give comparable sensitivity in the overlap energy range of about 40-60 keV. The instrument concept for **EXOSS** is based on reasonable extrapolations of current detector technology, such as developed by our balloon programs (Althouse et al. 1985, Grindlay et al. 1986, and Ramsey, Weisskopf and Joy 1989). An important advantage of the modular approach, with both phoswich and gas proportional counter detectors, is the ability to adapt the instrument size and envelope to a variety of different accommodation options for either a free flyer or the Space Station.

The baseline detector concept for **EXOSS** would have the sensitivity shown in Figure 2, which also shows the expected sensitivity compared to other planned space missions. In particular, it illustrates how **EXOSS** covers the 10-100 keV energy range between **AXAF** and **GRO** . **EXOSS** will be an excellent follow-on to the **GRO** , **XTE** , and **SIGMA** missions. With its advanced capabilities, **EXOSS** will satisfy the continuing need for the ability to make high-sensitivity observations of astrophysical objects in the 10-300 keV energy range. As such, it will address many of the objectives of the Hard X-ray Imaging Facility (HXIF) discussed in the report of the Space Science Board, "Space Science in the Twenty-First Century: Imperatives for the Decades 1995 to 2015".

As described in the next section, **EXOSS** will address fundamental problems in both extragalactic and galactic astrophysics:

Detailed studies of the hard X-ray and soft gamma-ray emission of quasars and active galactic nuclei (AGN) will determine the spectra and variability of these objects. The **EXOSS** observing program will include medium and deep surveys of selected fields of view to determine the spectra and luminosity functions for various classes of AGN as well as "starburst" galaxies.

Detailed studies of galactic hard X-ray emitters, including pulsars and compact X-ray binaries, will determine the spectra and variability of these objects. **EXOSS** observations will again include medium and deep surveys of selected regions in the galactic plane (e.g. the galactic bulge) to determine the luminosity functions and spatial distributions of various classes of compact objects (e.g. CVs, LMXBs and black hole systems) emitting hard X-ray spectral components. Black hole candidates (like Cyg X-1) as well as apparently super-Eddington sources will also be observable in nearby galaxies (e.g. M31).

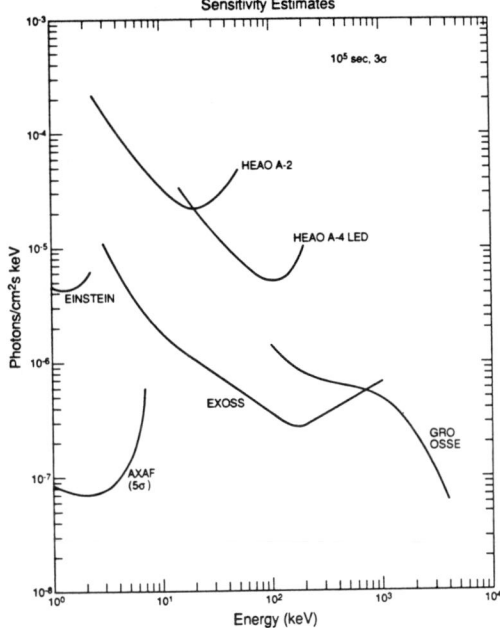

Figure 2. Sensitivity of EXOSS versus other missions

3 Scientific Objectives

3.1 Studies of Active Galactic Nuclei (AGN)

Because AGN typically have photon spectra flatter than $\alpha = 2$ (at energies above $\sim$3 keV), the emitted power per logarithmic energy interval increases with energy. As a consequence, AGN emit a significant fraction of their power at energies above 10 keV. The study of the hard X-ray and soft gamma-ray emission is thus crucial to the understanding of active galaxies.

At low-energies, the spectral indices measured by **HEAO 1** in the "medium" X-ray band (2-20 keV) are generally consistent with an index of about 1.5-1.7. This is not the case, however, for AGN spectra in the soft X-ray band (0.1 - 3 keV) observed with the **Einstein** Observatory. Here, a wide range of apparent spectral index values (1 - 3) is indicated (cf. Elvis et al. 1985). Thus it is particularly important to measure the hard X-ray spectra of a wide range of AGN types to determine whether the diversity of soft spectra is due to varying contributions from different emission components (e.g. a thermal accretion disk and/or a steeper non-thermal contribution) or whether it is due to evolutionary or luminosity effects (the **Einstein** objects are in general both more luminous and more distant than the predominantly Seyfert 1 galaxy sample measured thus far at higher energies). The hard X-ray spectrum is also likely to be modified in some cases by scattering from cold accretion disk material or clouds in the vicinity of the central source. Additionally, Fe line features have been detected in several sources by the **GINGA** instrument and these promise to be powerful diagnostics of the cold material near the AGN source. It is important to have detectors, like those on **EXOSS** which cover the entire band from Fe line features to Comptonized non-thermal photon spectra.

At higher energies, measured spectra are available (Rothschild et al. 1983) for only about a dozen objects up to about 100 keV. Above 100 keV, spectral data exist only for the few strongest sources (Bassani and Dean 1983). Estimates (Grindlay 1978, Rothschild et al. 1983) of the integrated soft gamma-ray flux from all AGN require that the average spectrum of AGN steepen, probably at energies below about 1 MeV; otherwise, the estimates exceed the measured diffuse background. It is therefore important to extend the measurements of AGN to the lowest possible flux limits at hard X-ray energies in order to better characterize their contribution to the diffuse background. Likewise, a measurement of a turnover in individual AGN spectra, or even a limit on the turnover energy is important for determining the total luminosity and particle spectra of AGN.

Definitive hard X-ray observations would enable sensitive tests of models of the central source. In particular, increasingly realistic synchrotron self-Compton (SSC) and inverse Compton (IC) models for the central sources have been developed (e.g., Band and Grindlay 1986, Band and Malkan 1989) that predict hard X-ray power-law spectra to emerge from the steeper soft X-ray spectrum that may arise from both thermal emission from the accretion disk and the extrapolation of a non-thermal synchrotron spectrum. The amplitude and spectral index of these hard X-ray components can constrain the high-energy particle spectra and energy loss vs. acceleration mechanisms in regions close to the central object where direct observation at radio wavelengths is probably not possible due to optical depth effects. Very different spectra are likely for the nuclei of "starburst" galaxies (e.g. M82) if LMXBs vs. BH candidates dominate, and these systems have been invoked to explain the X-ray background.

In summary, high-sensitivity observations of hard X-ray and soft gamma-ray emission from a large sample of AGN could lead to substantial progress on several important astrophysical questions:

1. What is the central power source in AGN? The shape of the photon spectrum and its time variability are crucial in determining the energetics and size scale of the central object.

2. What are the dominant emission mechanisms? Spectra from individual objects correlated with observations at other energies will aid in deciding among the various possibilities, such as Compton scattering, synchrotron-self-Compton emission, photon production near black holes, and relativistic beam models.

3. How do the various classes of AGN differ as hard X-ray and soft gamma-ray emitters? Luminosity functions are needed separately for quasars, Seyferts, BL Lac objects, and other classes of AGN. Spectral data from a large sample of objects will determine the general characteristics of these different classes of objects and determine if evolutionary effects are important.

4. What is the contribution of AGN to the diffuse hard X-ray and soft gamma-ray background? Estimates of the contribution of quasars and Seyferts to the diffuse gamma-ray background depend on the range of redshifts from which AGN contribute, and thus on the "turn-on" age for AGN. Comparisons of the spectra of individual AGN with the diffuse background should contribute to our understanding of the evolution of these objects.

5. Do "starburst" galaxies have the hard X-ray spectral signatures required if they are important for the X-ray background?

Figure 3 shows an estimate of the number of detectable AGN at 30 keV, derived from both the observed HEAO-A4 fluxes as well as extrapolations (with photon index 1.7) from the **Einstein** medium sensitivity survey as a function of minimum flux sensitivity. Also shown in Figure 3 are

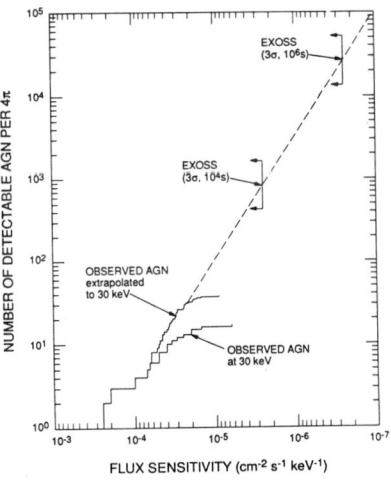

Figure 3. Number AGN detectable

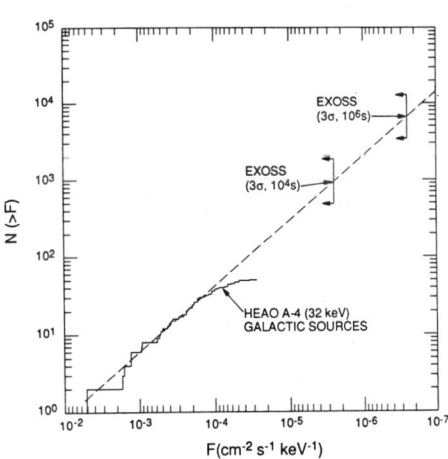

Figure 4. Number galactic sources

estimates of the flux sensitivities (3σ) at 30 keV that would be achieved by **EXOSS** pointed observations of 10^4 and 10^6 sec. It is clear that as many as 25,000 AGN would be detectable with **EXOSS** over the whole sky. This number could be even larger if luminosity evolution is important at high-energies. For the first time, a large number of AGN detected at other wavelengths will be accessible for study in the hard X-ray energy range (cf. survey objectives in section 3.4). Conversely, **EXOSS** will be able to obtain large samples of hard X-ray selected AGN which, by analogy with previous radio through soft X-ray surveys, virtually guarantees new results.

3.2 Studies of Galactic Sources

Compact X-ray binaries are relatively well studied at soft and medium X-ray energies (i.e., up to 20 keV) but only very partially explored at the still higher X-ray energies in the hard X-ray band open to **EXOSS** . Before commenting on the individual classes of galactic sources available for study with **EXOSS** , it is worth noting that a large number of galactic objects will be detectable and available for detailed study. In Figure 4 we show the log N - log S diagram expected for galactic sources detectable, based on extrapolations from the much lower sensitivity **HEAO-A4** survey. The large number of galactic sources expected motivates the galactic survey objectives (section 3.4).

Cataclysmic Variables (CVs). The fact that magnetic white dwarfs, the AM-Her type systems, have pronounced hard X-ray spectral components, make them also particularly attractive for detailed studies with **EXOSS** . Since the free-fall temperature in a standoff shock near the white dwarf surface is of order 100 keV, these objects can radiate a large part of their total energy in the hard X-ray band. A large number of hard X-ray emission CVs are expected to be detected in the galactic bulge as already suggested from the **Einstein** galactic plane survey (Hertz and Grindlay 1983, 1988). Thus **EXOSS** observations could help to constrain the spatial density and distribution of CVs in the Galaxy and the galactic bulge and thereby limit models for both stellar and binary evolution in the Galaxy. The precise (20 arcsec) positions derived from the low-medium energy experiment (gas counter) on **EXOSS** makes the required identification process no more difficult than for the **Einstein** survey.

X-ray Pulsars. Thus far, at least two pulsars, Her X-1 and 4U0115+62 (and possibly the Crab, A0535+26 and even Cyg X-3) have had cyclotron features identified in their spectra, at energies between about 20 keV and 100 keV. Several other pulsars have been searched (e.g., 4U0900-40) and only upper limits derived. It is likely that changes in the absorption and scattering properties of the pulsar accretion column will appear as changes in the cyclotron emission or absorption profile (and central energy) with pulse phase. Lower sensitivity experiments could miss bonafide features as they are smeared out (in time-averaged spectra) or have limited statistics (in attempted pulse-phased spectra). Because the line profiles are expected to be both broad and variable, high sensitivity and not high spectral resolution is of paramount importance. Pulse-phased cyclotron line spectroscopy could provide important new constraints on the origin and evolution of strong magnetic fields in neutron stars, the physics of radiative transfer in strong magnetic fields, and the physical processes operating (e.g., shock formation) in magnetized accretion columns.

Candidate Black Hole Systems. The black hole candidates such as Cyg X-1 and GX339-4 are known to have pronounced hard X-ray spectral components extending up to 200 keV. Black hole systems are also thought to be particularly time variable, down to the shortest time scales (commensurate with their Schwarzschild radii). The very large area detectors proposed for **EXOSS**, particularly the low-energy detector with an effective area larger than that for **XTE**, are especially well suited for short time variability studies of black hole candidates. The black hole candidates display both hard X-ray tails and steep soft (i.e., below 10 keV) X-ray spectra. Therefore it would be particularly interesting to observe the many additional sources that show soft excesses but have not yet been searched for hard X-ray components. The combined medium (3-30 keV) and hard (>30 keV) sensitivity of the two instruments proposed for **EXOSS** make these objects ideally suited for study. Many more soft-excess black hole candidates will doubtless be found with the all-sky survey conducted by **ROSAT**, and it will be important for theories of stellar and binary evolution to ascertain how many of these are probable black hole systems. Similarly, if black hole systems such as A0620-00, which undergo recurrent transient (soft) X-ray outbursts, have residual low level accretion during their quiescent periods between outbursts (as indicated by the optical disk colors detected in the quiescent A0620-00 system), then they may have residual hard X-ray spectra (Hameury, Lasota and King 1987). Finally, **EXOSS** could detect the hard X-ray spectrum of Cyg X-1 in M31, suggesting a "complete" inventory of BH candidates could be done in very deep surveys of nearby galaxies.

Sites of Particle Acceleration. Finally, it is becoming increasingly clear from recent radio (VLA) studies of luminous galactic bulge type X-ray sources such as Cyg X-3 (Molnar et al. 1988) and GX13+1 (Garcia et al. 1988; Grindlay and Seaquist 1986) that these objects are luminous sources of high-energy particles. In both these cases, and probably several others, the low-level radio flares typically observed imply the (periodic ?) injection of relativistic particles with total energies in excess of the observed soft X-ray luminosities. Thus particle acceleration and production may be a dominant energy loss mechanism in these intriguing objects, and hard X-ray studies are very much needed to probe this non-thermal behavior.

3.3 Studies of Diffuse Sources

The superior angular resolution and energy range of the **EXOSS** telescopes will allow a meaningful attack on several new problems in high energy astrophysics not accessible to earlier investigations: inverse Compton emission from galaxy clusters and hard X-ray emission from SNR.

Galaxy Clusters. The soft X-ray emission from galaxy clusters is decidedly thermal, with intense 6.7 keV iron line emission detected in most systems observed with sufficient sensitivity. In many clusters diffuse synchrotron radio emission is also detected from regions comparable in spatial extent (i.e. with cluster core radii of typically 0.3 mpc) to that of the diffuse thermal X-ray emission. The implied cluster magnetic field is possibly of major importance in determining the gas conductivity and thus cooling rate and could actually be measured by the detection of the inverse Compton hard X-ray emission that must arise from the scattering of the electrons on the ambient microwave background photons in the cluster. As worked out by Harris and

Grindlay (1979), the hard X-ray spectrum and flux ultimately depend on how the energy is shared between the electrons and field, and the combination of the radio spectrum and hard X-ray spectrum are enough to determine these two quantities.

The angular resolution of the gas counter telescope on **EXOSS**, which would be 1 arcmin, is more than sufficient to study this diffuse inverse Compton non-thermal emission in a variety of relatively nearby clusters. For example, Perseus has an apparent core radius of about 30 arcmin and contains a known central hard X-ray point source (NGC 1275). The **EXOSS** telescopes could clearly resolve the central source, measure its spectrum alone (for the first time) out to high energies, measure the spectrum of the expected surrounding diffuse emission, and (for the higher resolution gas counter telescope) map its structure for detailed comparison with the diffuse radio emission and thus measurement of the magnetic field.

Supernova Remnants. The spectrum of the Tycho SNR extends to beyond 20 keV from a remnant with apparent size of 8 arcmin and no central concentration or point source that would indicate a pulsar. The Cas A remnant is similar: with apparent diameter of some 4 arcmin, no pulsar (candidate), and yet a spectral component extending beyond 30 keV. Spatially resolved images of these, and other (larger), galactic supernova remnants at high energies are thus of vital importance to understand the origin of this emission. **AXAF** images, and spectra, may well enlighten us on the origin of this hard emission but again their limited energy range means the higher energy range of coded aperture imaging is needed.

3.4 Hard X-ray Survey Objectives

The survey observations of **EXOSS** will capitalize on the excellent sensitivity and angular resolution to perform medium and deep surveys of selected sky fields and portions of the galactic plane. Significant portions of the sky will be imaged for the first time at hard X-ray and soft gamma-ray energies. For example, it is likely that a large fraction of the galactic plane and galactic poles could be mapped in the process of carrying out the following important survey objectives:

Medium and deep surveys of portions of the galactic plane to determine the number, type, spectra, and variability of various classes of galactic hard X-ray emitting objects, for instance, LMXBs, cataclysmic variables, pulsars, and black hole systems. Given the 2.6° x 2.6° and 7.3° x 7.3° fields of view of the **EXOSS** telescopes, a substantial fraction of the galactic plane will be surveyed in medium (10^4-10^5 s) pointed observations on selected individual galactic objects. In addition, observing time should be allocated to "deep" pointings (>10^6 s) of selected fields, e.g. the rich galactic bulge region, to obtain hard X-ray selected samples of objects at the highest sensitivity. The **EXOSS** instruments, capable of simultaneous observations of several sources, are optimal for this type of study at 10-100 keV.

Medium and deep surveys of extragalactic objects to study the luminosity and distribution functions of AGN. As with galactic sources, some medium surveys can be made in conjunction with observations of individual objects, e.g. NGC1275 in the Perseus Cluster. In addition, **EXOSS** can perform deep-pointed (> 10^6 s) surveys of a small number of "blank" fields in order to probe the AGN luminosity and flux distribution to the deepest levels. Such deep pointings will make it possible to acquire unbiased samples of hard X-ray and gamma-ray selected objects down to the limits of instrumental sensitivity (see Figure 2.2). In particular, **EXOSS** can measure the log N vs. log F relation over three decades of flux, which together with measurements of a large sample of AGN spectra, will yield considerable information on the emission characteristics of classes of AGN and on the contribution of AGN to the diffuse background. As mentioned above, **EXOSS** could detect individual Cyg X-1 like sources in M31.

The last survey objective is particularly important for other surveys now planned for upcoming space missions. It would greatly complement the **ROSAT** all-sky soft X-ray survey by distinguishing soft vs. hard X-ray selected AGN. It would also complement the "medium surveys" to be carried out at soft and medium X-ray energies by **AXAF**.

4 Instrumental Requirements and Considerations

In order to achieve the basic scientific objectives outlined above, a mission such as **EXOSS** must have certain requirements for sensitivity, angular resolution and spectral/temporal resolution. These are described in the following sections.

4.1 Sensitivity

The instrumental flux sensitivity needed to make significant progress in the studies of AGN and galactic hard X-ray sources may be estimated using Figures 3 and 4. Figure 2 compares **EXOSS** to **AXAF**, **GRO** and other missions (note that the latest sensitivity estimates for OSSE on **GRO** (cf. Johnson et al 1989, *Proc. GRO Science Wkshp.*, GSFC, p. 2-22) are a factor of ∼1.5 less sensitive than plotted here). **AXAF** will not be sensitive above 10 keV and the **GRO** sensitivity is confusion limited below about 200 keV. What is required to fill this gap is a mission that lies between **AXAF** and **GRO** in both energy coverage and sensitivity. SIGMA and XTE cover the 20-200 keV energy range, but their sensitivity is at least an order of magnitude away from the 10^{-7}phot/cm^2-sec-keV levels at 100-200 keV needed to complement **AXAF** and **GRO**. In addition, medium and deep surveys of galactic and extragalactic sources require coverage of reasonable fields of view with a large number of angular resolution elements. Because **EXOSS** has an order of magnitude larger solid angle field of view, and it enjoys the multiplex imaging advantage, its sensitivity for medium and deep surveys will be at least a factor of 15 better than **XTE**.

Significant advances can be made if the number of objects available for study is increased by an order of magnitude and if the spectra and time variability of individual objects can be studied in greater detail. Figures 2 and 3 also show that **EXOSS** will have sufficient sensitivity to measure the spectrum of sources an order of magnitude fainter than 3C273 in 10^4 seconds out to 250 keV. Such observations, made frequently on selected objects, will determine the time-variability, and therefore the size scale of the emitting regions. Figure 4 shows that **EXOSS** would also significantly increase the number of galactic hard X-ray sources available for detailed study.

4.2 Angular Resolution

Good angular resolution was not critical for the initial, low-sensitivity observations of hard X-ray sources. However, with better flux sensitivity, the number of detectable hard X-ray sources will increase dramatically, making angular resolution essential to avoid source confusion and to limit the contribution of "AGN background fluctuation" errors to source flux measurements. **OSSE**, on **GRO**, will be confusion limited (given its non-imaging $\sim$40 square degree field of view) in sensitivity at a flux level of $\sim$2 x 10^{-6} photons/cm^2-sec-keV. Both the number of sources that may be detected and the accuracy with which their spectra may be measured will be limited if instrumental angular resolution is not sufficiently fine (Finger, 1987). For example, to reach a sensitivity of 10^{-7}phot/cm^2 -sec-keV at 100 keV requires an angular resolution (pixel size) of a few times 10^{-2} square degrees. **EXOSS** has a $\leq$12 arcmin, or 4 x 10^{-2} square degree cell size and a 10^6 s deep pointing sensitivity of 1.1 x 10^{-7}phot /cm^2-sec- keV and thus has comparable statistical and confusion limit sensitivity levels for deep pointed observations (as does **XTE** at reduced sensitivity). We note that at lower energies, although the **EXOSS** scintillation detectors will reach the confusion limit, the gas counters, with 1 arcmin resolution (3 x 10^{-4} square degrees) will be free of confusion, and should be effective in detecting possible cases of source confusion in the scintillators.

EXOSS will provide locations with accuracy significantly better than the 1 arcmin (at $\geq$100 keV) and $\sim$10 arcsec (at $\sim$3-60 keV) angular resolution by using inter-pixel interpolation. This will usually permit identification with objects at radio, optical, or soft X-ray wavelengths. Finally, the $\sim$1 and 12 arcmin resolution of the baseline **EXOSS** telescopes will allow images of extended sources such as galaxy clusters and SNR for the first time at high energies.

4.3 Spectral and Temporal Resolution

A coded-aperture telescope will measure a spectrum for any object in the field of view. If the field of view is sufficiently wide, a significant improvement in "operational" sensitivity results for observations of crowded galactic fields or deep survey observations of extragalactic objects. The spectral resolution for the **EXOSS** low-energy telescope (FWHM <6% up to 80 keV) and high-energy telescope (FWHM <12% above 100 keV) telescopes are sufficient to accurately measure continuum spectra and to detect the (typically broad) cyclotron lines expected in the hard X-ray band.

EXOSS will also carry out high time resolution studies of compact objects. We note that there has been a misconception in the past that it is difficult to measure the spectral and temporal characteristics of multiple sources in the field of view of a coded aperture telescope. This is not the case. In fact, unambiguous extraction of spectra and timing information for individual sources is straightforward and typically less computationally intensive than producing an image. Source confusion does not complicate time variability studies as sensitivity improves, unlike the non-imaging **GINGA** observations of SN1987A vs. LMC X-1.

5 Baseline Instruments for EXOSS

The relatively advanced status of scintillation detector, gas counter, and coded-aperture technology allows specification of a modular instrument with considerable latitude in the configuration details of a hard X-ray instrument complement (cf. Figure 1 and Table 1). Although the **EXOSS** concept shown is for the Space Station, it could be adapted readily for a free flyer implementation.

5.1 Imaging Phoswich Scintillation Detector (IPSD)

The baseline design of the Imaging Phoswich Scintillation Detector (IPSD) contains four identical square imaging phoswich modules (shown in Figure 5 and described below) which view the sky through a common coded-aperture array consisting of multiple repetitions of a basic 37 x 37 element pattern (Figure 6). The basic parameters for IPSD are given in Table 1. The size of the mask cell is 2.4 cm, chosen to yield good sensitivity given a detector position resolution of $\sim$ 9 mm FWHM at 100 keV. The mask to detector spacing of 7 meters then yields an angular resolution of approximately 12 arc minutes and an imaged field of view of 7.3° x 7.3°.

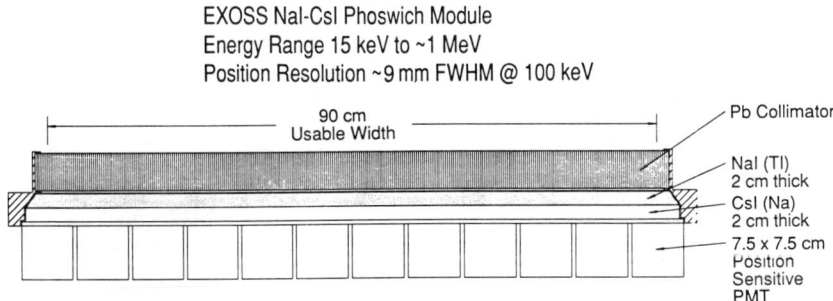

Figure 5. Schematic representation of the detector for the IPSD

The required sensitivity is obtained using NaI(Tl)-CsI(Na) "phoswich" style camera plate modules designed for both large area and low background. The concept, although new, is a straightforward evolution of existing technology. As illustrated schematically in Figure 5, the primary detector is a 2 cm thick layer of NaI(Tl). An adjacent 2 cm thick layer of CsI(Na) provides shielding from back-incident X-rays and rejection of penetrating charged particles. A Pb honeycomb collimator restricts the field of view to 3.6° FWHM. The total unobstructed, usable area of the four detector modules is approximately 27500 cm^2.

The camera plates are each viewed through a 0.5 cm thick optical window by a 12 x 12 array of 7.5 cm square-faced position sensitive multi-anode photomultiplier tubes (alternative readout schemes would include an array of still larger imaging PMTs, under development, or an array of smaller non-imaging PMTs). A thin stainless steel or fiberglass strongback grid (not shown in Figure 5) interleaves the PMTs and is bonded to the optical window to provide mechanical support. PMT anodes could be connected to form x and y strips which span the detector, with two strips per PMT width. Readout electronics would then incorporate pulse height analysis for each strip (24 x and 24 y strips imply 48 PHAs per module) and on-board calculation of event positions and energies. Depending on the available telemetry rate, data would either be transmitted event-by-event or recorded on-board the station with commercially available PCM encoders and VCR recorders. Detector count rate maps would also be accumulated in several energy bands on-board and telemetered to ground in real-time. The data rate for event-by-event information would be approximately 350 kb/s, while the data rate for accumulated count rate maps would be at least 10 times lower.

To minimize background, the coded-aperture mask is of graded-Z design in which the closed cells are formed by a sandwich of 2 cm lead, 1 mm tin, and 1 mm copper. The cells are supported on a light-weight carbon-composite structure (not shown in Figure 6) which can be made essentially transparent at hard X-ray energies. It is interesting to note that the 37 x 37 element coded-aperture pattern selected for the baseline design is not a conventional URA, but has the required delta function auto correlation, provided that the central cell of the unit pattern is "half-filled". The pattern is based on a two-dimensional Galois field. We could have as well selected standard URA patterns of 41 x 43 cells but the 37 x 37 square pattern is a better choice to optimize angular resolution, sensitivity, and field of view.

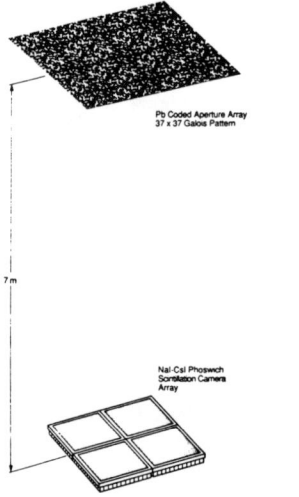
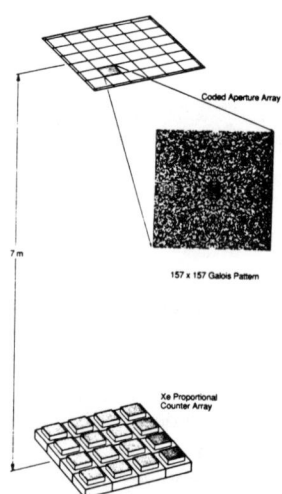

Figure 6. Schematic view of the IPSD   **Figure 7.** Schematic view of the IGPD

By keeping the necessary shielding immediately adjacent to the detector via the phoswich design and the Pb honeycomb collimator, the weight of the baseline instrument has been kept to a minimum. The approximate estimate of the weight of the IPSD is 1500 kg.

5.2 Imaging Gas Proportional Detector (IGPD)

The baseline design of the Imaging Gas Proportional Detector (IGPD, Figure 7) contains an array of 16 identical imaging proportional counter modules viewing a coded aperture consisting of repetitions of a 157 x 157 Galois pattern. The mask cell-size is 2 mm, chosen to yield good flux sensitivity given the spatial resolution of the proportional counters which ranges between 0.4 and 0.7 mm FWHM at energies from 3 to 60 keV. With the mask at a distance of 7 meters from the detector, the angular resolution is 1 arcmin and the imaged field is 2.7° x 2.7°. Each detector is fitted with a slat collimator which restricts the field of view to $\sim$ 1.3° x 1.3° FWHM both to minimize the background counting rate due to diffuse cosmic flux and to eliminate possible imaging artifacts due to sources just outside the imaged field. The basic parameters are summarized in Table 1.

The imaging proportional counters for the baseline instrument (see Figure 8 and Table 1) are of the "multistep" design developed at MSFC (Ramsey and Weisskopf 1986) to provide both

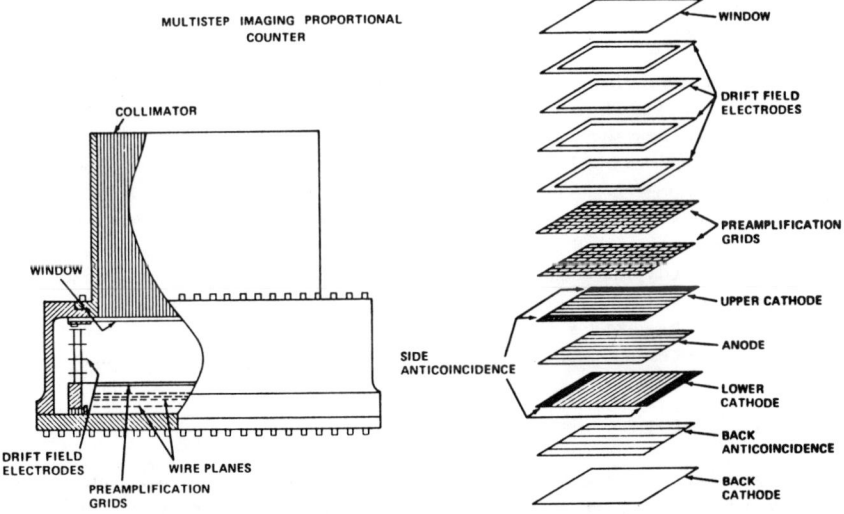

Figure 8. Schematic representation of the MWPC for the IGPD

good energy and position resolution, combined with high detection efficiency and background rejection. The detector consists of an absorption and drift region followed by two stages of charge amplification. The first of these, the preamplification region, is formed from a pair of stainless steel etched grids separated by 1 mm. It is from these grids that a primary energy signal is taken. Five percent of the charge produced here is transferred to a second stage consisting of a standard multiwire proportional counter (MWPC). This second stage forms the imaging section of the instrument and is also responsible for the rejection of charged particles.

It has been shown that by separating the detector into two sections the device can be simultaneously optimized for both energy and spatial resolution (Ramsey and Weisskopf 1987). The use of a parallel field first stage provides excellent energy resolution and leaves the second stage free to be optimized for spatial resolution. In addition, the presence of the first stage actually improves the energy resolution on the second stage by spreading the charge cloud and thereby reducing space charge effects (Ramsey and Weisskopf 1986).

The imaging second stage of the detector is read out via the induced charge on segmented x and y cathode wire planes above and below the anode wire plane. Each cathode plane is subdivided into eighty strips (each containing 4 cathode wires), and each strip is pulse-height analyzed. Event positions are determined by reconstructing the induced charge distribution on the two (orthogonal) sets of cathode strips. The principal advantage of such an imaging system over less complex techniques is the ability to resolve pairs of events which materialize simultaneously in the sensitive volume. This is crucial for the instrument's "escape-gated" mode, discussed below.

Charged particle rejection capability is vital for non-focusing instruments. The baseline detector uses standard 5-sided anticoincidence techniques coupled with risetime discrimination, and below 35 keV achieves rejection rates typical of these methods (Ramsey et al. 1989). Above 35 keV use is made of the fact that 75% of X-ray interactions lead to xenon K-shell fluorescence, a good fraction (0.4) of which are reabsorbed in the detector sensitive volume. This double event is a characteristic signature of a true X-ray event which can be used to discriminate against charged particles which interact primarily with outer shell electrons and hence do not produce K-shell fluorescence. The net result of this so called "escape gating" is an increase in sensitivity and, as a byproduct, an improvement in energy resolution due to a prior knowledge of the energy of the K-shell fluorescent photons. Event selection and on-board background rejection mean that typical telemetry rates for IGPD should be only 20 kbs, even for bright sources. An event-by-event timing and spectroscopy mode could be achieved with either 300 kbs telemetry or, more likely, a data storage and playback system.

The preliminary estimate for the weight of the baseline IGPD design is ~ 500 kg.

6 Conclusions

The need for a high sensitivity hard X-ray/soft gamma-ray imaging mission is acute. It would more than just fill in the gap between the high sensitivity provided by **AXAF** and **XMM** below 10-15 keV and **GRO** (and its follow-on missions) above 100-200 keV. It would open the entire domain of non-thermal astronomy and very high temperature astrophysics to detailed study for the first time. The technologies are at hand for proceeding with the development of a detailed mission concept. A platform on the Space Station for **EXOSS** would be a most important use of this national resource. However, the basic criteria for a hard X-ray imaging mission could also be met on a variety of other platforms and mision possibilities: from "moderate" missions (Explorer class) to future lunar base observatories. The time is ripe to begin development now so that an **EXOSS** -type mission can fly within the Great Observatory timeframe.

7 References

1. Althouse, W.E. et al, *Proc. 19th Intl. Cosmic Ray Conf. (La Jolla)*, 3, 299 (1985).
2. Band, D. and Grindlay, J., *Ap. J.*, 308, 576 (1986).
3. Band, D. and Malkan, M., *Ap. J.*, 345, 122 (1989).
4. Bassani, L. and Dean, A. J., *Space Science Rev.*, 35, 367-398 (1983).
5. Elvis, M., Wilkes, B. and Tanabaum, H., *Ap. J.*, 292, 357 (1985).
6. Fenimore, E. E. and Cannon, T. M., *Applied Optics*, 17, 337 (1978).
7. Finger, M. H., PhD Thesis, California Institute of Technology (1987).
8. Garcia, M. et al., *Ap. J.*, 328, 552 (1988).
9. Grindlay, J. et al., *IEEE Trans. Nucl. Sci.*, 33, 750 (1986).
10. Grindlay, J. and Seaquist, E., *Ap. J.*, 310, 172 (1986).
11. Grindlay, J. et al., *IAU Circular*, No. 4408 (1987).
12. Grindlay, J. *Nature*, 273, 211 (1978).
13. Hameury, J. M., King, A. R. and Lasota, J.P., *Astron. Astrophys.*, 162, 71 (1987).
14. Harris, D. and Grindlay, J., *Mon. Not. R. Astron. Soc.*, 188, 25 (1979).
15. Hertz, P. and Grindlay, J. E., *Ap. J.*, 278, 137 (1984).
16. Hertz, P. and Grindlay, J. E., *Astron. J.*, in press (1988).
17. Molnar, L., Reid, M. and Grindlay, J., *Ap. J.*, 331, 494 (1988).
18. Ramsey, B. D. and Weisskopf, M. C., *Nucl. Instr. Meth.*, A248, 550 (1986).
19. Ramsey, B. D., Weisskopf, M. and Elsner, R., *Proc. SPIE: X-ray Instrumentation in Astronomy*, 597, 213 (1986).
20. Ramsey, B. D. and Weisskopf, M. C., *IEEE Trans. Nucl. Sci.*, NS34, 3 (1987).
21. Ramsey, B., Weisskopf, M., and Joy, M. *Proc. SPIE Symp. 1159*, 1159, 246 (1989).
22. Rothschild, R. E. et al., *Ap. J.*, 269, 423 (1983).

DISCUSSION

Gerald Share

How many supernovae would you expect to see from the Virgo Cluster in continuum radiation near 100-200 eV during a 10^6 second observation?

Grindlay

With a continuum sensitivity of $\sim 10^{-7}$ photons cm^{-2} s^{-1} keV^{-1} at 100 keV (over 10^6 s), EXOSS could readily detect a Type Ia supernova (with intrinsic flux $10 - 100\times$ that of SN 1987A). Thus perhaps $1 - 2$ such events could be detected per year (recalling that as a probable rate for SN-Ia's in Virgo).

Hugh Hudson

One question about the optimization of the X-ray optics: wouldn't an array of coded-aperture resolutions (as we're doing on CASES) be better for a variety of diffuse-source angular scales? Such an arrangement wouldn't hurt the point-source sensitivity particularly.

Grindlay

Probably not, since ~ 1 arc minute resolution ($\sim 5 - 100$ keV) is already optimum for galaxy clusters (i.e. ~ 10 arc seconds would lose sensitivity, whereas ~ 10 arc minutes would not resolve most clusters). The main reason, however, is to allow for maximum throughput and not further subdivide the maximum available detector area.

THE NUCLEAR ASTROPHYSICS EXPLORER

J.L. Matteson
Center for Astrophysics and Space Sciences,
University of California, San Diego

B. J. Teegarden and N. Gehrels
Laboratory for High Energy Astrophysics, Goddard Space Flight Center

W. A. Mahoney
Jet Propulsion Laboratory, California Institute of Technology

ABSTRACT

The Nuclear Astrophysics Explorer (NAE) is a concept for a possible future NASA Explorer mission which would obtain high resolution, $E/\Delta E \sim 500$, observations of gamma-ray lines in order to study many fundamental problems in astrophysics. It operates from 15 keV to 10 MeV with a 3σ sensitivity of $\sim 3 \times 10^{-6}$ ph/cm^2-s in a 10^6 s observation. This is 100 times below the presently known gamma-ray line fluxes. The NAE uses a heavily shielded array of 9 cooled Ge detectors in a very low background configuration. Its 10° field of view contains a versatile coded mask system which provides 2-D imaging with 4° resolution, 1-D imaging with 2° resolution and efficient measurements of emission from diffuse and point sources. The late 1990's is the earliest the NAE mission could begin. The scientific motivation, instrument concept, mission concept and expected results, and status and plans for the NAE are presented.

I. INTRODUCTION

Continuing progress in gamma-ray astronomy requires new instruments which provide significant improvements in sensitivity, angular resolution and energy resolution. The Gamma-Ray Observatory (GRO), scheduled for launch in June 1990, will obtain substantial improvements in sensitivity and angular resolution over earlier space missions. However, its energy resolution in the MeV energy range, $E/\Delta E \sim 15$, will not be adequate to resolve gamma-ray lines, many of which are already known to be produced in a wide variety of astrophysical objects. Thus the fundamental information carried by line widths, profiles and energy shifts generally will not be obtained. In this paper we describe a mission concept which would overcome this problem. This is the Nuclear Astrophysics Explorer (NAE), a concept for a possible future NASA Explorer mission dedicated to high resolution gamma-ray spectroscopy. Its energy resolution of 2 keV at 1 MeV, i.e. $E/\Delta E = 500$, would allow most gamma-ray lines to be resolved and have their profiles measured. The NAE operates in the 15 keV to 10 MeV range with a sensitivity to narrow gamma-ray lines of $\sim 3 \times 10^{-6}$ ph/cm^2-s in a 10^6 s observation. This is 100 times below the known line fluxes and 10 times below the sensitivity of the GRO.

In July 1989 NASA's Explorer Concept Study Program completed 1-year feasibility studies of the NAE and three other mission concepts. The NAE was studied by a 28-scientist, 12-institution international collaboration lead by J. Matteson, Principal Investigator, B. Teegarden and W. Mahoney, Co-Principal Investigators, and N. Gehrels, Study Scientist. In October 1989 two of the four concepts were selected by NASA for development and spaceflight in the mid to late 1990's. Although the NAE was not selected, there is strong interest within NASA and the US and international scientific communities in performing the NAE mission. Indeed, the NAE is the highest priority new mission in the recent Report of the NASA Gamma Ray Program Working Group (Lingenfelter et al. 1988, pp. 1–16, 24, 25). The scientific merit of high resolution gamma-ray spectroscopy has been recognized by national scientific advisory committees, e.g. the Astronomy Survey Committee (Field et al. 1982, pp. 30, 64, 106) and the Task Group on Astronomy and Astrophysics (Burke et al. 1988, p. 61).

II. SCIENTIFIC MOTIVATION

Gamma-ray lines are the most direct probe of cosmic nuclear processes. High resolution spectroscopy of these lines can provide basic information on fundamental astrophysical problems such as nucleosynthesis, supernovae dynamics, neutron star and black hole physics, and particle acceleration and interactions. Gamma-ray line observations have, in fact, already made major contributions to our understanding of a wide range of astrophysical objects and phenomena. Table 1 contains a summary of these observations, indicating the line forming processes, energies, sources and line fluxes. All the observations were made with instruments that had relatively poor sensitivity, greater than $\sim 2 \times 10^{-4}$ ph/cm^2-s, and, in many cases, low energy resolution, $E/\Delta E \sim 10$. These results show that gamma-ray lines are produced in astrophysical objects by a variety of processes, i.e. electron-positron annihilation, radioactive decay, nuclear excitation, neutron capture and cyclotron processes in strong magnetic fields, i.e. transitions between quantized electron energy levels. Since these processes can only be directly observed in the gamma-ray band, the NAE would make available the study of unique astrophysics.

Gamma-ray lines can be produced when the temperature exceeds $\sim 10^8$ K, energy exceeds ~ 1 MeV/nucleon or magnetic field exceeds $\sim 10^{12}$ gauss. Although these are extreme conditions by terrestrial standards, they occur frequently in the cosmos. Unique astrophysical information is encoded in gamma-ray lines; not only do their energies indicate the presence of specific nuclei and excitation processes, electron-positron pairs or the magnetic field strength, but the line parameters, i.e. intensities, energy shifts from rest values, widths and profiles, carry information on abundances, bulk velocities, gravitational potentials, densities, temperatures, particle energy spectra and the magnetic field geometry. The literature contains many reviews of gamma-ray line astrophysics (Clayton 1973, Lingenfelter and Ramaty 1978, Ramaty and Lingenfelter 1979, Ramaty and Lingenfelter 1982, Matteson 1983, Lingenfelter and Ramaty 1985, Peterson 1988, Lingenfelter 1988, Chupp 1988).

TABLE 1. Astronomical Gamma Ray Line Observations

Process	Observed Energy	Source	Flux, ph/cm²-s	Ref
$e^{\pm}$ Annihilation	511	Galactic Center	0.6-1.8×10^{-3}	1
Radiation	511	Interstellar Gas	$\sim 2 \times 10^{-3}$/rad	2
	511	Solar Flares	up to ~ 0.1	3
(Redshifted)	400-460	Gamma Ray Bursters	up to 70	4
(Redshifted)	~ 400	CrabPulsar Transient	2-7×10^{-3}	5
(Redshifted)	~ 413	10June74 Transient	7×10^{-3}	6
	500-2000	Cygnus X-1	up to 2×10^{-2}	7
Radioactive Decay				
$^{56}Co(\epsilon\gamma, \beta^+\gamma)^{56}Fe$	847	Supernova 1987A	$\sim 10^{-3}$	8
	1238	" "	$\sim 10^{-3}$	8
	2598	" "	$\sim 10^{-3}$	9
$^{26}Al(\beta^+\gamma)^{26}Mg$	1809	Interstellar Gas	4.8×10^{-4}/rad	10
Nuclear Excitation				
$^{56}Fe\ (p,p'\gamma)$	847	Solar Flares	up to ~ 0.05	3
$^{24}Mg\ (p,p'\gamma)$	1369	" "	up to ~ 0.08	3
$^{20}Ne\ (p,p'\gamma)$	1634	" "	up to ~ 0.1	3
$^{28}Si\ (p,p'\gamma)$	1779	" "	up to ~ 0.08	3
$^{12}C\ (p,p'\gamma)$	4438	" "	up to ~ 0.1	3
$^{16}O\ (p,p'\gamma)$	6129	" "	up to ~ 0.1	3
Neutron Capture				
$^{1}H\ (n,\gamma)^{2}H$	2223	Solar Flares	up to ~ 1	3,11
$^{1}H\ (n,\gamma)^{2}H$	2223	10June74 Transient	1.5×10^{-2}	6
(Redshifted)	1790	" "	3×10^{-2}	6
$^{56}Fe\ (n,\gamma)^{57}Fe$	5947	" "	1.5×10^{-2}	6
(Redshifted)				
Cyclotron Emission	20-70	Gamma Ray Bursters	up to 3	12
& Absorption in	20-58	X-Ray Pulsators	1-3×10^{-3}	13
$\sim 10^{12}$ **gauss fields**	73-79	Crab Pulsar Transient	4×10^{-3}	14

References: 1. Haymes et al. 1975, Leventhal et al. 1978, 1989, Riegler et al. 1981, Riegler et al. 1985, Leventhal et al. 1989, Matteson et al. 1989; 2. Mahoney 1988, Share et al. 1988; 3. Chupp et al. 1973, Chupp 1984, Yoshimori et al. 1983; 4. Mazets et al. 1979, 1981, Teegarden and Cline 1980; 5. Leventhal et al. 1977, Ayre et al. 1983; 6. Jacobson et al. 1978, Ling et al. 1982; 7. Nolan and Matteson 1983, Ling et al. 1987, Ling and Wheaton 1989; 8. Cook et al. 1988, Mahoney et al. 1988, Matz et al. 1988, Sandie et al. 1988, Matteson et al. 1989, Rester et al. 1988, Teegarden et al. 1989, Tueller et al. 1989; 9. Matz et al. 1989, Tueller et al. 1989; 10. Mahoney et al. 1984, Share et al. 1985, v. Ballmoos et al. 1987; 11. Hudson et al. 1980, Prince et al. 1982; 12. Mazets et al. 1981, Dennis et al. 1982, Hueter 1984, Murakami et al. 1988; 13. Trümper et al. 1978, Wheaton et al. 1979, Gruber et al. 1980, Tueller et al. 1984, Maurer et al. 1982; 14. Ling et al. 1979, Strickman et al. 1982, Ayre et al. 1983.

The observations summarized in Table 1 have lead to many important conclusions and promise a wealth of new information when new instruments are flown in space. Some of these are discussed further below.

Electron-positron pair plasmas are expected to form at the extreme densities and temperatures of matter accreting onto black holes, e.g. Burns *et al.* (1983). Electron-positron annihilation radiation is the signature of this phenomenon, appearing as a broad line shifted above 511 keV. This feature has been observed in the spectra of Cyg X-1 and a variable source near the Galactic Center, both of which contain a narrow 511 keV annihilation line as well. These sources have been observed to vary on 6-month time scales, but the variability has not been resolved and its connection to underlying mechanisms is unknown. Electron-positron annihilation and the decay of radioactive ^{26}Al, which has a 740,000 year half life, have been observed to occur in the central 5 kpc of the Galaxy, producing gamma-rays at 511 and 1809 keV. The latter proves that nucleosynthesis is an ongoing process in the Galaxy and the former is thought to be primarily due to the decay of nucleosynthetic ^{56}Co produced in supernovae (Lingenfelter and Ramaty 1989). This produces positrons, some of which escape the remnant and have lifetimes of 10^5 to 10^6 years in the interstellar medium. These gamma-rays provide tracers of the sites and nature of galactic nucleosynthesis in the present epoch and the physical conditions of the interstellar medium (Lingenfelter and Ramaty 1990). The latter results from the dependence of the profile of the 511 keV line and associated positronium continuum on the physical conditions of the annihilation medium. Lingenfelter and Ramaty predicted that the annihilation line from the interstellar medium is very narrow, $\sim$ 1.6 keV FWHM, because the annihilations primarily occur in its warm, $\sim 10^4$ K, phase. This has been confirmed by the recent observations of the Galactic Center region by Matteson *et al.* (1989). They placed a limit of 1.8 keV on the 511 keV line width at a time when the line flux was low and consistent with the diffuse galactic flux. 7 months earlier observations by Leventhal *et al.* (1989) showed a larger flux, due to the compact source being active, and a broader line width, 3.5 keV. The latter requires a temperature of $\sim 10^5$ K. With these results line width measurements have become as important as line flux measurements.

Observations of gamma-rays from radioactive ^{56}Co synthesized in the recent Type II supernova SN1987A proved that explosive nucleosynthesis of heavy elements occurred in this supernova. The lack of red- or blueshift of the ^{56}Co lines and the need to obscure most of the ^{56}Co requires that the ejecta is fragmented (Teegarden *et al.* 1989, Tueller *et al.* 1989). Gamma-rays from radioactive material synthesized in supernovae and novae can be used as diagnostics of the yield, expansion and energetics of these events (Gehrels *et al.* 1987, Chan and Lingenfelter 1987,1988, Woosley and Pinto 1988). Assuming 0.5 $M_\odot$ of ^{56}Ni is produced in a Type I supernova (Woosley and Weaver 1986), the NAE should be able to observe the ^{56}Co gamma-rays from these events out to distances of 20 Mpc, i.e. the Virgo cluster, with $\sim$ 1 week time resolution for several months after the explosion. ^{44}Ti is expected to be produced in supernovae (Woosley *et al.* 1986) and its 54 year half-life allows its gamma-rays to be used as tracers of the undiscovered $\sim$ 10 galactic supernovae that have occurred in the past 500 years, $\sim$ 5 of which should be discovered by the NAE.

Gamma-ray lines due to nuclear excitation, neutron capture and electron-positron annihilation are diagnostics of particle acceleration to 10's to 100's of MeV, and their subsequent interactions. Solar flares are copious producers of these lines, which have been interpreted to determine the spectrum of the accelerated particles and the abundances in the solar atmosphere (Ramaty and Murphy 1987). With future high resolution measurements many more lines will be observed and they will be used to refine and extend the abundance determinations and study the geometry of the accelerated particles' confinement, transport and interactions. Nuclear excitation must also occur throughout the Galaxy due to cosmic ray interactions with the interstellar medium. The predicted fluxes (Ramaty et al. 1980) are below present detection capabilities, but the NAE should detect the stronger lines and use them to obtain new information on the abundances of the interstellar gas and dust, and the flux and spectrum of the low energy, <100 MeV, cosmic rays, which cannot penetrate the solar system.

Absorption and emission lines in the 20 to 70 keV range, interpreted as due to cyclotron processes in intense, $\sim 10^{12}$ gauss, magnetic fields have been observed in the spectra of many objects thought to be neutron stars, e.g. gamma-ray bursters, X-ray pulsators and the pulsar in the Crab Nebula. Since the transport of radiation in these conditions is strongly dependent on photon energy, propagation direction relative to the magnetic field, and electron temperature (Bussard 1980, Nagel 1981), the lines are sensitive diagnostics of the geometry of the magnetic field and the conditions in it.

Generally only the intensities and energies of gamma-ray lines have been determined to a useful accuracy and the information carried by the other line parameters has not yet been obtained. However, observations with the NAE would obtain detailed measurements of the parameters of known lines and many weaker lines as well. It would obtain 1) a hundred fold improvement in sensitivity to point sources of narrow lines, i.e. $\sim 3 \times 10^{-6}$ ph/cm^2-s, and a diffuse flux sensitivity of $\sim 2 \times 10^{-5}$ ph/cm^2-s-rad, both in 10^6 s observations, 2) an energy resolution that is less than most lines' known or predicted widths in order to determine energy shifts, widths and profiles, e.g. 2 keV resolution at 1 MeV, and 3) an angular resolution of a few degrees in order to map diffuse emission, resolve source complexes and determine source positions. With these capabilities the NAE could effectively pursue a wide range of objectives in astrophysics.

III. INSTRUMENT CONCEPT

The NAE instrument concept is shown in Figure 1 and its parameters are given in Table 2. It contains 9 large, $\sim$ 300 cm^3 each, Ge detectors in a heavily shielded 3 x 3 array. These are cooled by a combination of a Stirling Cycle mechanical refrigerator and a thermoelectric cooler to 85 K. The detectors have very low background, because of 4 essential features. 1) They use position sensitivity, obtained through axial segmentation and pulse shape discrimination, to discriminate against induced β-decay radioactivity in

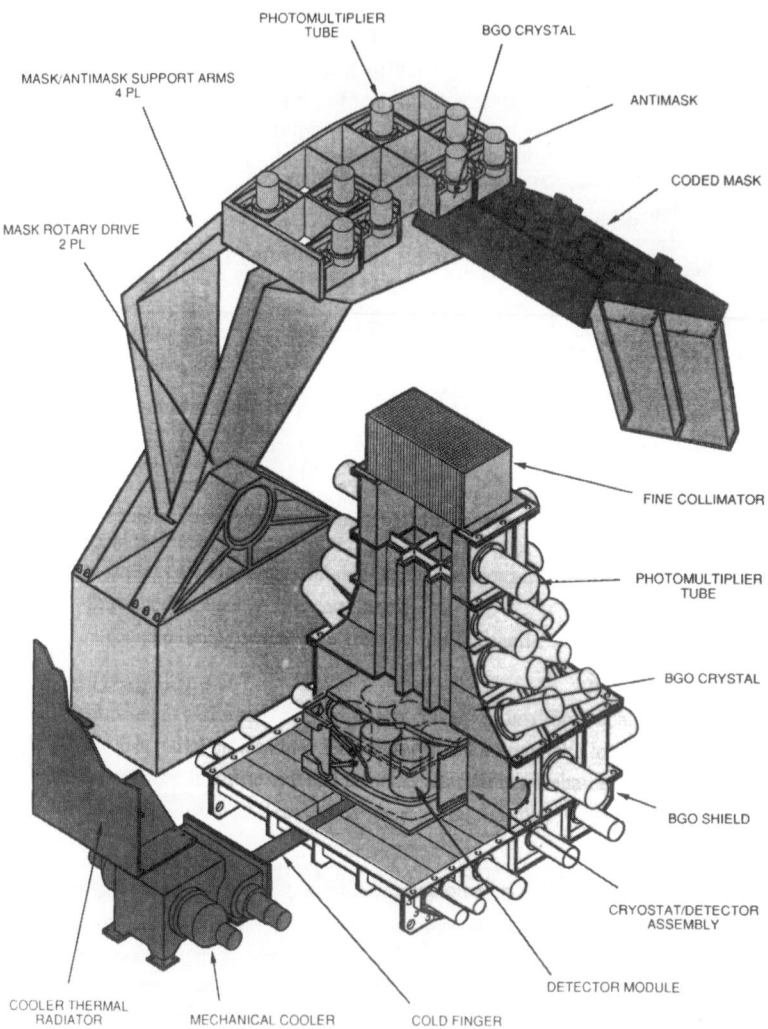

Figure 1. The Nuclear Astrophysics Explorer instrument concept. The BGO anticoincidence shield and cryostat are shown in a cutaway view to allow the germanium detector modules to be seen. The mask/antimask system is also shown cutaway. Its support arms extend to a second rotary drive (not shown) at the right side of the instrument.

TABLE 2.   NAE Instrument Concept Parameters

Energy Range	15 keV to 10 MeV
Energy Resolution	2 keV at 1 MeV
Sensitivity, 3σ	3×10^{-6} ph/cm^2-s in 10^6 sec
Detector System	9 cooled Ge detectors, 300 cm^3 each
Anticoincidence Shielding	10 cm thick bismuth germanate
Imaging and Aperture Modulation System	25 element mask/anti-mask system, 7 cm thick bismuth germanate elements
Field of View	10° FWHM (Image Mode, Knife-Edge Mode)
	20° FWOM (Image Mode, Knife-Edge Mode)
	1° FWHM (Fine Collimator Mode)
Angular Resolution	4° FWHM (Image Mode)
	2° FWHM (Knife-Edge Mode)
Instrument Size	1m(W) x 1.3m(L) x 1.8m(H)
Cryostat Thermal Load	0.5 W at 85 K
Cryostat Cooling System	Mechanical refrigerator, thermoelectric cooler
Power	242 W
Mass	1500 kg
Bit Rate	8 kbps (av), 25 kbps (peak)
Pointing Requirement	0.05°, unrestricted viewing in any direction
Orbit Requirement	low exposure to trapped radiation and cosmic rays, e.g. <10° incl. x 500 km alt.

the detectors themselves. Here the multiple-site signature of Compton scattered gamma-rays is distinguished from the single-site signature of β-decays, which are a major background component in heavily shielded instruments. 2) A 10 cm thick anticoincidence shield, made of 740 kg of bismuth germanate (BGO), greatly attenuates the background due to ambient gamma-rays. Its transmission is 1 percent at 1 MeV. The shield also eliminates prompt background from charged particles. 3) A 10° FWHM field of view, defined by apertures in the BGO shield, reduces the background near 1 MeV due to aperture gamma-rays to the level of the residual, non-rejected detector radioactivity. 4) A low inclination, low altitude orbit, i.e. $< 10°$ inclination and ~ 500 km altitude, is preferred for the NAE since it has the lowest possible fluences of the background producing trapped protons, cosmic rays and their secondaries.

Imaging and aperture chopping, for suppression of background systematics, are simultaneously obtained through the use of a 25-element coded mask/antimask system. The mask and antimask codes, or patterns, form complementary 5 x 5 arrays, which are produced by open elements and 7 cm thick BGO elements, which are nearly opaque to gamma-rays. They produce shadowgrams on the detector array which are deconvolved by matrix multiplication to produce a sidelobe-free 2-D image of point and point-like sources with 4° angular resolution. A simulated map produced by the mask/antimask system is shown in Figure 2. The mask and antimask are alternately placed in the aperture during imaging observations. This is performed with a few minute cycle in order to suppress the systematic effects of varying background caused by changing cosmic ray cutoff energies around the orbit. The mask/antimask combination can also be moved together in and out of the aperture to obtain either 1) a "knife-edge" which is smoothly moved over the aperture to produce a 1-D image with 2° angular resolution, in order to study complex source regions and better locate sources, or 2) a totally blocked or totally open aperture, in order to modulate and detect diffuse flux with high efficiency. A 1° FWHM collimator that is effective below 150 keV can be placed in the aperture to improve the sensitivity and angular resolution at low energies.

The 10° field of view and 4° and 2° angular resolution were selected in order to optimize the sensitivity to diffuse and point sources, while providing for mapping of diffuse emission and obtaining adequate resolution for point source studies. Finer resolution would improve the latter in some cases, but at the expense of losing sensitivity in mapping the diffuse emission. Important points in this regard are that the strongest steady gamma-ray lines are the diffuse galactic emission at 511 and 1809 keV, and that many important point sources, e.g. extragalactic supernovae, are at known positions and transient, so that source confusion will not be a problem.

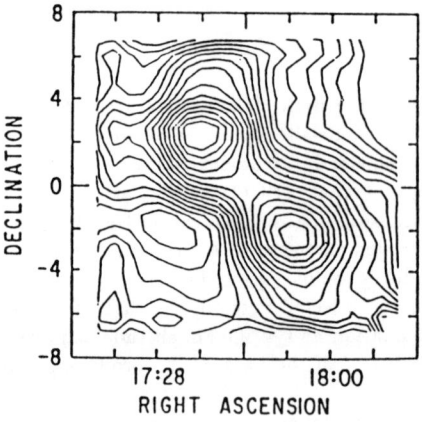

Figure 2. Simulated map from the NAE's mask/antimask system for 2 assumed sources in the vicinity of the Galactic Center separated by 7° and each detected at 60σ.

As a consequence of its very low background, the NAE would become background limited at very low flux levels, $\sim 1\times10^{-5}$ ph/cm²-s at 1 MeV. It would obtain sensitivity to larger fluxes very rapidly. Only 1/2 hour would be required to reach 3×10^{-4} ph/cm²-s, the limiting flux for previous instruments, and 1 day to reach 10^{-5} ph/cm²-s. With 10^6 s of data, a sensitivity of $\sim 3\times10^{-6}$ ph/cm²-s would be reached. Even better sensitivity could be obtained with longer observations, for example, of the galactic plane or extragalactic supernovae. The NAE's sensitivity versus energy is shown in Figure 3 along with those of the Gamma-Ray Spectrometer on the HEAO-3, which used high resolution Ge detectors, and the Oriented Scintillation Spectrometer Experiment (OSSE) on the GRO, which uses lower resolution NaI detectors, with $E/\Delta E \sim 15$.

At 1 MeV the NAE would have a sensitivity and energy resolution that are 10 and 30 times better than the OSSE. Its sensitivity would be 100 times better than the HEAO-3. The very good sensitivity predicted for the NAE is the result of careful consideration of the many factors which affect sensitivity. The most significant of these, in comparison with the HEAO-3, are: 6 times more detector volume, 6 times more observing time (due to

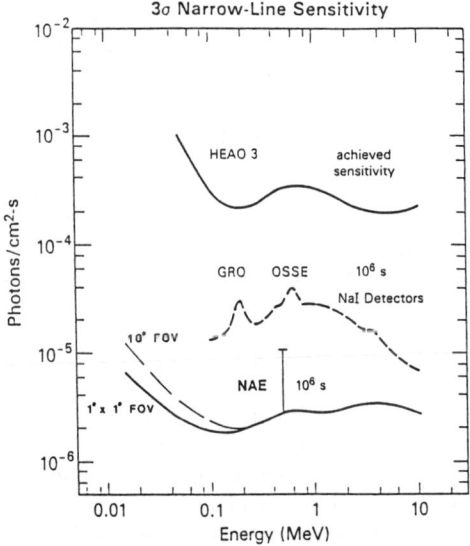

Figure 3. The predicted sensitivity of the Nuclear Astrophysics Explorer to narrow gamma-ray lines in a 10^6 sec observation. Shown for comparison are the predicted sensitivity of the OSSE instrument on the Gamma-Ray Observatory, in 10^6 sec, and the achieved sensitivity of the Gamma-Ray Spectrometer on the HEAO-3, which typically obtained $\sim 2\times10^5$ sec on a source. The NAE sensitivity is 100 times below the known gamma-ray line fluxes.

pointed observations), 2 times more useful data and ~ 100 times lower background per unit detector volume. The latter is primarily due to an orbit with much lower background, much thicker shielding, and the rejection of β-decay background.

The NAE concept in based on technology that is available or in development. Ge detectors with internal segmentation and pulse shape discrimination have been modeled (Roth et al. 1984), developed (Luke et al. 1984), tested in the laboratory (Smith et al. 1988) and in balloon flights (Feffer et al. 1989). A large gamma-ray spectrometer with an array of 12 of these detectors and a 5 cm thick BGO shield has been developed by a UCSD-led collaboration (Matteson et al. 1985). It had its first balloon flight in May 1989, obtaining the 1.8 keV limit on the Galactic Center region's 511 keV line width (Matteson et al. 1989). Another gamma-ray spectrometer incorporating 7 large, ~ 300 cm^3, Ge detectors and a mask/antimask system has been developed by a GSFC-led collaboration (Teegarden et al. 1985, Tueller et al. 1988). It made 2 balloon flights in 1988, measuring the flux and profile of the ^{56}Co gamma-rays from SN1987A (Teegarden et al. 1988, Tueller et al. 1989) and the broad, 3.5 keV, width of the Galactic Center region 511 keV line when the compact source near the Galactic Center was active (Leventhal et al. 1989). Ge detectors with external segmentation have been modeled (Gehrels 1985) and tested in the laboratory (Varnell et al. 1988) and are planned to be tested in balloon flights during the next year.

These developments and observations were performed primarily within NASA's High Resolution Gamma-Ray Spectroscopy Program. (There have also been major contributions to these programs from France and the Sandia National Laboratories.) Its purpose is the development of new techniques relevant to future space missions for high resolution gamma-ray spectroscopy and tests and scientific observations with these techniques on high altitude balloons. This program is continuing with major emphasis during the next few years on further detector and shield development, coded masks, mechanical refrigerators and scientific observations.

IV. MISSION CONCEPT AND EXPECTED RESULTS

In order to provide very effective observations of important transient gamma-ray line sources, e.g. extragalactic supernovae, and allow extended viewing of the sun, the NAE spacecraft would be able to view in any direction for any length of time. Due to its low altitude orbit, earth occultation could eliminate ~ 50 percent of the observing time. The spacecraft would rapidly slew to alternate targets to avoid these losses. The NAE's sensitivity and versatility will allow it to be used to pursue many astrophysical problems with relatively brief observations, e.g. a few hours to a few days. Therefore, the observing program would extend beyond the mission development team and involve a large number of scientists who would use the NAE as a facility. In a 2-year mission scenario the following observational program could be accomplished.

a) Galactic Plane Survey and Mapping (6 months)

A complete survey of the galactic plane would be performed with 5 days of observations at each 10° step in longitude. Mask/antimask and blocked/open modes would be used to detect and map the diffuse emission with 10° and 4° resolution, allowing the total galactic flux to be determined and separated into its diffuse and point-like, < 4°, components. The galactic 511 and 1809 keV gamma-rays would be detected at $\sim 50\sigma$ and $\sim 20\sigma$ significance in each 10° step. A sensitive, high-contrast map would result which could be used to determine (1) the sites and rates of nucleosynthesis in the past 10^6 years; longitude from the map, distance from the lines' Doppler velocities, determined to < 30 km/sec, and the galactic rotation model known from 21 cm observations, (2) the nature of the mixing of nucleosynthetic material into the Galaxy and temperature of the electron-positron annihilation regions from the line widths and profiles, (3) the sites and nature of $\sim$ 5 undiscovered galactic supernovae which have occurred in the past ~ 500 years from their ^{44}Ti gamma-ray emission and (4) the scale height and velocity dispersion of the postulated high-velocity plasmas in the galactic disk, bulge and corona.

b) Detailed Mapping of Selected Regions (2 months)

The knife-edge mode would be used to obtain 2° resolution mapping at multiple position angles to resolve source complexes and better locate point sources. The vicinity of the Galactic Center is already known to be a region where this will be required. The knife-edge mode would also be used to map the latitude distribution of galactic gamma-rays with high precision, in order to determine their scale height and study their possible association with various components of the Galaxy.

c) Extragalactic Supernovae (6 months)

About 4 Type I supernova/year are expected at distances out to the Virgo cluster, $\sim$ 20 Mpc, and at this distance the 847 keV line flux from ^{56}Co decay should be detectable for several months with $\sim$ 1 week time resolution. Measurements of line widths will test models of supernova explosions and measurements their intensities, profiles and time evolution will give information on the nucleosynthetic yield, and the energetics and mass distribution of the ejecta.

d) Galactic Novae (2 months)

Several galactic novae within 2 kpc are expected to be discovered during a 2 year mission. These will be close enough for sensitive tests of the predicted nucleosynthesis of ^{22}Na (2.6 year half-life) by searching for its 1275 keV gamma-ray. The predicted ^{22}Na yield is sensitive to the thermal history and dynamics of a nova outburst, so it is expected to vary greatly from one nova to the next. The NAE sensitivity corresponds to a yield of $\sim$ 6×10^{-9} $M_\odot$ for a nova at 2 kpc, which is 100 times below present limits.

e) Observations of Known Point Sources (8 months)

Many known galactic gamma-ray sources will require regular monitoring to observe unpredictable temporal changes that are known to occur but have not been resolved. The compact source near the Galactic Center and Cygnus X-1 are two of the best known in this class. Other objects, such as X-ray binaries, have their unique, predictable variability that requires observations at specific times. External galaxies which have energetic nuclei characterized by high energy, nonthermal radiation, i.e. the active galactic nuclei, are prime candidates for observations. Models predict a massive, accreting black hole at the nucleus with an associated electron-positron plasma or relativistic particle jets which transport energies over vast distances. In either scenario gamma-ray lines are expected (Matteson 1983) and their discovery would place the theoretical ideas on a much firmer footing. The sun is known to be a source of intense gamma-ray line fluxes during solar flares which often last for tens of minutes. The NAE would perform dedicated solar observations when a large flare is deemed likely.

f) Benefits of an Extended Mission

An extended mission lasting much longer than 2 years is very desirable scientifically and technically feasible. The latter follows from the lack of consumables in the instrument and spacecraft concepts and the detectors' > 10 year resistance to radiation damage. In a 5 to 10 year scenario much more extensive observations could be performed of objects where unpredictable variability will carry key information. We already know that this is true of active galaxies, and the Galactic Center and Cyg X-1. An extended mission will allow the galactic plane survey to be repeated several times in order to obtain information on source variability, with variable sources being selected for follow-up observations. A large number, ~ 20, of extragalactic Type I supernovae would be observed, providing definitive information on their range of nucleosynthetic yield and explosion mechanism(s). The likelihood of the mission overlapping solar maximum, the only time when gamma-ray producing flares are frequent, would be greatly increased and long periods, many months to years, of dedicated solar viewing could be obtained. In this scenario ~ 30 large flares would be expected, with hundreds to thousands of photons counted in the 15 strongest lines in each flare.

IV. STATUS AND PLANS

In October 1989 NASA announced that the NAE was not selected for development. However, strong interest was expressed in the NAE and the Astrophysics Division of NASA is planning to continue its support of technology and instrument development and observations in high resolution gamma-ray spectroscopy. The next round of future Explorer selections is planned to begin with a Spring 1991 request for proposals for 1 year feasibility studies. The NAE will be proposed again at that time and the earliest it could be launched would be the late 1990's. In November 1989 a European-US collaboration submitted a proposal to ESA for a "Blue Box" mission in the late 1990's. This would combine ESA and NASA Explorer resources in a mission named INTEGRAL. This joins the NAE instrument and the Gamma-Ray Imager from the ESA/GRASP mission concept, which was studied in 1987-88.

ACKNOWLEDGEMENTS

The development of the NAE concept was the result of the work of many people. We wish to acknowledge the contributions to this effort by R. Lingenfelter, R. Ramaty, J. Higdon, M. Leventhal, R. Muller, J. Tueller, M. Pelling, R. Pehl, W. Wheaton, P. Durouchoux and V. Schönfelder. This work was supported in part by NASA contract NAS5-30338 and NASA grant NAGW-449.

REFERENCES

Ayre, C.A. et al. 1983, *Mon. Not. R. Astr. Soc.*, **205**, 285.
Barthelmy, S. et al. 1988, *IAU Circular No. 4593*.
Burns, M., Harding, A., and Ramaty, R. eds. 1983, *Positron-Electron Pairs in Astrophysics* (AIP, NY).
Burke, B. et al. 1988, *Space Science in the Twenty-First Century: Imperatives for the Decades 1995 to 2015* (National Academy Press, Washington DC).
Bussard, R.W. 1980, *Ap.J.*, **237**, 970.
Chan, K. W. and Lingenfelter, R. E. 1987, *Ap.J. (Letters)*, **318**, L51.
Chan, K. W. and Lingenfelter, R. E. 1988, in *Nuclear Spectroscopy of Astrophysical Sources*, eds. N. Gehrels and G. H. Share (AIP, New York) p. 110.
Chupp, E.L. et al. 1973, *Nature*, **241**, 333.
Chupp, E.L. 1984, *Ann. Rev. Astr. Ap.*, **22**, 359.
Chupp, E.L. 1988, in *Nuclear Spectroscopy of Astrophysical Sources*, eds. N. Gehrels and G. H. Share (AIP, New York) p. 24.
Cook, W.R. et al. 1988, *IAU Circular No. 4527*.
Clayton, D.D. 1973, in *Gamma-Ray Astrophysics*, eds. F. W. Stecker and J. I. Trombka (NASA SP-339, Washington, DC) p. 263.
Dennis, B.R. et al. 1982, in *Gamma-Ray Transients and Related Astrophysical Phenomena*, eds. R. E. Lingenfelter, H. S. Hudson and D. M. Worrall (AIP, New York) p. 153.
Feffer, P. et al. 1989, in *Proceedings of the SPIE*, in press.
Field, G. B. et al. 1982, *Astronomy and Astrophysics for the 1990's, Volume 1: Report of the Astronomy Survey Committee* (National Academy Press, Washington DC).
Gehrels, N. 1985, *Nuc. Inst. Mech.*, **A239**, 324.
Gehrels, N. et al. 1987, *Ap.J.*, **322**, 215.
Gruber, D.E. et al. 1980, *Ap.J. (Letters)*, **240**, L127.

Haymes, R.C. et al. 1975, *Ap.J.*, **201**, 593.
Hudson, H.S. et al. 1980, *Ap.J. (Letters)*, **236**, L91.
Hueter, G.J. 1984, in *High Energy Transients in Astrophysics*, ed. S. E. Woosley (AIP, New York) p. 373.
Jacobson, A.S. et al. 1978, in *Gamma-Ray Spectroscopy in Astrophysics*, eds. T. L. Cline and R. Ramaty (NASA TM 79619, Greenbelt, MD) p. 228.
Leventhal, M. et al. 1977, *Ap.J.*, **216**, 491.
Leventhal, M. et al. 1978, *Ap.J. (Letters)*, **225**, L11.
Leventhal, M. et al. 1989, *Nature*, **339**, 36.
Ling, J.C. et al. 1979, *Ap.J.*, **231**, 896.
Ling, J.C. 1982, in *Gamma-Ray Transients and Related Astrophysical Phenomena*, eds. R. E. Lingenfelter, H. S. Hudson and D. M. Worrall (AIP, New York) p. 143.
Ling, J.C. et al. 1987, *Ap.J. (Letters)*, **321**, L117. (1987).
Ling, J.C. 1988, in *Nuclear Spectroscopy of Astrophysical Sources*, eds. N. Gehrels and G.H. Share, (AIP, New York) p. 315.
Ling, J.C. and Wheaton, W.A. 1989, *Ap. J. (Letters)*, **343** L57.
Lingenfelter, R.E., and Ramaty, R. 1978, *Physics Today*, **31**, No. 3, 40.
Lingenfelter, R.E., and Ramaty, R. 1985, in *Conference Papers of 19th International Cosmic Ray Conference*, **5**, p. 19.
Lingenfelter, R.E. 1988, in *Nuclear Spectroscopy of Astrophysical Sources*, eds. N. Gehrels and G. H. Share (AIP, New York) p. 17.
Lingenfelter, R. E. et al. 1988, *Gamma Ray Astrophysics to the Year 2000, Report of the NASA Gamma Ray Program Working Group* (NASA, Washington DC).
Lingenfelter, R.E., and Ramaty, R. 1989, in *High Resolution Gamma-Ray Cosmology*, eds. D.B. Cline and E. Fenyves, Nuclear Physics B Proc. Suppl. 10B (North Holland, Amsterdam) p. 67.
Lingenfelter, R.E., and Ramaty, R. 1990, in *Proceedings of the 21st International Cosmic Ray Conference*, paper OG7.2-6, in press.
Luke, P.N. et al. 1984, *IEEE Trans. Nuc. Sci.*, **NS-31**.
Mahoney, W.A. et al. 1984, *Ap.J.*, **286**, 578.
Mahoney, W.A. 1988, in *Nuclear Spectroscopy of Astrophysical Sources*, eds. N. Gehrels and G.H. Share (AIP, New York) p. 149.
Mahoney, W.A. et al. 1988, *IAU Circular No. 4584*.
Matteson, J.L. 1983, *Adv. Space Res.*, **3**, No. 4, 135.
Matteson, J.L. 1983, in *Electron-Positron Pairs in Astrophysics*, eds. M.L. Burns, A.K. Harding and R. Ramaty (AIP, New York) p. 292.
Matteson, J.L. et al. 1985 in *Proc. 19th ICRC*, **3**, 326.
Matteson, J.L. et al. 1989, *IAU Circular No. 4889*.
Matz, M. et al. 1988, *IAU Circular No. 4568*.
Maurer, G.S. et al. 1982, *Ap.J.*, **254**, 271.

Mazets, E.P. et al. 1979, Nature, bf 282, 587.
Mazets, E.P. et al. 1981, Nature, **290**, 378.
Murakami, T. et al. 1988, Nature, **355**, 234.
Nagel, E. 1981, Ap.J., **251**, 288.
Nolan, P.L., and Matteson, J.L. 1983, Ap.J., **265**, 389.
Peterson, L.E. 1988, in *Nuclear Spectroscopy of Astrophysical Sources*, eds. N. Gehrels and G. H. Share (AIP, New York) p.1.
Prince, T. et al. 1982, Ap.J. (Letters), **255**, L81.
Ramaty, R. and Lingenfelter, R.E. 1979, Nature, **278**, 127.
Ramaty, R., Kozlovsky, B., and Lingenfelter, R.E. 1979, Ap.J. Supp., **40**, 487.
Ramaty, R. and Lingenfelter, R.E. 1982, Ann. Rev. Nucl. Part. Sci., **32**, 235.
Ramaty, R. and Murphy, R. 1987, Space Sci. Rev., **45**, 213.
Rester, A.C. et al. *IAU Circular No. 4535*.
Riegler, G.R. et al. 1981, Ap.J. (Letters), **248**, L13.
Riegler, G.R. et al. 1985, Ap.J. (Letters), **294**, L13.
Roth, J. et al. 1984, IEEE Trans. Nuc. Sci., **NS 31**, 367.
Sandie, W. et al. 1988, *IAU Circular No. 4526*.
Share, G.H. et al. 1985, Ap.J. (Letters), **292**, L61.
Share, G.H. et al. 1988, Ap.J. **326**, 717.
Smith, D.M. et al. 1988, in *Nucl. Spectroscopy of Astrophysical Sources*, eds. N. Gehrels and G.H. Share (AIP, New York), p. 484.
Strickman, M.S. et al. 1982, Ap.J. (Letters), **253**, L23.
Teegarden, B.J., and Cline. T.L. 1980, Ap.J. (Letters), **236**, L67.
Teegarden, B.J. et al. 1985, in *Conference Papers of the 19th International Cosmic Ray Conference*, **3**, 307.
Teegarden, B.J. et al. 1989, Nature, **339**, 122.
Trümper, J.R. et al. 1978, Ap.J. (Letters), **219**, L105.
Tueller, J. et al. 1984, Ap.J., **279**, 177.
Tueller, J. et al. 1988, *Nuclear Spectroscopy of Astrophysical Sources*, eds. N. Gehrels and G.H. Share (AIP, New York), p. 439.
Tueller, J. et al. 1989, in *Proceedings of the Gamma-Ray Observatory Workshop*, ed. N. Johnson p. 4–258.
Wheaton, W.A. et al. 1979, Nature, **282**, 240.
Woosley, S.E. et al. 1986, Ap.J., **301**, 601.
Woosley, S. E. and Pinto, P. A. 1988, in *Nuclear Spectroscopy of Astrophysical Sources*, eds. N. Gehrels and G.H. Share (AIP, New York), p. 98.
Yoshimori, M. et al. 1983, Solar Physics, **86**, 375.

DISCUSSION

Steven Kahn

How does INTEGRAL differ from NAE?

Matteson

INTEGRAL is a just-proposed joint ESA-NASA mission. It includes the NAE as conceived in the NAE Phase A Study plus the Gamma-Ray Imager as conceived in the ESA-GRASP Study done in 1988. The Gamma-Ray Imager has > 10 m Crab sensitivity in the +0.3 − 30 MeV range and uses a coded mask to obtain 12 arcminute resolution over a 6' field. Its spectral resolution, $E/\Delta E$, is ~ 10.

Kevin Hurley

The Soviets and the Europeans seem to prefer highly eccentric (~ 4 day) orbits for many of their high-energy astrophysics missions. Could you discuss the orbit tradeoff studies which were done for NAE?

Matteson

Thirteen orbits at various altitudes and ranging from near equatorial to outside the magnetosphere were considered in order to assess their particle and gamma-ray fluxes and residual radioactivity in the instrument. The conclusions were: (1) The South Atlantic Anomaly must be avoided, it raises the background and causes radioactivity which contaminates up to ~ half the mission time. (2) Two options are available: (a) Low inclination (> 10°), low altitude (~ 500 km) has the lowest possible background, but requires slewing to alternate targets to avoid ~ 50% observing time less due to occultation. (b) Orbits which are at high altitude, and avoid the trapped protons, are at the interplanetary cosmic-ray flux and will have higher background than (a), but do not have earth occultation to contend with. Since the NAE's goal is the best sensitivity, the orbit (a) was considered to be the most favorable. However, this is an issue that requries further assessment. An important consideration is the payload mass that can be put into orbit (b), since lower mass leads to less shielding and higher background.

LOW-COST SMALL SATELLITES FOR ASTROPHYSICAL MISSIONS

William C. Priedhorsky

Space Science Division
Los Alamos National Laboratory

1. Introduction: What is a Miniature Satellite?

A miniature satellite is a low-cost platform to support a small space experiment. Space astrophysics has been hindered by decades-long delays in important experiments. With miniature satellites, one hopes to reduce both experiment cost and lead time to an affordable level.

Miniature satellites are not a new idea. The first scientific satellites, including Explorer I, were small and developed on a timescale of months. Important science was done by these pioneer missions. Though the easy discoveries have been made, important missions in exploration and follow-up can still be carried out from small platforms.

Successful small satellite programs continue to this day. These include the OSCAR amateur radio satellite program, in which 12 small satellites, built by amateurs, have been flown over 25 years with no satellite failures (Fleeter, 1988). Two small free-flyers, GLOMAR and NUSAT, were ejected from the Shuttle in 1985. GLOMAR, a radio-relay experiment, was built in less than a year for under $1 million, and operated over a year in orbit. Small satellite projects continue to this day. Approaching launch are the Air Force STACKSAT array of 3 small satellites (P87-2), a number of other small satellites under Department of Defense auspices. The Air Force Space Test Program is developing a standard small experiment platform called STEP (Space Test Experiment Platform). NASA has started a small explorer program, beginning with SAMPEX, a solar and magnetospheric particle explorer, FAST, a fast auroral snapshot experiment, and SWAS, a submillimeter astronomy experiment.

The definition of a miniature satellite is subjective. In my opinion, miniature satellites to date fall within the following approximate bounds.

Table I: Miniature Satellite Parameters

Power	≤ 50 Watts
Payload Mass	≤ 50 kg
Design Lifetime	≤ 1 year
Project Lifetime	≤ 3 years
Spacecraft Cost	≤ $1-4 million

The modest experiment scope allow an expedited project organization; the modest investment allows a certain amount of risk. Larger projects can in principle be run as miniature satellites, but risk budget and schedule runaway in overmanagement, quality control, and documentation.

2. Commercial Miniature Satellites

Now is a good time to buy a miniature satellite: competition is intense, and satellite buyers are eagerly sought. In the case of ALEXIS, Los Alamos National Laboratory issued a request for quotation in November 1988. Nine companies responded. In the final round, we received four strong proposals. The final selection was made on the basis of cost and technical merit.

The ALEXIS spacecraft discussed below could be upgraded at modest cost. The spacecraft could point in 3 axes with the addition of a momentum wheel and, possibly, a small star tracker. More data storage could be added by replacing the present 256K SRAM array with expected 1M components. The extended ALEXIS bus could thus accept and store approximately 100 kilobits/second of telemetry (time averaged), and point to celestial targets to within ~0.1°.

Ball Aerospace Systems Division, under their TECHSTARS program, is developing a miniature satellite capability. A bus massing less than 120 lbm could provide > 50 Watts to a payload, with 5 MIPS of processor power, data storage of 1 Gbit, and 3-axis attitude control to 0.2° and knowledge to 0.1°. Such a bus would cost ~$4M for the first vehicle, and ~$2M for copies (T. Higbee, private communication).

Fairchild Space Company is also active in the small satellite field, and presently building SPINSAT for the Office of Naval Research. Fairchild, also, find that small satellites with significant capabilities can be built for a few million dollars. Their smallest "Class 1" design study would support a payload of < 100 lbm with a platform of < 100 lbm, providing < 15 Watts to the payload for a recurring platform cost less than $1 million. At a slightly larger scale, their "Class 2" design study would support a payload of < 750 lbm with a platform of < 250 lbm, providing < 100 Watts to the payload for a recurring platform cost less than $3 million (Bartlett 1989).

Other potential vendors include Globesat, Inc., of Logan, Utah, Intraspace of Salt Lake City, Space Data Corporation of Tempe, Arizona, and Defense Systems, Inc. of McLean, Virginia. With such intense competition, several companies are offering attractively-priced high performance spacecraft.

3. The ALEXIS Project

ALEXIS is an example of a miniature satellite for astrophysics. ALEXIS (Array of Low-Energy X-Ray Imaging Sensors) is an ultrasoft X-ray monitor experiment that consists of six compact normal-incidence telescopes operating in narrow bands centered on 66, 72, and 93 eV. The Air Force Space Test Program is scheduled to launch ALEXIS via a Pegasus air-launched vehicle into a 400 nm polar orbit in April 1991. The ALEXIS satellite and experiment are funded by DOE, and are being built by a collaboration of Los Alamos and Sandia National Laboratories, and Space Sciences Laboratory/UC Berkeley. The technological objectives of ALEXIS are to 1) demonstrate the feasibility of a wide field-of-view, normal incidence ultrasoft X-ray telescope system and 2) to determine ultrasoft X-ray backgrounds in the space environment. In addition, ALEXIS will pursue a number of scientific objectives including: mapping the diffuse background in three emission line bands with the highest angular resolution to date, performing a narrow-band survey of point sources, searching for transient phenomena in the ultrasoft X-ray band, and providing synoptic monitoring of variable ultrasoft X-ray sources such as cataclysmic variables and flare stars.

Each ALEXIS telescope consists of a layered synthetic microstructure (LSM) spherical mirror, a prime-focus curved microchannel plate detector with a wedge and strip readout, background-rejecting filters and magnets, and image processing readout electronics. Figure 1 shows a cutaway of one ALEXIS telescope. The multilayer optics are discussed in more detail in Smith, Bloch, and Roussel-Dupré (1989, 1990). The ALEXIS telescopes are aligned in pairs, each pair covering a 33° field-of-view. The geometric area of each telescope will be about 25 cm^2, with spherical aberration limited resolution of about 1/2°. The 5σ survey sensitivity is estimated to be several x 10^{-3} photons cm^{-2} s^{-1} for line emission at the center of the bandpass. Other aspects of the experiment are discussed in Priedhorsky et al. 1989. ALEXIS, with its wide fields-of-view and well-defined wavelength bands, will complement the upcoming NASA Extreme Ultraviolet Explorer and ROSAT EUV Wide Field Camera, which are sensitive broad-band survey experiments. Unlike these experiments, ALEXIS is tuned to emission in the ~70eV Fe IX-XII complex, can observe transient emission over half the sky, and monitors variable sources continuously for 6 months of each year.

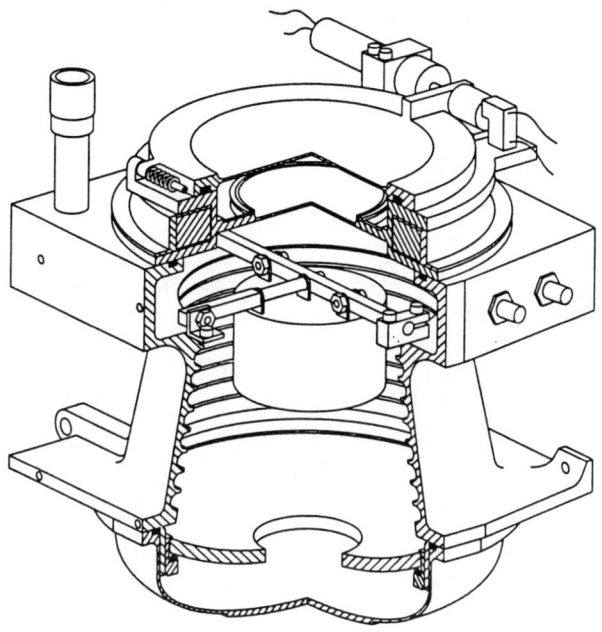

Figure1. Cutaway view of one of the six ALEXIS ultrasoft X-ray telescopes.

Position and time of arrival are recorded for each detected photon. The ALEXIS telescopes are aligned to sweep over the entire anti-Sun hemisphere each 30-second spin; the whole sky is therefore mapped out in one year. ALEXIS will always be in a survey-monitor mode, with no individual source pointings. Because it sweeps the entire anti-Sun hemisphere, it is well-suited for simultaneous observations with ground-based observers who prefer to observe sources at opposition. Coordinated observations need not be arranged before the fact, because all sources in the anti-Sun hemisphere will be observed and archived.

ALEXIS and a piggyback experiment mass 125 lbm, draw 45 Watts, and produce 10 kilobits/second of data. The supporting spacecraft, described in the next section, has less mass (100 lbm) and draws less power (13 W) than the experiments.

4. The ALEXIS Spacecraft

The ALEXIS spacecraft is a spinning vehicle which provides power, telemetry, data storage, attitude control and determination, and other services to ALEXIS and its piggyback experiment. The spacecraft is being built by Aero-Astro, Inc., of Reston, Virginia, (formerly Astronautics) under contract to Los Alamos National Laboratory. The contract price for the ALEXIS spacecraft is less than $2 million.

The complete ALEXIS satellite masses 225 lbm at launch. The satellite consists of the Aero-Astro spacecraft and the Sandia-Los Alamos-Berkeley payload. In its stowed configuration, the satellite is 24 inches in diameter by 30 inches high, exclusive of antennae, with a spacecraft section 9 inches high. The payload is on the anti-Sun side of the spacecraft. Figure 2 shows the spacecraft.

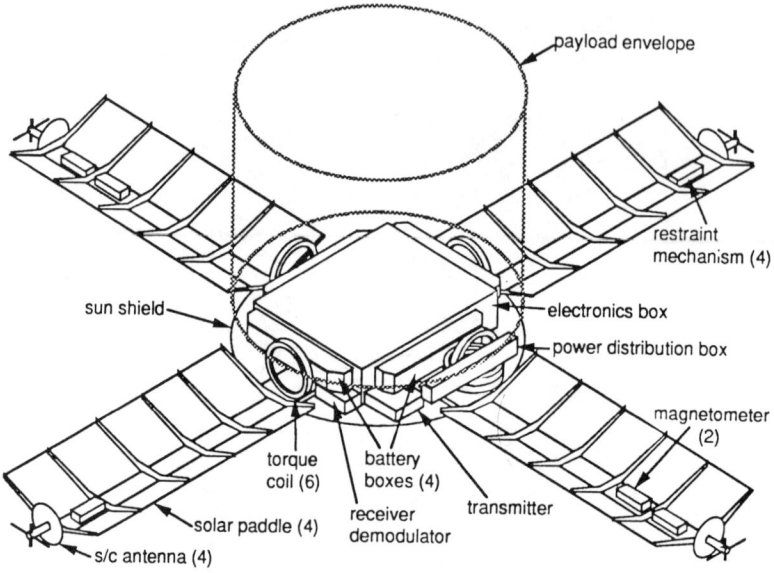

Figure 2. ALEXIS spacecraft with subsystems

The spacecraft spins about its longitudinal axis, which is pointed to within 2° of the Sun at all times. Attitude control is by means of magnetic torque coils in 3 axes. Spin axis determination is by a fine sun sensor. Spin phase angle is from an infrared horizon crossing indicator and a 3-axis magnetometer. Ground algorithms are used to reconstruct attitude to ±0.25° (3 sigma worst case). Upon insertion into orbit, the magnetic torquers damp out tumble rates of up to 10 rpm, using magnetometer data, then spin up the spacecraft to a nominal 2 rpm. Using coarse Sun sensor data, the spin axis is precessed towards to Sun. The solar panels are then deployed; once deployed, the moments of inertia are such that the spacecraft is stable for spin about its longitudinal axis. For the rest of the mission, the satellite spins at 2 rpm and keeps station on the Sun.

Spacecraft power originates in a silicon solar cell array, provided by Israeli Aircraft Industries, which supplies 114 Watts at the beginning of mission. During orbital nighttime, power is supplied by 4 redundant strings of commercial nickel-cadmium cells. Exclusive of spacecraft systems power, the power system provides to the payload a minimum of 56 W of unregulated 28 V power at beginning of life, declining to 51 W after 1 year on orbit.

Experiment data are stored continuously and transmitted during passes over a Los Alamos ground station. The data storage and telemetry system is sized to accept experiment data at a mean rate of

10 kilobits/second. Due to the polar orbit, the satellite is visible from Los Alamos during one or two passes every 12 hours. ALEXIS thus carries 448 Mbits of static RAM mass memory, net of error detection and correction. The radio frequency downlink carries 750 kilobits/second at 2260 MHz, with an estimated error rate less than 2×10^{-9}/bit. Command data are uplinked at up to 5 Megabits/day, with a calculated error rate less than 1.2×10^{-9}/bit.

The ground station is based on a 1.8-meter steerable dish installed on a roof at Los Alamos. The ground RF circuitry is custom built. Immediate control of the data flow is via a Macintosh II computer, with archiving and first pass processing on a Sun workstation. Data will be archived on an optical disk, and transmitted to collaborators at Berkeley on VCR-type magnetic tapes.

Given the schedule and budget of ALEXIS, it is impractical to require S-level electronic parts. We ensure reliability with a mixture of military (883-B) and industrial-grade parts, selected based on budget and reliability considerations. Parts will receive a dose of ~5 krad if the mission lasts 3 years. We have therefore chosen and tested parts for radiation tolerance. Again, schedule and budget prohibit the use of explicitly radiation-hard parts. Typical parts are, for the mass memory, Dense Pak memory modules built from Hitachi 256K SRAM chips, and for logic circuitry, Texas Instruments HCMOS.

To be cost-effective, ALEXIS must be built as quickly as we can, while meeting project technical and quality requirements. The ALEXIS team in place represents a fixed cost per month; to first order, the project cost is proportional to the time elapsed between program start and launch. We are therefore building ALEXIS to the following schedule:

Table 2: ALEXIS Schedule

- January 1988   Start
- May 1988   Apply to Space Test Program for Launch
- August 1988   System Design Review
- April 1989   Preliminary Design Review
- May 1989   Memorandum of Agreement for Launch
- October 1989   Critical Design Review
- April 1990   Prototype Integration
- November 1990   Payload/Spacecraft Integration
- April 1991   Launch

5. Conclusion

Though ALEXIS is hardly complete, we have learned a lot about miniature satellites already. The following advice is proffered to those contemplating similar projects:

A miniature satellite should carry one experiment. Multiple experiments lead to increased spacecraft requirements and more costly integration. The ALEXIS spacecraft, which carries a main experiment and a piggyback, is more complicated than it would have been for the ALEXIS experiment alone.

A miniature satellite project must accept risk. Ultimate reliability is not consistent with a foreshortened schedule and modest budget. In the case of ALEXIS, we accept risk in the use of some industrial parts and a substantially single-string design.

Minimize paper deliverables. The ALEXIS project is built around a zero-based paperwork requirement; a proposed document is rejected unless absolutely necessary. We have found that preliminary and critical design reviews, interface control documents, software specifications, and test plans cannot be done without.

Minimize weight and volume. Even for a miniature satellite, each additional payload pound adds complexity and support requirements. Kilograms cost dollars.

Move fast. Time costs dollars.

Fit-test interfaces early. Interfacing systems should be brought together and tested at the earliest possible phase of the project. The goal is not a successful fit, but identification of problems.

Thoroughly test at the satellite level. Ultimately, our confidence in the ALEXIS system will derive from vibration, functional, and thermal vacuum tests of the integrated satellite. Some subsystems are thermally cycled and vibration tested at the box level.

Scrounge vigorously. Miniature satellites and their components are a buyer's market. Some suppliers will provide components below cost to enter the market.

Keep the project team tightly integrated. Two half-timers do not make the equivalent of one full-timer for project progress.

Control the satellite from a local ground station. Ground station hardware and software is now cheaper than interfacing to a system such as TDRSS.

Be flexible about launch opportunities. In the case of ALEXIS, we began to design the satellite before we knew the launch vehicle. The design was in compatible with both Scout and Pegasus boosters. Because we had an ongoing program, it was relatively easy to obtain a launch.

The ALEXIS mission, including launch, vehicle integration, and flight operations, but excluding the piggyback experiment, represents an investment of approximately $15 million. It is therefore a very cost-effective approach to a scientific goal, as modern scientific satellites typically cost $100 million and up. However, ALEXIS is not a particular bargain on a per-pound basis ($150K/payload pound). Astrophysics missions for miniature satellites should be those that yield significant science in a small package. Particular missions that might be suitable include 1) small X-ray, ultraviolet, and solar telescopes. These can provide a continuous observational capability between major missions, provide additional observing time beyond that provided by major missions, and support multiwavelength observational campaigns; 2) transient monitors, sensitive to X-ray stars, gamma-ray bursts, etc.; 3) transient response, dedicated to continuous monitoring of identified transients such as X-ray transients and supernovae, and 4) technology development. Other missions will not doubt arise from the ingenuity of the astrophysical community.

Acknowledgements: I would like to acknowledge contributions from Bob Bartlett of Fairchild, Terry Higbee of Ball, and the entire ALEXIS team. In particular, figure 1 is the work of Jim Miller and Lee Morrison, figure 2 is from Aero-Astro, and Jeff Bloch and Steve Wallin made helpful suggestions on the manuscript.

REFERENCES

Bartlett, R. O. 1989, "Small Satellite Mission Requirements and Concepts", presentation at *Low Cost Access to Space Conference*, Paris.

Fleeter, R., 1988, "Miniature Satellite Technologies for Space Astronomy", in *SPIE Proceedings* **982**, *X-Ray Instrumentation in Astronomy II*, 173.

Priedhorsky, W., et al. 1989, "ALEXIS: A Narrow-Band Survey/Monitor of the Ultrasoft X-Ray Sky", proceedings, Berkeley Colloquium on Extreme Ultraviolet Astronomy, Berkeley, CA.

Smith, B. W., Bloch, J. J., and Roussel-Dupré, D. 1989, "Metal Multilayer Mirrors for EUV Wide Field Telescopes", in *SPIE Proceedings* **1160**, *EUV Optics for Astronomy and Microscopy*, 171.

Smith, B. W., Bloch, J. J., and Roussel-Dupré, D. 1990, *Optical Engineering*, in press.

THE ENERGETIC TRANSIENT ARRAY

— ETA —

A NETWORK OF "SPACE BUOYS" IN SOLAR ORBIT FOR OBSERVATIONS OF GAMMA-RAY BURSTS

GEORGE R. RICKER

Center for Space Research
Massachusetts Institute of Technology
Cambridge, Massachusetts 02139

ABSTRACT

The Energetic Transient Array (ETA) is a concept for a dedicated interplanetary network of ~40 microsatellites ("space buoys") deployed in an ~1AU radius solar orbit for the observation of cosmic gamma ray bursts (GRBs). Such a network is essential for the determination of highly accurate (~0.1 arc sec) error boxes for GRBs. For each of ~100 bursts which would be detectable per year of observation by such a network, high resolution ($\Delta E/E$ ~0.2% at 1 MeV) spectra could be obtained through the use of passively-cooled Ge gamma-ray detectors. Stabilization of each microsatellite would be achieved by a novel technique based on the radiation pressure exerted on "featherable" solar paddles. Because of the simplicity of the microsats, as well as the economics of mass production and the failure tolerance of such a network of *independent* satellites, a unit cost of ~$250 K per microsat can be anticipated. Should such a project be undertaken in the mid 1990's, possibly as an International mission, it should be possible to have a fully functional array of satellites in place before the end of the decade for a total cost of ~$20M, exclusive of launcher fees.

I. INTRODUCTION

It has become increasingly apparent in recent years that the origin of cosmic gamma-ray bursts (GRBs) can only be determined by identifying burst source counterparts in other energy ranges. To date, about a half dozen GRBs have positions determined to 10 arcminutes or better (Sagdeev and Zacharov 1989; Atteia et al 1987). Infrared (Schaefer et al 1987), optical (Schaefer et al 1987; Ricker et al 1986; Pederson et al 1983; Schaefer et al 1983), and soft X-ray (Boer et al 1988; Pizzichini et al 1986) searches have not revealed *any* convincing candidates for GRBs. From these observations, it is known that the quiescent counterparts for GRBs are intrinsically very faint, B ≥ +24. Because of the high spatial density of such faint optical sources, it is likely that GRB error boxes in the 0."1 – 1."0 range will be required to establish identifications based on positional coincidences alone. Certainly, gamma-ray burst-derived positions with such accuracies will be essential to justify the expenditure of any significant amount of observing time with the major new space-borne observatories with high angular resolution which are coming on-line during this decade —HST (Launch in 1990), AXAF (launch in 1997), and SIRTF (launch in 1998). (The Gamma Ray Observatory, GRO, will not be able to provide GRB locations any more accurate than ~1 arc minute.) Of recent great interest are the cyclotron and annihilation radiation features reported in GRB spectra (Fenimore et al 1988; Murikami et al 1988; Hueter et al 1982; Mazets et al 1981; Dennis et al 1981). Thus, it is important that high resolution spectra be obtained simultaneously with such position determinations, in order that the geometry of the emission regions might be studied (Refs. in Harding , Petrosian, and Teegarden 1984).

In the following sections, we describe a new instrumental technique which provides highly accurate measurements of both positions and spectra of gamma ray bursts in the 0.01 – 2 MeV energy range. Results from such measurements are essential to conclusively identify the cosmic sources of GRBs. Our technique relies critically upon a new, potentially low cost observational platform – the *microsatellite*. With an "armada" of such microsats, deployed within the inner solar system, a large baseline array for GRB position measurements can be achieved using compact solid state detectors with excellent spectral resolution.

II. SCIENTIFIC AND TECHNICAL OBJECTIVES

The importance of establishing small (~0."1) error boxes for GRBs cannot be overemphasized. The history of establishing optical counterparts to "empty field" radio and "crowded field" X-ray sources has lead to the general criterion that, in order to assure that a putative identification be secure, there be no more than ~1 accidental source per ~40 "beams" (error box) for the non-optical observing technique which seeks to establish an optical counterpart. Thus, if we set as a goal the establishment of useful *positional* identifications of GRBs based on quiescent optical counterparts, then we must produce error boxes for which there would be only about 1 B~ +31 optical source per 40 error boxes. This limit is an appropriate one for the powerful techniques developed for use on the new generation of 10 meter class ground-based telescopes (T. Tyson, private communication), as well as for observations at the limiting magnitudes which the WF/PC and FOC on the Hubble Space Telescope will achieve. These limits *require* 0.3 arc sec (3 σ) level of confidence GRB error boxes. Furthermore, many of the current models for GRBs which associate them with faint degenerate stars (e.g., old neutron stars) also predict quiescent counterparts with +30 ≤B≤ +35, depending on source distances.

Given a goal of deriving such error boxes, it is useful to consider what methods might be utilized. The concept of measuring GRBs positions from times-of-arrival at independent spacecraft separated by interplanetary distances dates back to the discovery of the phenomenon in the late 1960s, and is an obvious choice. However, this technique has not been as fruitful as one might have expected because of two major problems:

1) Lack of a sufficiently dense array of detectors (too few spacecraft, with non optimum spacings); usually, these were "hitchhiker" experiments on planetary science missions.

2) Lack of uniformity of the *detection* and *signal processing techniques*, since the instruments were usually developed independently without regard to precisely how they would function with other GRB detectors in the network. In general, the political and bureaucratic climate made the *a priori* design coordination between American, European, Japanese, and Soviet instruments difficult. Also, the mounting of the instruments on the carrier spacecraft was not often optimal, because they were regarded as "secondary science."

The approach outlined in Table 1 should correct these two problems, and permit the achievement of the 0.1 arcsec positions we wish to establish.

> **TABLE 1: KEYS TO ACCURACY OF ETA-DERIVED POSITIONS**
> - Long baseline (2.2 AU) between 20 pairs of identical, mass-produced μsats
> - Template matching of burst signals from 40 identical satellites:
>
>   Relations: $\Delta\Theta < (N/2)^{-0.5} \tau_m/\tau_c$
>
>   where $\Delta\Theta$ is the absolute accuracy of the GRB position;
>   N is the number of μsats in the "constellation";
> τ_m is the burst template match timing uncertainty;
> τ_c is the μsat pair separation (light-sec).
>
> - Template Fidelity 40 μsats assure that there are always ~3 μsats within ~0.2 A.U. of earth, so that high fidelity templates can be rapidly (<5 minutes) transmitted to a dedicated earth-based station for selection of the burst (and its template) for which a corrected arrival time will be transmitted to all other μsats in the network as the "burst-to-keep" for the observing day in question.

If we estimate the value of $\Delta\Theta$, the absolute accuracy of locating a "typical" gamma ray burst, we find:

For an $S > 5 \times 10^{-6}$ ergs cm^{-2} s^{-1} gamma-ray burst,

$\Delta\Theta < (N/2)^{-0.5} \tau_m/\tau_c = (40/2)^{0.5} (10^{-3} \text{ s} / 10^{+3} \text{ s}) \cong 0.2$ μradian (1σ),

or 0.05 arc sec (1σ).

> THUS, ≤ 0.3 ARC SEC ERROR BOXES (3σ) COULD BE ACHIEVED FOR ~100 GAMMA-RAY BURSTS PER YEAR OF OPERATION FOR A FULL 2.2 A.U. BASELINE OF IDENTICAL DETECTORS.

The next question which arises is the nature of an instrument design and supporting system which could provide the above positioning data, as well as simultaneous high resolution spectroscopy. These issues are discussed in the following section.

III. KEY IDEAS AND CHARACTERISTICS

The key ideas and characteristics of the ETA microsatellite system are given in Table 2 and Figure 1. Discussion of the deployment scenario is given below (Figure 2 and Table 3). As indicated in Table 2, the ETA is a tightly integrated system, with the detector, bus, and ground stations all being optimized to support the ETA mission. The number of microsats, (~40) is determined by both the need to tolerate failure rates of ~10%, as well as the need to have at least 1 satellite within

~0.2 AU of the earth at all times. The detectors are conventional high purity, N type Ge solid state sensors, which are radiatively cooled to 80K. Their energy resolution is $\Delta E/E < 0.2\%$ at 1 MeV. The detectors and their radiators continuously face anti-Sun, greatly simplifying the thermal design. Also, the power system is relatively straightforward, since there are no eclipse periods to consider. A key technological element in making the ETA concept feasible is a compact, low power, low cost stabilization system which will permit the use of an earth-directed microsatellite transmitting antenna. The enabling technology for this system was developed at MIT more than 20 years ago for the SUNBLAZER project (Colombo 1966; Falcowitz 1966; and Baker et al 1969). Essentially, this technique utilized the radially directed light pressure from the Sun to orient one face of a small (~20 kg) microsatellite toward the Sun, by means of the force directed on dihedrally-oriented solar paddles. By "feathering" the solar paddles at their attachment points to the satellite body, the roll rate of the satellite could also be adjusted. Detailed stability analyses of these questions were carried out for a deep space mission using a 20 kg microsatellite using 1960's technology (see above references). With 1990's microtechnology, most of the mission drivers foreseen on the original studies in the '60's are no longer areas of difficulty. Thus, spacecraft parameters, as well as cost and schedule issues, can be estimated with some degree of confidence, based on the body of experience from such programs as the ALEXIS (Priedhorsky et al 1988) and HETE (Ricker et al 1988) "minisats".

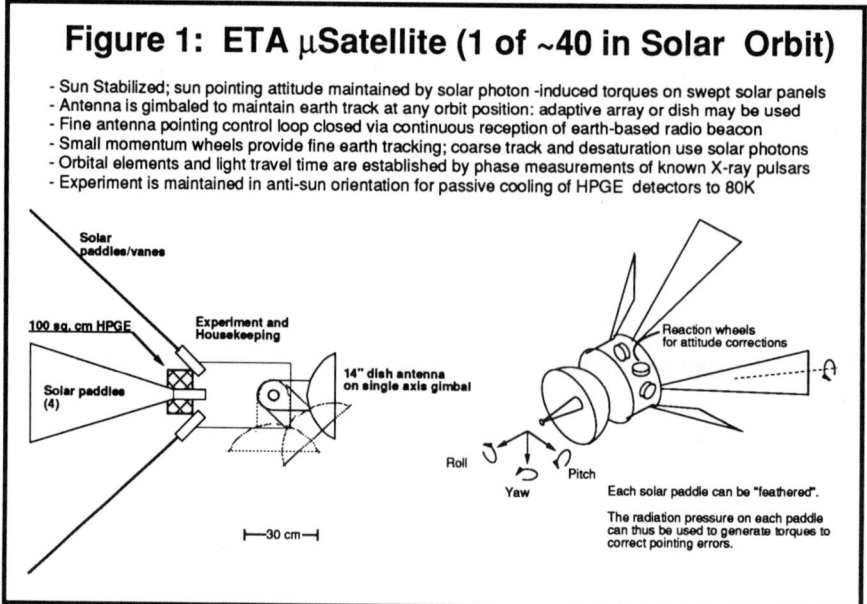

Figure 1: ETA μSatellite (1 of ~40 in Solar Orbit)

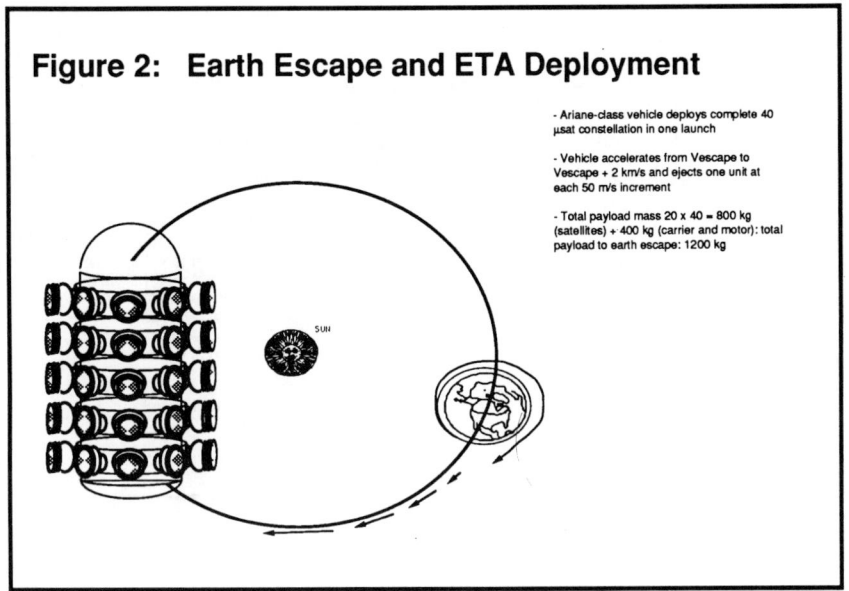

Figure 2: Earth Escape and ETA Deployment

- Ariane-class vehicle deploys complete 40 μsat constellation in one launch
- Vehicle accelerates from Vescape to Vescape + 2 km/s and ejects one unit at each 50 m/s increment
- Total payload mass 20 x 40 = 800 kg (satellites) + 400 kg (carrier and motor): total payload to earth escape: 1200 kg

TABLE 2: SUMMARY OF THE ETA SYSTEM CHARACTERISTICS

PARAMETER	VALUE
Number of μSats	40 in total
Deployment	Solar Orbit (2.2 A.U. major axis)
Detectors	
Type	HPGe (coaxial n-type)
Energy range	5 keV-10 MeV
Area	25 sq. cm. (5.5 cm D x 5 cm H)
Number	4 per μSat
Dimensions per μSat	35cm x 35cm x 55 cm
Total Mass per μSat	20 kg
Power per μSat	19 W
Stabilization	Solar radiation pressure ("SunBlazer")
Transmitted power	1 W (10 GHz)
Minimum bit rate	6.5 bit/s @ 2.2 A.U.
Ground Stations	3 (dedicated to the ETA)
Ground-based Antennas	4 meter diameter

Table 3: Summary of Deployment Options

Deployment Option	Launch Vehicle	# of Launches	Upper Stage Required	Pros / Cons
Direct Insertion from LEO	STS	10	35 kg Star-class: 1 per satellite	+ Gradual deployment - upgrade potential + Potential low launch cost (~$4M) - Shuttle safety, high cost of multiple Stars - Each satellite has significant G&C Ovhd
Direct Insertion from LEO	Scout / Pegasus	8	35 kg Star-class: 1 per satellite	+ Gradual deployment - upgrade potential - High launch cost (~$80M) - Each satellite has significant G&C Ovhd
Insertion from Geosynch Xfer	Ariane or Delta (secondary payload)	5	20 kg Star-class: 1 per satellite	+ Gradual deployment - upgrade potential - Moderate launch cost (~$6M to $15M) - Each satellite has significant G&C Ovhd
Dispensing from Accelerating Vehicle	Ariane or Delta (primary payload), or shared Proton to solar orbit	1	Dispenser stage mass ~400 kg to insert 40 units	+ Single Launch+Single insertion stage=lower cost, higher reliability + No satellite G&C overhead + Compatible with Ariane, Delta, Proton with performance margin

* G&C = Guidance and Control

The ETA μsat bit rates are such that data from ~1 GRB/day can be telemetered from each μsat even for those μsats 2 AU from earth; we assume that each μsat in the constellation telemeters data from the *same GRB* by providing each satellite with semiconductor memory storage for a full day of data, and by using high rate telemetry data from a μsat within ~0.2 AU of earth to choose *which stored GRB event* that all ~40 μsats will be commanded to downlink to earth. To support the ETA, a modest, dedicated "mini-DSN" is required and would be costed within the program (See Conclusions and Program Plan discussion below).

In Figure 1, we depict one of the "mass-produced" 20 kg ETA microsats. The comments within the figure give additional details on some of the spacecraft systems. A 14 inch diameter parabolic X-band telemetry antenna dish is specified. However, a phased-array antenna could also be used, although the geometry of the spacecraft and its roll degree of freedom seem to make a single-axis gimballed dish a natural choice.

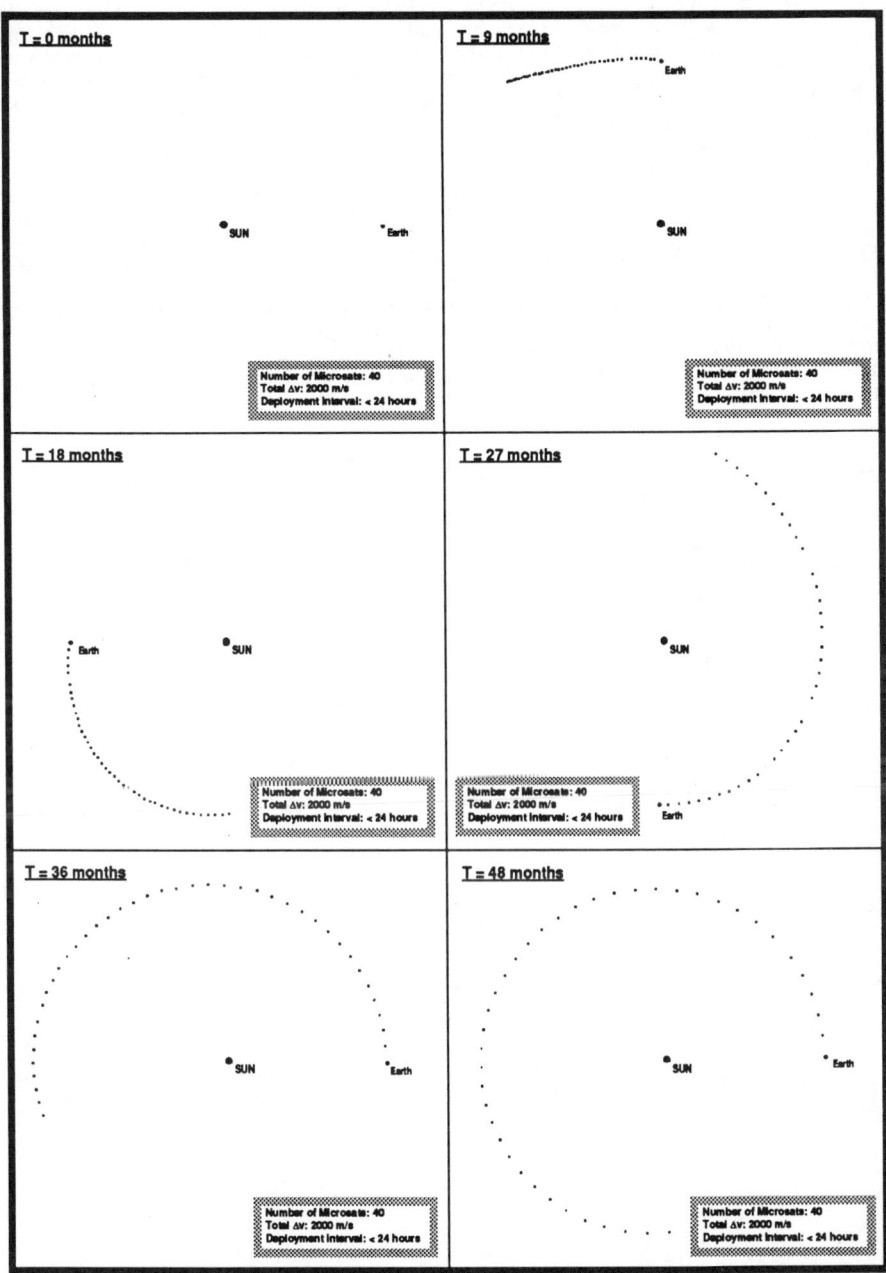

Figure 3: Evolution in time of a constellation of 40 microsatellites comprising the Energetic Transient Array (ETA), as viewed from the North eliptic pole. Successively later "snapshots" are shown left-to-right, top-to-bottom. By 27 months following launch, the full 2.2 A. U. baseline has been established for a satellite-to-satellite dispersion velocity of 50 m/s.

To deploy the ETA microsats, we have considered several possible launch vehicles. In Figure 2, we show the simplest one, which would require an Ariane-class vehicle with a spin stabilized "pinecone" transtage which ejects the microsats one-by-one as it accelerates smoothly up to 2000 m/s beyond earth escape. In Table 3, we show three other possible deployment options. Obviously, because the μsats operate autonomously once launched, programmatic and cost considerations would be important in choosing a particular launch and deployment scheme. However, the single launch on an Ariane-class vehicle would be the simplest approach.

In Figure 3, we show the results of an N-body calculation showing the development of the ETA constellation of 40 microsats in the months following launch. Perturbations by the earth-moon system have been ignored, a simplification which will be connected in work in progress. **There are no propulsion units on the microsats** - -the evolution seen with time is strictly a function of the fact that each has initially undergone a Hohmann-transfer into a separate Keplerian orbit, due to a spread (50 m/s increments) in the initial ejection velocity imparted by the transtage bus. In Table 4, we summarize the baseline achieved by the constellation as a function of time. As early as 6 months post-launch, the ETA should begin producing <1 arc second GRB position measurements, and by 24 months a full 2 AU baseline will be achieved, resulting in the ultimate absolute accuracy of ~0.1 arcsec (1 sigma) for the ETA constellation.

TABLE 4: μSAT CONSTELLATION SEPARATION DISTANCES

Time	Baseline
Launch (=L)	0.0 A.U.
L + 6 months	0.5 A.U.
L + 9 months	1.0 A.U.
L + 12 months	1.4 A.U.
L + 24 months	2.0 A.U.
L + 32 months*	2.2 A.U.*

*Constellation fully spread around a 180° arc in less than 3 years.

In Figure 4, we present the results of a preliminary system analysis for the ETA, including a downlink budget, a mass budget, and a power budget. No "show stoppers" are evident in this preliminary assessment.

Figure 4: ETA Solar μSat System Analysis

- **Mass Budget:** Total mass 20 kg. Battery mass is minimized because there is no eclipse operation.
- **Power Budget:** Solar panel area driven by control authority - therefore beginning of life margin is over 35%, end of life margin is over 15%. Keep alive time is for only CPU and actuators on.
- **Downlink Budget:** 1 Watt transmitter with 14" dish or array yields adequate link margin assuming 20,000 bits per event, 1 event per day per satellite plus 10% data overhead for status and state of health.

Mass Budget for an ETA μSat

item	mass (kg)
payload	10.0
batteries	0.5
Tx	0.5
Rx	1.0
antenna	1.5
solar cells	3.4
CPU	1.0
structure	1.8
attach clamp	0.3
total	20.0

Power Budget for an ETA μSat

ITEM	POWER (W)	
payload	5	
Tx	7	
Rx	3	
CPU	3	
actuators	1	
total	19	
power collected	24	BOL
margin	5	
batteries energy	10	W-hrs
operating time	32	minutes
survival time	150	minutes

Downlink Budget for ETA Sun Orbiting μSats

based on 1 x 20,000 bit event per day per satellite

SC Tx power	30.0	dBm
SC ant. gain	30.0	dB
Carrier freq.	10.0	GHz
Rx noise temp	60.0	Kelvin
Sky noise temp	50.0	Kelvin
GS ant. dia.	4.0	meters
GS ant. G/T	29.0	dB
range	299.0	x10^9 meters
path loss	282.0	dB
CNR at GS	5.7	dB-Hz
data rate	6.54	bits/sec.
GS Rx Bw	0.2042	Hz
SNR	12.6	dB
SNR req'd 10^-5 BER	11.0	dB
margin	1.6	dB

FIGURE 5: ETA PROGRAM DEVELOPMENT PLAN

Phase I: Test Flight:
A single prototype ETA microsat is flown to geosynchronous orbit (piggyback with commercial launch). The single satellite will demonstrate:

- Operation of the solar radiation pressure stabilization system
- RF link at low bandwidth including satellite antenna tracking
- Thermal standoff of passively cooled detectors

Phase II: Operational Flight
The 40 μsat armada is launched aboard either Delta, Ariane, or a shared Proton

Phase I:	Phase II:
- Prototype development	- Design for mass production
- Launch of a single test μsat to geosynchronous orbit and mission data collection	- 40 unit production
	- Launch and Operations of full 40 μsat armada

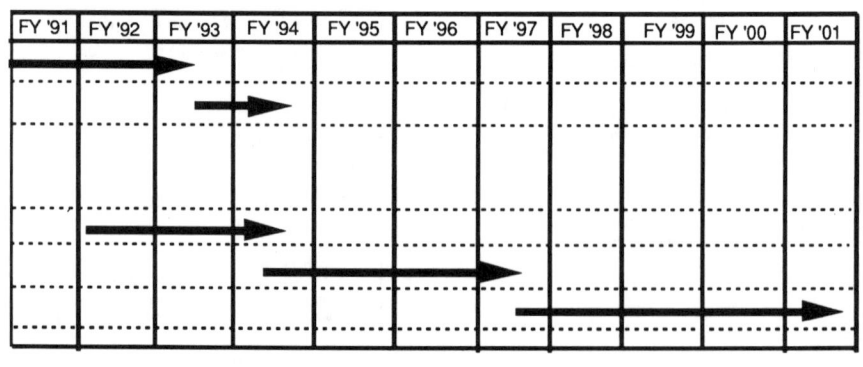

IV. CONCLUSIONS AND PROGRAM PLAN

The promise of the ETA, and the pressing scientific need to identify the origin of GRBs leads us to consider the practicality of proposing such a program within the near future. In Figure 5, we provide a time line for one such possible program, which would occur in 2 phases, as indicated in the figure. The Phase I, a demonstration test flight of a single ETA microsat, would test the key technologies, including the solar radiation pressure stabilization system, the RF link and X-ray pulsar synchronization technique, and the passive detector cooling design. Since a single microsat would only weigh 20 kg, it should be possible to carry out this phase as a secondary "hitchhiker" payload on a rocket transporting a primary satellite to geosynchronous orbit. Such a launch could occur as early as mid FY 1993 (see Figure 5), with the in-orbit testing phase lasting ~12 months. Should this effort be successful, a Phase II would consist of a "design-for-manufacturability" effort (concurrent with the latter half of Phase I), followed by a 3-year mass production effort, in which a production rate of 1 microsat/month would be achieved. This manufacturing period could extend from mid FY 1994 through mid FY 1997. Launch of the 40 µsat armada on a single rocket could then take place in FY 1997. Mission operations would extend from FY 1997 to FY 2001. We would expect failures of individual µsats to occur during the operations phase, since a reliability level of 10 - 20% per µsat would be specified. However, since the *overall constellation* is sufficiently robust to tolerate such failures, there would be no significant impact on the scientific return from the ETA.

We have attempted to estimate the costs of a "full-up" ETA, exclusive of launch charges. This information is given in Table 5. All estimates except for the µsat costs are based on industry quotes. For the microsats, a parametric estimation method was used, based on recent DARPA-sponsored communication microsat programs.

TABLE 5: PRE-LAUNCH COST ESTIMATE FOR ETA PROJECT

Item	Amount
µsats (40@$250K)	$10M
Ge Detectors (40@$100K)	$4M
Upper Stage Bus (including ejectors)	$3M
Ground Stations (3@$600K)	$2M
TOTAL	**$19M**

In conclusion, the ETA could be a conceptually-simple, yet state-of-the-art mission which would revolutionize the study of gamma-ray burst sources by providing the firm identifications which are essential to bring this area of study into the fold of the maturing discipline of high energy astrophysics. The technology required for the ETA has only recently come of age, and is fortunately readily accessible because of the enormous advances made in micro-circuitry and commercial communications technology over the past decades. Should it prove possible to initiate the ETA in the near future, it could be operational when AXAF and SIRTF are launched toward the end of this decade. Because of the distributed, yet autonomous nature of the ETA, it also lends itself to formulation as an international mission, which would be entirely in keeping with the fact that gamma-ray burst astronomy has been an international discipline since its inception, perhaps more so than has any other area of study in high energy astrophysics, with vigorous groups in Europe, Japan, the Soviet Union, and the United States (e.g., see References).

V. ACKNOWLEDGMENTS

The ideas described in this paper were developed in two periods: 1965–1970, and more recently during 1988–1989. Conversations with John Doty of MIT were especially helpful in clarifying the concepts for burst synchronization and data relay from the network to the dedicated earth stations, as well as for a general understanding of "system" issues. Rick Fleeter of Aero-Astro, Inc. contributed many ideas which permitted realistic assessments of the practicality of assembling such an armada of microsatellites, including the link margin estimates (with Greg Huffman), the mass/power estimates (with Bob Dil), and evaluations of launch options (with Richard Warner). Richard Warner of Aero-Astro, Inc., and Eliot Young of MIT wrote the orbital simulation programs which were essential in estimating both the constellation deployment time as well as the minimum number of microsatellites in the armada that would assure that a high fidelity burst template could be relayed to earth for retransmission to more distant microsats to enable precise burst profile matching. Kevin Hurley of University of California-Berkeley helped estimate the number of bursts detectable by the ETA and provided useful information on radiation damage in germanium detectors. Stan Woosley of the University of California-Santa Cruz and Hugh Hudson of the University of California-San Diego asked challenging questions on the limitations of the technique, as well as its possible application to studies of solar flares, a topic which will be discussed in a later paper. Peter Tappan of MIT conceived a compact, low mass radiator suitable for passively cooling the germanium detectors. Finally, I thank the many members of the original MIT SunBlazer team who not only conceived, designed, and breadboarded the first solar radiation pressure-stabilized microsatellite more than 20 years ago, but also documented their design (which, alas, was never flown) in sufficient detail to support the revival of the microsat idea presented here. Ironically, it is the relentless march of *commercially-driven* data communication and computer high technology, rather than aerospace or defense-related project research, that has made the approach advocated by these early pioneers a viable technique today. Support for the work presented here was provided by the NASA Office of Space Science and Applications.

REFERENCES

Atteia, J.-L., Barat, C., Hurley, K., Niel, M., Vedrenne, G., Evans, W.D., Fenimore, E.E., Klebesadel, R.W., Laros, J.G., Cline, T., Desai, U., Teegarden, B., Estulin, I.V., Zenchenko, V.M., Kuznetsov, A.V., and Kurt, V.G., 1987 *Ap. J. Supp.* **64**, 305.

Baker, R.H., et al, 1969, "Conceptual Design of a Small Solar Probe (SUNBLAZER)", MIT/CSR Internal Report, TR-69-1.

Boer, M., Atteia, J.-L., Gottardi, M., Hurley, K., Niel, M., Barat, C., Pizzichini, G., Mason, K., Branduardi-Raymont, G., Cordova, F., Laros, J.G., Evans, W.D., Fenimore, E.E., Klebesadel, R.W., Sims, M.R., and Martin, C., 1988, *Astron. Astrophys.* **202**, 117.

Colombo, G., 1966, "Passive Stabilization of a Sunblazer Probe by Means of Radiation Torque", MIT/CSR Internal Report, TR-66-5.

Dennis, B.R., Frost, K.J., Kiplinger, A.L., Orwig, L.E., Desai, U., Cline, T.L., 1981, *AIP Conference Proceedings*, **77**, 153.

Falcovitz, J., 1966, "Attitude Control of a Spinning Sun-Orbiting Spacecraft by Means of a Grated Solar Sail", MIT/CSR Internal Report, TR-66-17.

Fenimore, E.E., Conner, J.P., Epstein, R.I., Klebesedel, R.W., Laros, J.G., Yoshida, A., Fujii, M., Hayashida, K., Itoh, M., Murikami, T., Nishimura, J., Yamagani, T., Kondo, I., Kawai, N., 1988, *Ap. J.*, **335**, L71.

Harding, A.K., Petrosian, V., and Teegarden, B.J., 1984, in Gamma Ray Bursts (ed. E.P. Liang and V. Petrosian), A.I.P. (New York), p. 75.

Hueter, G.J., and Gruber, D.E., 1982, in Accreting Neutron Stars, (eds. W. Brinkmann and J. Trümper), MPI Garching-bei-München, 213.

Mazets, E.P., Golenetskii, S.V., Aptekar', R.L., Gur'yan, Yu.A., Ilinskii, V.N., 1981, *Nature*, **290**, 378.

Motch, C., Pedersen, H., Ilovaisky, S.A., Chevalier, C., Hurley, K., and Pizzichini, G., 1985, *Astron. Astrophys.* **145**, 201.

Murikami, T., Fujii, M., Hayashida, K., Itoh, M., Nishimura, J., Yamagani, T., Conner, J.P., Evans, W.D., Fenimore, E.E., Klebesedal, R.W., Yoshida, A., Kondo, I., Kawai, N., 1988, *Nature*, **335**, 234.

Pedersen, H., Motch, C., Tarenghi, M., Danziger, Pizzichini, G., and Lewin, W., 1983, *Ap. J. Lett.* **270**, L43.

Pizzichini, G., Gottardi, M., Atteia, J.-L., Barat, C., Hurley, K., Niel, M., Vedrenne, G., Laros, J.G., Evans, W.D., Fenimore, E.E., Klebesadel, R.W., Cline, T.L., Desai, U.D., Kurt, V.G., Kuznetsov, A.V., and Zenchenko, V.M., 1986, *Ap. J.*, **301**, 641.

Priedhorsky, W.C., Block, J.J., Smith, B.W., Strobel, K., Ulibari, M., Chavez, J., Evans, E., Siegmund, O.H.W., Marshall, H., Vallerga, J., and Vedder, P., 1988, in X-ray Instrumentation in Astronomy II (ed. L. Golub), *SPIE*, **982**, 188.

Ricker, G.R., Vanderspek, R.K., and Ajhar, E.A., 1986, *Adv. in Space Res.*, **6**, 75.

Ricker, G.R., Doty, J.P., Rappaport, S., Hurley, K., Fenimore, E., Roussell-Dupre, D, Niel, M., Vedrenne, G., Lamb, D., and Woosley, S., 1988, in Nuclear Spectroscopy of Astrophysical Sources, (ed. N. Gehrels and G. Share), A.I.D., (New York), 407.

Sagdeev, R.Z., and Zacharov, A.V., 1989, *Nature*, **341**, 581.

Schaefer, B., Seitzer, P., and Bradt, H., 1983, *Ap. J. Lett.* **270**, L49.

Schaefer, B.E., Cline, T.L., Desai, U., Teegarden, B.J.,Atteia, J.-L., Barat, C., Hurley, K., Niel, M., Vedrenne, G., Evans, W.D., Fenimore, E.E., Klebesadel, Laros, J.G., Estulin, I.V., and Kuznetsov, A.V., 1987 *Ap. J.* **313**, 226.

VI. Rapporteurs' Reports

Alan N. Bunner, Chairman

X-RAY ASTRONOMY IN THE 21st CENTURY
RAPPORTEUR'S REPORT

Claude R. Canizares

Department of Physics and Center for Space Research
Massachusetts Institute of Technology

I. The Question of Long Range Planning

The astronomical community is now engaged in a plethora of planning exercises. In X-ray Astronomy alone we have this workshop and the associated HEAMOWG deliberations, the X-ray Program Working Group chaired by Steve Murray, and the High Energy from Space Panel chaired by Bruce Margon. The last of these deals with both X- and Gamma-ray astronomy and reports to the Astronomy and Astrophysics Survey Committee, which, under the leadership of John Bahcall, will formulate a strategy for the 1990s and beyond. As the discipline's representative on the Bahcall committee and as a member of several of the others, I have had lots of time to sit on airplanes and ask myself how worthwhile it all might be. That is, putting aside the question of political expedience, how well can we expect to do in forecasting the future directions of our field and in choosing fruitful areas for technological development?

My own belief is that X-ray Astronomy has now matured to a degree that makes it quite amenable to such planning exercises. Let us look briefly at the history. Jeff McClintock recently sent me the transcript of a fascinating meeting held at SAO in 1960 and attended by Bruno Rossi, Riccardo Giacconi, Alber Baez, Paul Kirkpatrick, Bill Liller, Stan Olbert and others. The participants speculated on the possible sources of X-rays. It should not surprise you that such a creative and imaginative group made some specific predictions about possibly fruitful areas of investigation which later proved to be correct. For example, they discussed both the Crab and M87 as likely sources of X-rays, and they were right. But most of the great discoveries of *Uhuru* could not have been foreseen, and it would have been difficult for these pioneers to make compelling arguments for pushing one particular kind of investigation over another, such as high angular vs. high time resolution.

By the time *Einstein* was launched, I think the situation was quite different. The community at least knew what it wanted to look at with *Einstein* and to a great extent it knew what kinds of instrumental capabilities would be most effective for each objective. With a few important exceptions like the hot gas in elliptical galaxies and particular classes of stellar coronae, the predictions were correct. We certainly did *not* know specifically what we would find nor could we predict the astrophysical implications to be derived from our studies -- both of which are manifest in Einstein's undeniable impact on astronomy. But the accuracy of our pre-launch planning exercises was probably closer to ~75% than to the few percent that one might estimate for the group at AS&E. This is partly because *Einstein* and the various European and Japanese X-ray missions of the 1980's were largely making detailed surveys of a territory whose gross features had already been identified by the exploratory missions of the 1970's. We still have lots more of that to do, particularly in the area of spectroscopy. My own guess is that we are far from saturation (notwithstanding Martin Harwitt's compelling studies of saturation effects in the rate of discovery in his book *Cosmic Discovery*), so that we are unlikely to do better than ~75-80% accuracy in predicting the most fruitful areas of research for AXAF (again, this does not mean that we predict the results, just the problems to work on).

I conclude that we are not wasting our time in planning exercises such as this, which attempt to define scientific drivers for NASA policy. At the same time few would deny that there remains a substantial unexplored realm beyond the observed boundaries of our knowledge in which surprising and important discoveries are waiting to be made; so the unpredictable ~20-25% may well be the most exciting. We must keep in mind Martin Harwitt's convincing evidence for the strong correlation between order-of-magnitude improvements in observational capabilities and the discovery of new phenomena.

Another aspect of the planning exercise deals with timescales. Here again we must take account of the present state of our discipline and its history. I believe that we must work to shorten the timescales rather than accept the anomalous conditions of the past decade as the inevitable model of the future.

Micro-missions and mini-missions should have timescales of 3-5 years. There is simply no reason for them to take longer than that. In fact, if they take longer than that, the technology goes out of date and the costs grow out of hand.

Macro-missions and even mega-missions should take 5-10 years, not 15-20. We need only look at *Einstein*, which took less than a decade from initial proposals to launch -- although to be sure the planning and political process stretched over a longer period. Remember too that *Einstein* required the development of a large number of detectors as well as the high resolution telescope itself whose characteristics far exceeded any that had gone before.

Nevertheless, we must also have at all times a 10-20 year plan, no matter how difficult it might be to formulate. That is in large part the charge of the Bahcall committee, since in our discipline much of the next ten years will be devoted to working through the backlog of missions from the previous decade. The process, of which this workshop is a part, is far from closure. I can report that so far the Margon panel has found little enthusiasm in our community for designating a specific mega-mission beyond the Great Observatories. In contrast there is strong sentiment in favor of a balanced program, in which large missions like AXAF are accompanied by several small and medium sized missions. Still, I believe it is very important for us to define one or two "strawman" mega-missions for the early 21st century, to focus our minds and to help us plan the intervening activities of the discipline. Of course, in sketching such missions we must also guard against the danger that those naive plans get taken too seriously a few years later. Long-range plans are helpful only to the extent that they can be revised and updated to reflect advances in technology and understanding.

II. Scientific Themes in 21st Century X-ray Astronomy

I will now turn to my role as a "rapporteur" for X-ray Astronomy. Saul Rappaport has given an excellent summary of the scientific topics that were discussed by the invited speakers. I do note that several major topics were inevitably left out by these speakers, such as stellar coronal spectroscopy or the physics of intracluster gas, so their presentations should not be taken as a comprehensive review of the field. I have tried to account for some of these.

From my notes, I have extracted seven overarching themes that span several subdisciplines and may help clarify new technology needs. There is no method to the order in which I present them.

1) *Large Samples:* It is clear that large samples and surveys are of great importance in many areas. We have heard calls for extensive surveys of stars, supernovae, galaxies, clusters, AGN, etc. The ROSAT all-sky survey will contribute, but it has limitations of bandwidth, spectral resolution and depth.

2) *High Throughput*: X-ray astronomers seem always to be doing battle at the limit of statistics. Most speakers noted that real progress in the future will require higher throughput so statistical accuracies of a few percent in imaging and spectroscopy can be achieved. In the case of imaging or temporal studies, a major reason to push for higher S/N is to permit correlated spectroscopy.

3) *Spectral Resolving Powers of 1,000 - 10,000:*. There seems to be a consensus that spectrometers with resolving powers of 1000 - 10,000 are what is needed to attack most identifiable problems in X-ray astronomy. This is roughly a factor 2-10 beyond what has been achieved so far or what will be achieved on AXAF. It is many factors of ten beyond what has been achieved for most most sources, because of the relatively poor sensitivity of past high-resolution spectrometers. It will be important to achieve high spectral resolution for both point and extended sources, and ideally one would like to obtain spatially resolved spectra for the latter. So far there seems to be little justification for pushing spectral resolution beyond this range.

4) *Opening the 10-100 keV Window:* Another theme is the need to go beyond the 0.1-10 keV range that has received most of the attention during the 70's and 80's and will still be our primary window in the 90's. XTE will provide significant new capabilities, but further improvements in sensitivity are required to obtain higher quality data and reach larger samples of sources. We can confidently predict some of the phenomena that will be studied through this window, such as cyclotron lines in X-ray binaries, spectral signatures of black holes, and non-thermal tails in AGN.

5) *Continuous Surveillance.* The existence of time variable phenomena such as transient X-ray sources, long-term cycles of binaries, and the outbursts of AGN, argue for continuous surveillance of the X-ray sky. Continuous *presence* of X-ray detectors in space, to respond to the next nearby supernova for example, is a distinct but related topic. It is likely that the series of planned missions starting with ROSAT and leading to AXAF and XMM will give us near continuous presence for the next two decades. But continuous surveillance, such as that provided by the all-sky-monitors on XTE and Spectrum X-Gamma, is not presently planned beyond those missions.

6) *Opening Unexplored Territory.* Finally there are some areas in which we can identify some scientific drivers, but which are more compellingly justified as territory for new exploration made possible by advances in technology of several orders of magnitude or more ("Harwitt space"). Here I include imaging on the scale of milliarcseconds, high signal to noise polarimetry and the sub-millisecond timing on large numbers of sources.

7) *Second Generation Instruments for AXAF*: This is one theme that has not been discussed at this meeting but which is, I believe, an important aspect of the

21st century in our discipline. AXAF will be a high performance telescope in orbit and will be accessible on at least one or two occasions during its lifetime (which on the current schedule lasts until 2012), and it will provide an extraordinary opportunity for pursuing many of the scientific objectives in X-ray Astronomy. Surely second generation AXAF instrumentation is an important goal of some of the detector development efforts described at this meeting, and mentioned below

III. Basic Technology and Enabling Technology

I will try to summarize the main areas of technology development that I have identified. Where appropriate, I have listed for each topic two distinct categories: *basic technology* development means the kind of laboratory research that is carried out by the attendees of this meeting; *enabling technology* development might be pursued as industrial research or by an academic group such as the Space Engineering Research Center (SERC) at MIT.

I want to emphasize at the outset that it is essential for us to establish and maintain close ties with the people and institutions that are developing the enabling technology. These "arms-length" research activities must have clear goals and direction from our community if are to be useful to us. I am happy to say that the MIT SERC is one model for this. Prof. Ed Crawley and colleagues from our Aeronautics and Astronautics department are working on precision controlled structures. They have gone far out of their way to establish ties to the scientific community. The very existence of this workshop and the Astrotech 21 program shows that NASA is moving in the right direction. We should applaud and encourage this, and furthermore we should work to establish the needed mechanisms to make sure it comes out right.

1) *Large Area Telescopes (0.1-10 keV):* This topic has received considerable attention and good publicity here. It seems to me that the goal must be to achieve large area with angular resolutions well below 10 arcsec, possibly at the ~2-3 arcsecond level. This is to avoid source confusion problems and adequately to resolve galaxy clusters and even galaxies at large redshift. That is a difficult goal for large-scale optics, but it is one we must reach for. We would like to go beyond 1 square meter to the 10 square meter range, and we would like to do it at reasonable cost. Clearly the primary goal is to achieve these parameters in the 0.1-10 keV range, but I think we should also continue to explore ways to push to 20 or 30 keV. Then one could image regions of non-thermal emission and obtain high S/N for cyclotron lines, etc.

<u>Basic Technology</u>: We have heard about various techniques (e.g. foils, replication, electroforming, flat plates) each of which holds some promise. I am no expert here but most likely most or all these should continue, because this development is of crucial importance to the field and it is not clear where progress will be made. Normal incidence optics may also play a role even in non-solar X-ray astronomy, especially if the multilayer technique can be pushed to shorter wavelengths.

<u>Enabling Technology</u>: Possible items in this category include optical benches and deployable structures, although it is unlikely that the requirements will push the state-of-the-art too hard until we get to very large areas or very long focal lengths. Once a viable optic and fabrication technique are chosen, another possible area for enabling technology will be the mass production of large numbers of elements at modest cost.

2) *Focal Plane Detectors I-- CCDs:*. There was not too much discussion at this meeting about CCD development, but it is well known that there is a lot yet to be achieved.

Basic Technology : Large area telescopes will inevitably require larger focal plane detectors -- we are pushing present technology already on AXAF. We heard about the need for CCDs as big as ping-pong tables -- even now we could use CCDs as big as ping-pong paddles. Present CCDs are roughly the size of a (flattened) ping-pong ball. Present CCDs all have pixels in the ~15-25 micron range, and I have heard calls for both larger and smaller pixels depending on the application. Deeper depletion regions are also needed to enhance the high energy efficiency, and there is a continuing concern about CCD susceptibility to long term radiation damage.

Enabling Technology: I found it difficult to identify enabling technology for this area, although one might list large scale fabrication of the customized devices we are likely to require. It is very unlikely that the astronomical community will be able to have a strong influence on industry in this area, which is driven by commercial and defense interests.

3) *Focal Plane Detectors II- Cryogenic Spectrometers:* This is an extremely exciting area and one in which we are sure to see startling developments occur over the next 5 years. I would list as a goal the achievement of a spectral resolution of 0.5 eV, so that one could achieve resolving powers better than 1000 at the oxygen K alpha line and the iron L lines. Ideally these detectors should be position sensitive, at least over a limited field. Logistics of the space program in the foreseeable future require detector lifetimes of at least five years, which puts strong demands on the cryogenic systems.

Basic Technology: There seem to be several promising technologies, such as micro-calorimeters, superconducting tunnel junctions, dielectrics, etc., and it is surely premature to anticipate which one will ultimately prove most useful.

Enabling Technology: The obvious issue is cryogenics at the milli-Kelvin level. The Astrotech 21 initiative already includes a cryogenics program, but it is important to insure that it meets the special needs of these detectors

4) *Other Basic Technology (Supporting Research and Technology):* In identifying major areas of basic technology as I have, it is important that we also emphasize the importance of continuing a variety of other, more specific development activities. In effect, we must maintain a balance between large and small research programs just as we must maintain a balance between large and small missions. A major message from this workshop is that remarkable returns have been achieved from modest investments in the SR&T program. During the long dry-spell of the 1980's in terms of flight opportunities, the SR&T program has delivered an impressive number of successes in technology development. That effort has to be continued while we proceed with the missions of the 1990's. A very incomplete list of areas for further investigation includes:
- high pressure gas counters
- synthetic multilayers (the goal should be achieving 2d spacings of ~10 A, both as Bragg diffractors and normal incidence mirrors)
- ultra-thin windows (these were not discussed here, but ~0.1-0.3 micron leak-tight membranes are within reach).
- improved reflection gratings
- improved transmission gratings
- various types of polarimeters
- etc...

5) *Other Enabling Technology:* For completeness I want to mention other kinds of enabling technology that are not unique to X-ray astronomy but on which we are very dependent. All our missions rely on gyros, tape recorders, batteries and solar cells, and I have the unsubstantiated impression that these electromechanical elements are often the weakest links in the reliability chain. *Einstein's* biggest problems came from gyros, and as many of you know, COBE suffered a gyro failure shortly after launch. I know that NASA is very aware of this issue, but we should be encouraging efforts to improve those components. At the same time, as part of an assessment of reliability that I will come to again, we might ask whether we are overdesigning the experiments given the reliability of these critical spacecraft subsystems. This is provocative and the answer may be a resounding no, but it seems to me the question is worth asking.

IV. Enabling Programmatics

Despite the fact that this is a technology workshop, I think it is not only appropriate but mandatory for us to address relevant programmatic issues. In terms of the planning process, the Bahcall committee has a Policy Panel, chaired by Dick McCray, but in the Margon panel we also intend to address policy issues relevant to our discipline.

My own opinion is that we can achieve real improvements in the quality and quantity of the science we do by addressing programmatic issues that affect the costs of our missions. Let me address two topics. First, the issue of a balanced program, and second, the issue of reducing mission costs.

(i) Need for a Balanced Program

The phrase "balanced program" has been used so frequently that it almost sounds trite. But there is no doubt that balance is essential for the scientific vitality of this or any discipline. I have already addressed basic instrument development and the SR&T program. Those, together with data analysis and theory, which I will not discuss, are important components of a balanced program. The other elements, which I will mention briefly, are the missions. I will consider them generically, in logarithmically ascending order of size and cost. How to allocate resources among the various categories and which specific missions to choose are difficult questions that must be sorted out through the various NASA and NAS advisory committees.

• $1-5M -- Sub-Orbital Program & Attached Payload

A vigorous albeit modest suborbital program can still play an important role in instrument development, graduate education, and for science (as the SN1987a campaign so clearly demonstrated). If and when the Space Station becomes reality, it might also provide some of these capabilities. One possibility was raised by Steve Kahn, who suggested that one of the LAMAR modules might become a test bed for new focal plane detectors. This is similar to the way a modest sized ground-based telescope is used to test new instruments that later might be used on a 4m. In a rational world, it should also be possible to have a standard pointer which could accommodate small, low-cost instruments for specialized investigations. While our experience with Spacelab, SPARTAN etc. has taught us to be very skeptical of any comparable promises about Space Station, is it hopeless to try again?

- $5-25M -- Foreign Missions & Micro-Missions

Opportunities to develop U.S. instruments on foreign missions have become an important and invigorating element of our program. The leverage on modest investments is substantial, in terms of technology development, frequency of flight opportunities and access to data. I think we owe our friends at NASA Code EZ a debt of gratitude for the considerable efforts they have expended to make these collaborations possible. One hopes that future joint ventures will flow more naturally now that precedents have been established with Japan, Europe, and the USSR.

We have heard heard some exciting new ideas about micro-missions from Bill Priedhorsky and George Ricker. These are small, fast-track, low-cost free flyers that in some sense could become in the 1990's what the sub-orbital programs were in the 1960's and early 70's. I do want to point out that although their total costs are very low, the micro-sats have a cost-per-pound several times higher than that of larger missions (I get numbers like $120K/lb vs. $30K/lb say for AXAF). This alone will limit the application of this concept. Furthermore, some of the management lessons learned here might well be applicable to larger missions.

- $20-100M -- Mini-Missions or Small Explorers

This is the cost range that used to be associated with Explorers, but which seems to be rather sparsely populated these days, as Explorers have grown in scope and cost. One of the difficulties of doing mini-missions, which might include Delta-class payloads, is the relatively high cost of three-axis stabilized and TDRSS compatible spacecraft. There are two ways of getting around this. The one which most people seem to prefer at present, is to emulate the micro-mission model of making the spacecraft part of the instrument, putting it together from commercially available modules. This is an attractive approach, but its viability and sufficiently low cost are not yet established to my knowledge. Another way is the old "holy grail" of a standardized spacecraft -- an attempt to follow the model of the rocket program. For example, I was quite impressed by the capabilities of ASTROSPAS, a standardized SPARTAN pointer that might have rescued that program had other events not intervened. Is it conceivable to develop a comparable spacecraft as a free flyer? I must admit that this idea is usually shouted down by my colleagues because of the long list of failed past attempts. But it seems worth raising the challenge, to ourselves to try and define modest and reasonable specifications, and to NASA to break all the old molds and get on with it, either in-house or through industry. I gather that even recent attempts at doing this in the context of Scout-class explorers have been frustrating.

As I stated above, missions of this size should have development times of 3-5 years. If they take much longer than that, their cost will inevitably exceed the guideline.

- $100-200M -- Macro-Missions or Large Explorers

The growth in the cost of Explorers has been driven in part by the selection process, which inevitably rewards proposers who put in more and more capabilities. But it is also a natural result of the maturation of our discipline, which requires more sophisticated instrumentation to address more sophisticated questions. Some of the general comments on cost control made below might be applied to keep these Explorers from pricing themselves out of the market altogether. Again, one method of controlling costs is to control schedule -- it should not take much more than 5 years for one of these.

- $200M & up -- Mega-Missions

Most of us would agree that the centerpieces of our program must be missions capable of addressing a broad range of questions at the frontiers of the discipline, such as AXAF. These will be costly and will require a considerable investment of political energy, but they will also be the instruments that provide the greatest scientific yield and serve the largest community. In fact, given the utility of an AXAF over 15 years, its amortized cost is comfortably in the Explorer range. I have already argued in favor of sketching possible mega-missions to follow AXAF and noted that their development timescales ought to be 8-10 years rather than 15-20, which in itself will help control the cost.

(ii) Improving the Cost/Benefit Ratio

There is a widespread perception in the community, which I share, that our missions are simply too expensive. Even if we develop new concepts for micro- and mini-missions, the present budget is too constipated to accommodate more than a few of them. Hard as it is to be sanguine about this, we simply must examine the possibility of cheaper ways to do business. I cannot help thinking that an improvement in efficiency might be easier to get than a comparable increase in the budget, and it would be morally satisfying. This is a very contentious issue, which quickly drives people into their foxholes to take up defensive positions. If we are going to make any progress here, it will require considerable diplomacy.

First, let me remind you of the current cost per pound of developing scientific payloads for launch:

Instrument Costs:	~$30-100K/lb
AXAF Development Costs:	~$30K/lb

(the AXAF costs refer to the entire payload: ~$1B for ~30,000 lbs). These costs are ten to a hundred times those of comparable instruments built for use in our laboratories.

This leads me to pose several questions:

1. Why are our missions so costly? What are the relative contributions of scientific requirements, man-rating, high reliability, quality assurance, systems engineering approach, etc.?

Part of the extra expense of space hardware is designing for survivability and lightness but these are not the only nor even the primary drivers. For example, the instruments we supply for foreign missions generally are cheaper than those we build for our own (i.e., there is a very significant difference in cost between the CCD cameras for Astro-D and AXAF, although the cost per pound is not very different), and I am told that some DOD payloads have also been much less costly.

Man-rating, associated with Shuttle launches and/or servicing, is another factor. My own opinion is that at the present time, astronaut safety factors, while significant, are *not* the major cost drivers. This is based on our experience at MIT with a variety of payloads including instrumentation for life sciences experiments on Spacelab (for which we built hardware that the astronauts actually *wear*). Meeting the safety requirements was not a major cost element. I believe that the same is true of our XTE and AXAF instruments. I realize that some would disagree with this conclusion. I am unaware of any quantitative analysis of this question.

I believe that the main cost drivers are programmatic and that these in turn are driven largely by the present approach to reliability and quality assurance (R&QA) and to systems engineering. What has evolved, at least partly as an accident of history, is a passionate phobia of failure combined with a particular method of systems engineering and R&QA that rely heavily on design studies, analyses, plans, reviews and reporting, all of which are expensive and stretch out the schedule, which is itself expensive. The cost-effectiveness of all this deserves reexamination,.

II. If reliability is a significant contributor to the high cost of space hardware, are we designing our instruments and spacecraft to the optimum level?

The need for high reliability derives largely from the perception that things in orbit must work because it costs a lot to put them there. But compare the development costs listed above to the following launch costs per pound of payload weight:

Launch Costs*:	Scout	~$20K/lb
	Atlas,Delta	~ $4K/lb
	Titan III,IV	~$4-7K/lb
	STS	~$5-10K/lb

The numbers show that excepting Scouts, launch costs are only 10-20% of the development costs. I have always found this a curious paradox. It means we are spending $10 extra to be sure that a payload launched for $1 is sure to work. Say we could halve the development cost and still achieve say 80% reliability for our instruments. We would end up with 1.7 times the number of launches and 1.4 times as many working spacecraft (assuming 100% reliability under the present system). Of course, for very visible, expensive and long-lived missions like AXAF, high reliability is laden with political significance, so failure is less tolerable. But this should be less true for more modest payloads, especially if the cost/benefit trade-off could be demonstrated.

We should examine the cost/benefit ratio of the present system with the goal of tailoring the R&QA requirements to the actual launch costs. Other factors like the political environment or the frequency of launch opportunities must be considered as well.

III. Whatever the conclusion about the optimum level of reliability, is the present management system optimally suited to achieve the desired level?

We now rely very much on paper studies to achieve reliability where we once used engineering test models and prototypes. For example, at MIT we built

---

* The first three are deduced from information in the draft Astrotech 21 Handbook assuming low earth orbit. The STS number is a WAG which I have heard quoted.

the HEAO focal plane instruments with a Data Procurement Document listing about 15 required documents. I understand that for HST instruments there were about 40 required documents and we have about 80 for our AXAF instruments. It is not the number of documents themselves that is at issue, but the management system they represent. The protracted schedules derive in part from this. Is this exponential growth actually the most efficient way to achieve the reliability we desire? Is it the most cost effective? Should NASA manage the procurement of a one-of-a-kind scientific payload with less than precisely defined specifications in basically the same way it does for a 24 V power supply? Or, could we find ways to shorten the schedules and reduce the costs without even lowering the level of reliability?

Obtaining believable answers to these questions will be extraordinarily difficult, and it will be still harder to implement changes. *We should urge NASA to undertake a serious review of these questions.* To be effective, this review would probably have to occur at the highest levels of NASA -- at least at Len Fisk's level and possibly with the blessing of Admiral Truly. Certainly it must include participation by the Comptroller, by various centers, by scientists like us and by industry. If improvements are to be made, all elements of the program would have to change their normal ways of doing business. That means that in the end it must be a "top down" rather than a "bottom up" initiative. At the same time we should applaud NASAs move toward low-cost approaches for small explorers, instruments for foreign missions and recently HETE -- it will be a victory if these work well, but it is unlikely that these small tails will wag the whole dog, at least not during our lifetimes.

Charlie Pellerin once quipped that this problem will be solved when NASA starts rewarding its managers for cheap failures. Maybe that is not such a bad idea, assuming that there would also be enough money available for several cheap successes as well.

These have been my own personal opinions. I have benefitted from numerous conversations with many colleagues, but I will not incriminate them here by name.

Rapporteur's Report Presented at the NASA Workshop on
"High-Energy Astrophysics in the 21st Century"

S. Rappaport
Department of Physics and Center for Space Research
Massachusetts Institute of Technology

A highly informative series of review talks on six broad areas of high-energy astrophysics was presented on the first day of the Workshop. The speakers reviewed many of the scientific highlights that emerged from previous space missions and looked ahead to what might be expected from the upcoming missions, including ROSAT, GRO, EUVE, XTE, and AXAF. In some cases, areas critical to the discipline that would not be covered by current or planned missions were identified. Bob Rosner reviewed the status of X-ray studies of normal stars. He emphasized the need for a more comprehensive study, even with Einstein-level capabilities, to obtain more definitive statistics on all stellar populations. This was felt to be essential, for example, in understanding the relation between stellar rotation and X-ray activity, at least for late-type stars. Other reporters referred to the importance of studying coronal plasma diagnostics with improved spectral resolution and sensitivity. Still others, who were even more ambitious, encouraged the development of instrumentation capable of milli-arc second imaging of stellar coronae. Don Lamb reported some of the highlights of what has been learned about neutron stars, white dwarfs, and black holes, in and out of binary systems, as well as γ-ray burst sources, from X-ray observations. He stressed the need for continuous coverage of these sources over the long term. Other speakers emphasized the need for improved temporal resolution down to the sub-millisecond level, and improved spectral resolution for the study of the cyclotron line features in γ-ray burst sources. Roger Chevalier looked at our understanding of supernovae and supernova remnants. He emphasized what could be learned from following the evolution of the radioactive lines, their intensities, centroid energies, and asymmetries. He also stressed the importance of searches for the breakout of the initial shock through the stellar photosphere of new supernovae. Continued spectroscopic and imaging studies of supernova remnants will, of course, be of great importance.

In the area of extragalactic studies, Pepi Fabbiano reviewed some of the things we have learned from the study of X-ray emission from normal galaxies. Observations of discrete X-ray sources in external galaxies provide invaluable information on the statistical properties of all classes of galactic-type X-ray sources (e.g., LMXRB, Be-star systems, globular cluster sources, supernova remnants, etc.). Advantages of studying this type of source in an external galaxy include the following: (i) the distance to the sources is known, (ii) there is often little absorption in the 1-10 keV band, and (iii) the relation of each X-ray source to the local stellar population in the host galaxy can be directly observed. Also of great interest are the luminous ($> 10^{39}$ ergs s^{-1}) X-ray sources observed in the nuclei of some normal galaxies (e.g., M81 & M101) and their relation to the more luminous AGN. Richard Mushotzky recalled some of the exciting new developments in X-ray studies of active galactic nuclei. Studies of rapid temporal variability and the use of Fe-K fluorescent line diagnostics (most recently highlighted with EXOSAT and Ginga) are two of the most promising techniques for understanding these objects. Margaret Geller described how the large scale structure in the universe could be probed with X-ray observations as a complement to the optical redshift surveys. She suggested a mission that could carry out a sensitive and comprehensive (X-ray) redshift survey of gas associated with clusters of galaxies as a tracer of large-scale structure in the universe.

In listening to the review talks, other scientific presentations, and subsequent discussions, I was struck by two general considerations which have no simple answers, but must be considered when deciding on the limited number of possible future missions. The first is the question of narrowly focused science versus projects that tend to involve large segments of the astrophysics community. Studies of X-ray and γ-ray bursts tend to fall into the former category, while the Einstein mission falls into the latter. Only if the narrowly focused science is sufficiently compelling (e.g., that of the COBE mission) and/or such missions can be made small and inexpensive, can they compete with missions addressing a broader range of scientific issues. A second general issue that emerged is related to the "weather vs. physics" question. Each astrophysicist has his or her own view of where the level of specific detail becomes sufficiently high that one is effectively studying "weather" rather than breaking fundamental new astrophysical ground. This will always remain a matter of personal taste, but it seems that such an issue raises strong emotions in debates over how to allocate limited space science resources.

It has been argued in many quarters that any substantial increase in instrumental sensitivity introduced by a new technology will invariably lead to significant breakthroughs in astrophysics (as well as other sciences). It could then be argued that future missions should be decided upon, at least in part, by the promise of greater sensitivity. There are a number of directions in instrumental parameter space where enhanced capabilities would be highly desirable. These include: (i) collecting area, (ii) spectral resolution and range, (iii) angular resolution and field of view, (iv) temporal resolution and extended coverage, and (v) polarization. The following points elaborate on these in somewhat more detail: (i) In the presentations made at this Workshop there were proposals for imaging and non-imaging instruments with areas of >1 m^2 and >100 m^2, respectively. Such large instruments would, of course, be very powerful but will require a clear demonstration that their costs can be contained. (ii) Observers were in agreement that spectral resolutions of 10^4 to 10^5 are all that will be needed for the foreseeable future. Extension of the X-ray energy range to several hundred keV will be especially important in the area of an all-sky survey. In this regard, it was claimed that such an experiment should lead to the discovery of more than 10^4 AGN as well as 10^4 galactic sources. Moreover, it is worth noting that, historically in high-energy astrophysics, the higher the energy the better the "filter" for detecting exciting new phenomena. (iii) Proposals were also put forth for improving the angular resolution of X-ray instruments. These ranged all the way from sub-arcsecond resolution (using an arc segment of an ordinary Wolter telescope) to milli-arcsecond resolution from a lunar-based X-ray interferometer. A complementary parameter to angular resolution is field of view and/or sky coverage. A number of innovative proposals for improving instrumentation in this direction were also put forth. (iv) In the area of temporal resolution, the XLA experiment (with an effective collecting area of ~100 m^2) was discussed. A related issue is one of long-term temporal coverage, i.e., sampling of low temporal frequencies (inverse hours to decades). This would require either a dedicated long-term mission or a patchwork of shorter missions. (v) Finally, the prospects for polarization studies have improved greatly with the advent of a technique that measures polarization via the photoelectric effect.

Clearly, any major enhancement in instrumental capability will lead to new discoveries. The selection of missions must be based on a combination of predetermined scientific objectives and instrumental capabilities. The Einstein observatory was a good example of both. The scientific objectives were well laid out in advance and the areas in which the observatory would enhance X-ray astronomy were, in most cases, correctly anticipated. On the other hand, its sheer increase in sensitivity over all previous missions virtually assured that it would make major new discoveries. Ubiquitous emission from virtually all types of stars and cooling flows in elliptical galaxies are but two examples of the latter.

In the area of improvements in instrumental capability, however, two important questions come to mind. First, how many decades of improvement in a given instrumental parameter are worth how much enhanced performance in another area? Second, how much of our limited resources are we willing to bring to bear to accomplish any of the improved instrumental capabilities discussed above? These can be difficult issues to judge, especially when one's favorite mission may be at stake.

When considering all of the outstanding science discussed during the first day of the Workshop, and all of the proposed new directions in instrumentation during the following two days, three future missions seem particularly important to me. These are:
1. A sensitive all-sky imaging survey extending up through the Fe-K energy range, conducted with > 1 m^2 of effective area and an angular resolution of ~ 10 arc sec.
2. A sensitive high-energy (up to ~ several hundred keV) X-ray imaging sky survey using a coded aperture telescope.
3. An all-sky transient monitor experiment covering a broad range of wavelengths from the visible through the γ-ray region.

Of all the technologies discussed during the Workshop, several seem particularly impressive and worthy of pursuit: (i) The development of low-temperature X-ray detectors, all the way from the well-studied microcalorimeters to the newly investigated superconducting devices, is worthy of much more research. (ii) Improvements in techniques for producing large areas of reflecting surfaces are extremely important. These include foils, bent glass, epoxy replication, and electroforming for producing large-area X-ray mirrors with moderate angular resolution. Work with silicon flats for producing flat-segment high-resolution imaging devices also appears to be worth pursuing. (iii) The future promise of high-pressure gas scintillation detectors seems excellent.

Finally, I report on a consensus of the meeting that seemed quite strong. It is important to note that this sentiment was expressed both by active participants in the large observatories as well as by those who prefer to work on smaller experiments. Simply stated, there is a clear need to increase the opportunities for quicker access to space. Specifically, there is a need to have a better balance between large and small missions. Such a balanced approach should be established as a clear policy at NASA that will not be jeopardized during periods when some of the larger missions get into trouble. More open access to space opportunities will (i) lead to the more rapid utilization of newly developing technologies, (ii) aid in the training of graduate students and experimenters in high-energy astrophysics to replace the current generation of scientists, (iii) maintain the welfare and interest of the current experimenters and theorists alike as they await future large missions, and (iv) provide all-important support for university activities in these areas.

Gamma-Ray Astronomy – Progress & Instruments

Rapporteur talk presented at
"High Energy Astrophysics in the 21st Century"
Taos, New Mexico
Dec. 11–14, 1989

L.E. Peterson
University of California, San Diego
La Jolla, CA   92093

INTRODUCTION

The hard X-ray and gamma-ray region of the astronomical electromagnetic spectrum concerns itself primarily with photons produced by fundamental processes occurring in compact, highly non-thermal sources, extended sources containing gases at extreme temperatures or with non-thermal particle populations, and situations where nuclear excitations and reactions occur. Research in this spectral region of high energy astrophysics is at a lesser level of maturity than that of X-ray astronomy. Although the exploratory phase is over, work here is characterized more by survey and discovery, and therefore the observational emphasis typically is on sensitivity, rather than on extensive observations of large classes of objects. Because of the transient nature of many of the phenomena in the energy range, serendipidity is important, and simply "being there" with an instrument of greater sensitivity or with some other parametric improvement, can result in a new discovery.

This paper is concerned with the observational objectives and instrumental techniques which seem appropriate for the first decade of the twenty-first century. By that time missions now operating or under development, such as GRANAT, XTE and GRO, will have produced a wealth of new data and results. Although at this time we can only predict scientific objectives based on present theoretical ideas and observational extrapolations, one can confidentially expect that progess in this field will occur primarily with improved instruments. The scientific objectives and instrumental concepts discussed here are summarized from the detailed papers presented at this conference.

In order to make a coherent presentation, the results are presented as a series of charts, ordered according to energy range, which naturally includes different phenomena and instrumental techniques.

Hard X-Rays

The processes occurring in the $\sim$ 10–100 keV range include radiation from hot gases (T $\sim 10^8$–10^9 K), cyclotron emission from neutron stars and X-ray pulsars, and non-thermal effects in active galactic nuclei. This is in the range where the spectral signature of thermal processes emerges into the signature of a non thermal process, and is therefore an important region for understanding energy transfer in these sources. Other objects where important emissions occur are in X-ray emitting cataclysmic variables, black hole systems, supernova remnants, and clusters of galaxies. The signature of a thermal plasma dominated by Compton scattering should be evident in this range.

Table I summarizes the objectives, techniques, some of the instrument parametric goals and the source of the information from these proceedings.

It would be extremely advantageous to extend focussing techniques to 40 keV, which requires development of grazing incidence mirror technology to high energies. However, the only technique which gives simultaneously a large area, large aperture and small angular resolution over the entire higher energy range seems to be a coded mask in conjunction with an imaging detector. The various detector techniques include the Imaging Gas Proportional Counter (IGPC) and its variants,

CsI(Na) imaging spectrometers, high pressure gas scintillation counters, and the Liquid Xenon counter (LXe). A new method which requires added development is the superconducting transition detector.

All these techniques are being utilized in balloon or space experiments or are under development as concept studies at the present time.

Soft gamma-rays

The 50 keV to 10 MeV range is the region where, in addition to continuum processes, the direct signature of nuclear processes in the form of gamma-ray lines occurs. These processes include the 0.511 MeV line from positron annihilation and positron-electron plasmas, radioactive decay from products of explosive nucleosynthesis, and neutron capture and inelastic scattering. Lines were first observed from solar flares, and since have been observed from both diffuse and compact astrophysical sources. The sources include the recent supernovae SN1987A, the galactic center and disk, gamma ray bursts, and the black hole candidate Cyg X-1. Recent results on the 0.511 MeV line and continuum from the galactic center have been interpreted in terms of the dynamics of a positron-electron plasma.

With present understanding, the energy resolution obtainable with cooled Ge counters appears to be adequate; the need is increased sensitivity. Combinations of coded mask apertures and Ge detectors with various shielding techniques are being developed to provide the total instrument configuration. The proposed Nuclear Astrophysics Explorer (NAE) utilizes these techniques. Further requirements for progress are in the areas of larger volume segmented Ge detectors, and improved shielding techniques. The use of the moon with the lunar soil as a shield has been suggested as a possibility for a low background cosmic gamma-ray spectrometer.

The Compton telescope is presently being utilized on the GRO to provide sensitivity over the 1–30 MeV range, particularly for diffuse components. Further development of this method requires large area detectors with position-sensitive Ge counters or a LXe telescope. An important advance in the Compton telescope technique would be the ability to determine the direction of the first scattered electron, which would eliminate one ambiguity in the interpretation of the data from the device.

High Energy Gamma-Rays

Process producing gamma-rays in the 30 MeV – 1 TeV range, which were not extensively discussed at this conference, include π° decays from cosmic-ray collisions, energetic Bremsstrahlung, exotic processes including dark matter, and TeV gamma-rays from cosmic rays interacting in high latitude galactic clouds. Already energetic gamma rays have been observed from the Crab pulsar, the Vela pulsar, the galactic disk, and the Geminga region. Extremely high energy TeV gamma-rays have been observed at the 4.8 hr period of Cyg X-3 with ground based techniques.

The range from 10 MeV – 30 GeV can be studied following the GRO with large volume drift chambers under development at GSFC.

High energies (to ~1 TeV) can be studied with a pair conversion telescope utilizing the large superconducting cosmic-ray magnets, ASTROMAG, planned for the Space Station. Even higher sensitivity can be obtained with an H_2O Cerenkov detector which has been suggested for the lunar surface.

Gamma Ray Bursts

These transient phenomena have not been included explicity in the previous discussions because the requirements and instrumentation are sufficiently different from those previously discussed. A simultaneous, multi-spectal approach is demanded for progress in certain problems, particularly identifications, since with one or two possible exceptions no gamma-ray burst has been identified with a known optical or radio object. Hopefully this can be accomplished by observing simultaneously in the gamma-ray, X-ray, and UV/IR ranges. A further requirement for progress is high resolution time-resolved spectroscopy of the gamma-ray portion of the burst itself. This requires large area Ge detectors.

The various missions and instruments proposed for use after the High Energy Transient Experiment (HETE) are shown in Table II. The Energetic Transient Array (ETA) consists of a number ($\sim$ 50) of simple spacecraft in interplanetary orbit designed to provide a large number of baselines for precise timing and triangulation.

TECHNOLOGY AND MISSION REQUIREMENTS

The technology requirements resulting from the conference are summarized in Table III. These can be classified as follows:

- Developing gas or LXe counters to operate at higher energies with larger area.

- Development of large volume position-sensitive Ge or equivalent detectors for gamma rays.

- Large stable structures to support focusing X-ray mirrors and coded masks for gamma rays.

- Large volume chambers of gamma-rays.

Mission requirements include a standard, inexpensive spacecraft for Explorers, on-orbit assembly of larger instruments, and possible utilization of the moon. To meet these requirements, international cooperation and simplified management techniques seem to be important issues.

TABLE I. OBJECTIVES AND INSTRUMENTS

Energy	Objective	Technique(s)	Area/ Sensitivity	E/ΔE	FOV	ΔΘ	Mission/ Proposer
10–40 keV	Cyclotron Lines in pulsars Non-Th. Processes AGN's	Grazing Incidence Telescope			1°	30	
10–100 keV	Cataclysmic Variables Black Hole Systems SNR's	Coded Mask/IGPC Coded Mask/IGPC Imaging Ge Spectrometer CsI(Na) Imaging Spectro. High Pressure Gas Scin. Counter	1.4 m² 0.5m²	14 40 8 80 @ 122 keV	2.6°	1.5 π 1ʳ 1ʳ	CASES/Hudson Grindlay/Prince Hailey Ziock Edberg/ Weisskopf
10–300 keV	Comptonized Plasmas	Coded Mask/ Imaging Phoswich Scintillation Counters LXe Detectors Super Conducting Transition Detectors	2.8m²	40 @ 600 keV	7.3°	12"	Grindlay/Prince Aprile Kurfess
50 keV–10 MeV	Pair Plasmas 0.511 and other γ-ray lines in Galaxy, Gal Center. Black Holes Systems AGN's, SNR's, Bursts	Coded Mask/Ge Dets/ BGO Shield " Coded Mask/Ge dets. Lunar Soil Shield Positron-Sensitive Ge Counters	3×10⁻⁶ ph/cm²-sec ~ 10⁻⁶ 3×10⁻⁷	500 @ 1 MeV " 500 @ 1 MeV	~ 12°	~ 0.1° ~ 0.1°	Matteson Nakano Gehrels/Mahoney

TABLE I. CONTINUED

Energy	Objective	Technique(s)	Area/ Sensitivity	E/ΔE	FOV	ΔΘ	Mission/ Proposer
1–30 MeV	Diffuse Background Nuclear Lines Pair Plasmas Non-T Bremms	Compton Telescope	2×10^{-7} ph/ cm^2-sec	200–1000		~ 0.1°	Johnson
10 MeV–1 GeV	Compton Scattering π^0 decays Pulsars Cos B Sources	LXe Compton Telescope Drift Chamber/ Anti/TOF/Calorimeter (to 30 GeV)	~ 1 m^2	7			Aprile Hunter Hunter
100 MeV – 1 TeV	Dark Matter	Pair Telescope/Magnet H$_2$O Cerenkov/Anti	7000 cm^2-s	100	70°	20′	ASTROGAM/Adams Lunar/Svoboda

TABLE II. GAMMA RAY BURST INSTRUMENTS

Name	Proposer	Description
Total Throughput Transient Spectrometer (TTTS)	Hurley	12 keV–16 MeV, 12 Ge dets 50 keV–100 MeV, BGO shield FOV 135° 400 cm² $\Delta t \sim 1$ ms $E/\Delta E \sim 500$
Lunar Tranient Observatory (LTO)	Ricker/Woosley	Sens 10× HETE (Lunar Surface) Ge det $\rightarrow E/\Delta E = 500$ Opt UV + IR + soft X-ray $\Delta 0 \sim 1"$
Energetic Transient	Ricker	~ 50 small s/c's in interplanetary orbit

TABLE III. TECHNOLOGIES

Technique	Technology Developments
Hard X-Ray Telescopes	Long Focal Lengths → Structures, Stable Highly Nested Mirrors Deformed Highly Polished Sheets Optics Design
Gamma-Ray Imaging (Coded Masks)	Mechanisms Position Sensitive Detectors Long Stable Booms → Structures, Deployable
Ge Counters	Larger Volumes Segmented Counters; Position Sensitive Isotope Enrichment Mechanical Coolers Be Cryostats
LXe	Background Rejection Modes Compton Telescope Chamber
Superconducting Transition Detectors	Granule Size Uniformity Fast Low Power Electronics
Compton Telescopes	Det arrays with better ΔE, ΔX image 1st scatter
Drift Chambers	Low power electronics Low outgas chambers
Lunar Outpost	Communications Transportation Lunar Construction Use of regolith as shield Data compression → transient monitor

NASA Workshop on "High-Energy Astrophysics
in the 21st Century"

Rapporteur's presentation by Richard Epstein

High-energy missions to date have been successful in defining many of the properties of astrophysical systems such as black holes, neutron stars, supernovae, active galactic nuclei, and clusters of galaxies. While our knowledge of these systems has expanded with these missions, so has the realization that we have just glimpsed some of their fundamental properties. The next generation of high-energy experiments with their greater sensitivity, broader range of energy response and improved spectral, temporal and angular resolution will further elucidate the nature of these systems. Here we illustrate some of the scientific issues that will be attacked.

X-ray and γ-ray observations of putative stellar mass black holes will study the creation and evolution of relativistic plasmas. The ratio of the gamma-ray luminosity L_γ of a source to its linear dimension R defines its *compactness*. Theoretical work on the black hole candidate Cyg X-1 suggests that when the source is compact (L_γ is high or R is small), the emission region is optically thick to gamma-rays. In this case the γ-ray luminosity is suppressed, and the X-ray emission is relatively strong. In contrast, when the compactness is lower, gamma-rays escape and the X-ray radiation is weaker. This anti-correlation between strong gamma-ray emission at $\gtrsim 1$ MeV and strong X-ray emission is what is observed in low-resolution studies of Cyg X-1. Future broad-band and high temporal and energy resolution observations of Cyg X-1 and other black hole candidates will clarify the radiation and plasma physics of black holes.

Active galactic nuclei (AGN) are powerful gamma-ray sources. In fact, for the sources which have been detected at high energies, more power is detected in the energy decade near 100 keV than any other decade. Observations are needed at $\gtrsim$ MeV to see how this emission falls off at higher energies. This falloff can reveal the energy distribution of the radiating electrons. Measurements of high-energy polarization and of the correlation of X-ray and gamma-ray luminosities will help determine how the radiation is generated; is it by synchrotron emission, Compton scattering, a combination or neither? The radiation from massive black holes which may exist in active galactic nuclei shares some attributes with the stellar mass black holes. In particular, an anti-correlation between hard and soft emission is expected. However for AGN's with their cool disks, the soft emission may come out in the UV or very soft X-ray range. High-resolution spectral observations of the ~ 7 keV iron line from near the center of AGN's will test whether AGN's contain massive black holes. The photo-excited iron emission from gas swirling around a massive black hole appears blue shifted and amplified when the gas is moving toward the observer, and it appears red-shifted and diminished when the gas is receding from the observer. These relativistic effects give the iron emission a characteristic spectral shape from which the mass of the central object can be deduced.

On a larger scale, iron-line observations of clusters of galaxies appear to be the most efficient and effective technique for measuring the large scale structure of the universe. Redshift surveys of individual galaxies are useful and feasible for studying thousands or tens of thousands of the nearest galaxies, but this approach becomes increasing impractical at larger distance scales. At great distances it is hard to identify clusters or groups of galaxies by their optical images; the high density of foreground objects makes it hard to recognize the enhancements in the surface number density of galaxies in a cluster objects, and it is hard to separate cluster objects from foreground and background galaxies. To determine the redshift of a single cluster, one must measure the redshift of many individual galaxies to ensure that the results are not contaminated by chance interlopers. Furthermore, many clusters would be missed since they have low apparent surface density contrast and it is not possible to measure the redshifts of all the background objects. In X-rays detecting and identifying clusters is much less confusing. Hot gas accumulates in the center of cluster, and the iron emission for from the gas in each cluster has a single well-defined redshift. A high-throughput x-ray spectrometer would survey the large scale structure of the universe to a much greater distance than would be feasible by optical spectrometry.

The only unambiguous observations of gamma-ray bursts, which are thought to arise from explosions on neutron stars, are in the X- and γ-ray energy bands. The study of gamma-ray bursts holds the promise of revealing much about the dynamics and evolution of neutron stars. Known gamma-ray bursters are more numerous that the sum of all other neutron star sources. It is likely that these bursters are older than radio pulsars; they can, therefore, provide clues to the evolution of neutron star magnetic fields and internal dynamics. The radiation properties of gamma-ray bursters are unique: the common or 'classical' burster radiates most power at $\sim$ 1 MeV with relatively little X-ray emission below $\sim$ 30 keV, and the 'soft gamma-ray repeaters' radiate mainly near 40 keV. Basic questions remain about how gamma-ray bursters fit into the framework of neutron stars: Are these two classes of gamma-ray bursters closely related? Are either or both akin to X-ray bursters, rotation-powered (radio) pulsars, or some other known type of system? To resolve these issues gamma-ray bursts have to be observed over broad spectral ranges with high temporal, spectral and angular resolution. The greatest problem to understanding gamma-ray bursts is that we do not know the distance to these sources. This question may be resolved statistically by more sensitive surveys in which the angular distribution of gamma-ray bursts might exhibit some anisotropy associated with the galactic plane. Direct observations of counterparts in other wavelength bands could be definitive. Multi-wavelength studies in which γ-ray observations are supplemented by searches in the soft X-ray, UV and optical bands hold promise; the planned HETE mission is based on this concept. Reported multiple cyclotron lines at $\sim$ 10-100 keV are compelling evidence for strong, uniform magnetic fields in the emission regions of these gamma-ray bursters. Sensitive, high-resolution studies of these spectral features can disclose how the magnetic field structure, gas density and geometry evolve during a burst due to the explosion or to stellar rotation.. The correlations of magnetic fields with other burst properties in different bursts could indicate how magnetic fields evolve with neutron star age. Since gamma-ray bursts are compact sources with high magnetic fields, the highest energy photons can be destroyed by interactions with the magnetic field or by two-photon pair production unless the radiation is highly collimated or much weaker than normally assumed. While no evidence for high-energy photon attenuation has yet been observed, spectral observation beyond ~ 100 MeV correlated with cyclotron line observation would constrain the nature of the radiation mechanism. Measurements of the γ-ray polarization would help distinguish between thermal and non-thermal processes.

The most tantalizing high-energy observations of neutron stars are the ground-based Cherenkov and air-shower measurements of TeV and PeV quanta from Her X-1, Cygnus X-3 and other X-ray pulsars. If verified, these observations will have profound implications for the particle acceleration and radiation processes in accreting neutron star binaries. In addition, there are indications that the observed quanta produce muon-rich air showers, unlike what is expected for high-energy photons; these data could have a huge impact on particle physics. High-energy space experiments that have sensitivity at ~ 1 TeV are well suited for investigating these sources Although the inferred photon flux is low at ~ 1 Tev energies, spacecraft-borne instruments can eliminate much of the background that confuse the signal for ground-based observers, and they would be able to directly determine whether the observed quanta from these pulsars are very high-energy gamma-rays.

The recent, nearby supernova SN1987A has given some indication of the power of X-ray and γ-ray observations for probing the structure of presupernova objects, the dynamics of supernova explosions, and the nucleosynthesis yield of such events. We cannot expect to have the good fortune in the immediate future of having another nearby supernova. High-energy observations of supernova should concentrate on supernova at distances comparable to that of the Virgo Cluster. The initial breakout of the supernova shock wave to low optical depth produces an intense soft x-ray burst. Observations of this burst would provide a useful diagnostic of the stellar structure and the burst energy and would be a trigger for launching a multiwavelength campaign for studying the early stages of the supernova outburst. Later, as the ejecta becomes transparent to gamma-rays, high energy resolution measurements of line emission from the decay chains of ^{56}Ni and ^{57}Ni would determine the amount of newly synthesized matter in the event, kinematics of the explosion, and the degree and depth of mixing within the ejecta.

The scientific productivity of many of NASA's high-energy missions could be greatly enhanced by correlated ground-based measurements. Simultaneous ground-based optical, IR or radio observations in conjunction with satellite X-ray or γ-ray measurements of black-hole candidates, AGN's or neutron star sources would be extremely helpful for deducing the nature of these objects. Since ground-based observations are so much less expensive than space missions, a relatively minor amount of NASA support for correlated ground-based measurements of high-energy sources, would increase the scientific productivity of some space missions.

One of the highest priorities in high-energy astrophysics is ensuring that younger scientists, both experimentalists and theorists, are being trained in this field. New people would be encouraged to enter this field if there were a reasonable number of space experiments whose development period is comparable to that of a graduate student's education.

AIP Conference Proceedings

		L.C. Number	ISBN
No. 124	Neutron-Nucleus Collisions – A Probe of Nuclear Structure (Burr Oak State Park - 1984)	84-73216	0-88318-323-4
No. 125	Capture Gamma-Ray Spectroscopy and Related Topics – 1984 (Internat. Symposium, Knoxville)	84-73303	0-88318-324-2
No. 126	Solar Neutrinos and Neutrino Astronomy (Homestake, 1984)	84-63143	0-88318-325-0
No. 127	Physics of High Energy Particle Accelerators (BNL/SUNY Summer School, 1983)	85-70057	0-88318-326-9
No. 128	Nuclear Physics with Stored, Cooled Beams (McCormick's Creek State Park, Indiana, 1984)	85-71167	0-88318-327-7
No. 129	Radiofrequency Plasma Heating (Sixth Topical Conference, Callaway Gardens, GA, 1985)	85-48027	0-88318-328-5
No. 130	Laser Acceleration of Particles (Malibu, California, 1985)	85-48028	0-88318-329-3
No. 131	Workshop on Polarized ^{3}He Beams and Targets (Princeton, New Jersey, 1984)	85-48026	0-88318-330-7
No. 132	Hadron Spectroscopy–1985 (International Conference, Univ. of Maryland)	85-72537	0-88318-331-5
No. 133	Hadronic Probes and Nuclear Interactions (Arizona State University, 1985)	85-72638	0-88318-332-3
No. 134	The State of High Energy Physics (BNL/SUNY Summer School, 1983)	85-73170	0-88318-333-1
No. 135	Energy Sources: Conservation and Renewables (APS, Washington, DC, 1985)	85-73019	0-88318-334-X
No. 136	Atomic Theory Workshop on Relativistic and QED Effects in Heavy Atoms	85-73790	0-88318-335-8
No. 137	Polymer-Flow Interaction (La Jolla Institute, 1985)	85-73915	0-88318-336-6
No. 138	Frontiers in Electronic Materials and Processing (Houston, TX, 1985)	86-70108	0-88318-337-4
No. 139	High-Current, High-Brightness, and High-Duty Factor Ion Injectors (La Jolla Institute, 1985)	86-70245	0-88318-338-2
No. 140	Boron-Rich Solids (Albuquerque, NM, 1985)	86-70246	0-88318-339-0
No. 141	Gamma-Ray Bursts (Stanford, CA, 1984)	86-70761	0-88318-340-4
No. 142	Nuclear Structure at High Spin, Excitation, and Momentum Transfer (Indiana University, 1985)	86-70837	0-88318-341-2
No. 143	Mexican School of Particles and Fields (Oaxtepec, México, 1984)	86-81187	0-88318-342-0

No. 144	Magnetospheric Phenomena in Astrophysics (Los Alamos, 1984)	86-71149	0-88318-343-9
No. 145	Polarized Beams at SSC & Polarized Antiprotons (Ann Arbor, MI & Bodega Bay, CA, 1985)	86-71343	0-88318-344-7
No. 146	Advances in Laser Science–I (Dallas, TX, 1985)	86-71536	0-88318-345-5
No. 147	Short Wavelength Coherent Radiation: Generation and Applications (Monterey, CA, 1986)	86-71674	0-88318-346-3
No. 148	Space Colonization: Technology and The Liberal Arts (Geneva, NY, 1985)	86-71675	0-88318-347-1
No. 149	Physics and Chemistry of Protective Coatings (Universal City, CA, 1985)	86-72019	0-88318-348-X
No. 150	Intersections Between Particle and Nuclear Physics (Lake Louise, Canada, 1986)	86-72018	0-88318-349-8
No. 151	Neural Networks for Computing (Snowbird, UT, 1986)	86-72481	0-88318-351-X
No. 152	Heavy Ion Inertial Fusion (Washington, DC, 1986)	86-73185	0-88318-352-8
No. 153	Physics of Particle Accelerators (SLAC Summer School, 1985) (Fermilab Summer School, 1984)	87-70103	0-88318-353-6
No. 154	Physics and Chemistry of Porous Media—II (Ridge Field, CT, 1986)	83-73640	0-88318-354-4
No. 155	The Galactic Center: Proceedings of the Symposium Honoring C. H. Townes (Berkeley, CA, 1986)	86-73186	0-88318-355-2
No. 156	Advanced Accelerator Concepts (Madison, WI, 1986)	87-70635	0-88318-358-0
No. 157	Stability of Amorphous Silicon Alloy Materials and Devices (Palo Alto, CA, 1987)	87-70990	0-88318-359-9
No. 158	Production and Neutralization of Negative Ions and Beams (Brookhaven, NY, 1986)	87-71695	0-88318-358-7
No. 159	Applications of Radio-Frequency Power to Plasma: Seventh Topical Conference (Kissimmee, FL, 1987)	87-71812	0-88318-359-5
No. 160	Advances in Laser Science–II (Seattle, WA, 1986)	87-71962	0-88318-360-9
No. 161	Electron Scattering in Nuclear and Particle Science: In Commemoration of the 35th Anniversary of the Lyman-Hanson-Scott Experiment (Urbana, IL, 1986)	87-72403	0-88318-361-7
No. 162	Few-Body Systems and Multiparticle Dynamics (Crystal City, VA, 1987)	87-72594	0-88318-362-5

No. 163	Pion–Nucleus Physics: Future Directions and New Facilities at LAMPF (Los Alamos, NM, 1987)	87-72961	0-88318-363-3
No. 164	Nuclei Far from Stability: Fifth International Conference (Rosseau Lake, ON, 1987)	87-73214	0-88318-364-1
No. 165	Thin Film Processing and Characterization of High-Temperature Superconductors	87-73420	0-88318-365-X
No. 166	Photovoltaic Safety (Denver, CO, 1988)	88-42854	0-88318-366-8
No. 167	Deposition and Growth: Limits for Microelectronics (Anaheim, CA, 1987)	88-71432	0-88318-367-6
No. 168	Atomic Processes in Plasmas (Santa Fe, NM, 1987)	88-71273	0-88318-368-4
No. 169	Modern Physics in America: A Michelson-Morley Centennial Symposium (Cleveland, OH, 1987)	88-71348	0-88318-369-2
No. 170	Nuclear Spectroscopy of Astrophysical Sources (Washington, D.C., 1987)	88-71625	0-88318-370-6
No. 171	Vacuum Design of Advanced and Compact Synchrotron Light Sources (Upton, NY, 1988)	88-71824	0-88318-371-4
No. 172	Advances in Laser Science–III: Proceedings of the International Laser Science Conference (Atlantic City, NJ, 1987)	88-71879	0-88318-372-2
No. 173	Cooperative Networks in Physics Education (Oaxtepec, Mexico 1987)	88-72091	0-88318-373-0
No. 174	Radio Wave Scattering in the Interstellar Medium (San Diego, CA 1988)	88-72092	0-88318-374-9
No. 175	Non-neutral Plasma Physics (Washington, DC 1988)	88-72275	0-88318-375-7
No. 176	Intersections Between Particle Land Nuclear Physics (Third International Conference) (Rockport, ME 1988)	88-62535	0-88318-376-5
No. 177	Linear Accelerator and Beam Optics Codes (La Jolla, CA 1988)	88-46074	0-88318-377-3
No. 178	Nuclear Arms Technologies in the 1990s (Washington, DC 1988)	88-83262	0-88318-378-1
No. 179	The Michelson Era in American Science: 1870–1930 (Cleveland, OH 1987)	88-83369	0-88318-379-X
No. 180	Frontiers in Science: International Symposium (Urbana, IL, 1987)	88-83526	0-88318-380-3

No. 181	Muon-Catalyzed Fusion (Sanibel Island, FL, 1988)	88-83636	0-88318-381-1
No. 176	Intersections Between Particle and Nuclear Physics (Third International Conference) (Rockport, ME 1988)	88-62535	0-88318-376-5
No. 177	Linear Accelerator and Beam Optics Codes (La Jolla, CA 1988)	88-46074	0-88318-377-3
No. 178	Nuclear Arms Technologies in the 1990s (Washington, DC 1988)	88-83262	0-88318-378-1
No. 179	The Michelson Era in American Science: 1870–190 (Cleveland, OH 1987)	88-83369	0-88318-379-X
No. 180	Frontiers in Science: International Symposium (Urbana, IL 1987)	88-83526	0-88318-380-3
No. 181	Muon-Catalyzed Fusion (Sanibel Island, FL 1988)	88-83636	0-88318-381-1
No. 182	High T_c Superconducting Thin Films, Devices, and Application (Atlanta, GA 1988)	88-03947	0-88318-382-X
No. 183	Cosmic Abundances of Matter (Minneapolis, MN 1988)	89-80147	0-88318-383-8
No. 184	Physics of Particle Accelerators (Ithaca, NY 1988)	89-83575	0-88318-384-6
No. 185	Glueballs, Hybrids, and Exotic Hadrons (Upton, NY 1988)	89-83513	0-88318-385-4
No. 186	High-Energy Radiation Background in Space (Sanibel Island, FL 1987)	89-83833	0-88318-386-2
No. 187	High-Energy Spin Physics (Minneapolis, MN 1988)	89-83948	0-88318-387-0
No. 188	International Symposium on Electron Beam Ion Sources and their Applications (Upton, NY 1988)	89-84343	0-88318-388-9
No. 189	Relativistic, Quantum Electrodynamic, and Weak Interaction Effects in Atoms (Santa Barbara, CA 1988)	89-84431	0-88318-389-7
No. 190	Radio-frequency Power in Plasmas (Irvine, CA 1989)	89-45805	0-88318-397-8
No. 191	Advances in Laser Science–IV (Atlanta, GA 1988)	89-85595	0-88318-391-9
No. 192	Vacuum Mechatronics (First International Workshop) (Santa Barbara, CA 1989)	89-45905	0-88318-394-3
No. 193	Advanced Accelerator Concepts (Lake Arrowhead, CA 1989)	89-45914	0-88318-393-5

No. 194	Quantum Fluids and Solids—1989 (Gainesville, FL, 1989)	89-81079	0-88318-395-1
No. 195	Dense Z-Pinches (Laguna Beach, CA, 1989)	89-46212	0-88318-396-X
No. 196	Heavy Quark Physics (Ithaca, NY, 1989)	89-81583	0-88318-644-6
No. 197	Drops and Bubbles (Monterey, CA, 1988)	89-46360	0-88318-392-7
No. 198	Astrophysics in Antarctica (Newark, DE, 1989)	89-46421	0-88318-398-6
No. 199	Surface Conditioning of Vacuum Systems (Los Angeles, CA, 1989)	89-82542	0-88318-756-6
No. 200	High T_c Superconducting Thin Films: Processing, Characterization, and Applications (Boston, MA, 1989)	90-80006	0-88318-759-0
No. 201	QED Stucture Functions (Ann Arbor, MI, 1989)	90-80229	0-88318-671-3
No. 202	NASA Workshop on Physics From a Lunar Base (Stanford, CA, 1989)	90-55073	0-88318-646-2
No. 203	Particle Astrophysics: The NASA Cosmic Ray Program for the 1990s and Beyond (Greenbelt, MD, 1989)	90-55077	0-88318-763-9
No. 204	Aspects of Electron-Molecule Scattering and Photoionization (New Haven, CT, 1989)	90-55175	0-88318-764-7
No. 205	The Physics of Electronic and Atomic Collisions (XVI International Conference) (New York, NY, 1989)	90-53183	0-88318-390-0
No. 206	Atomic Processes in Plasmas (Gaithersburg, MD, 1989)	90-55265	0-88318-769-8
No. 207	Astrophysics from the Moon (Annapolis, MD, 1990)	90-55582	0-88318-770-1
No. 208	Current Topics in Shock Waves (Bethlehem, PA, 1989)	90-55617	0-88318-776-0
No. 209	Computing for High Luminosity and High Intensity Facilities (Santa Fe, NM, 1990)	90-55634	0-88318-786-8
No. 210	Production and Neutralization of Negative Ions and Beams (Brookhaven, NY, 1990)	90-55316	0-88318-786-8

APR 16 1991